ENVIRONMENT AND SOCIETY

Environment and Society

A Reader

Edited by
Christopher Schlottmann,
Dale Jamieson,
Colin Jerolmack, *and*
Anne Rademacher

With Maria Damon

NEW YORK UNIVERSITY PRESS
New York

NEW YORK UNIVERSITY PRESS
New York
www.nyupress.org

© 2017 by New York University
All rights reserved

References to Internet websites (URLs) were accurate at the time of writing.
Neither the author nor New York University Press is responsible for URLs
that may have expired or changed since the manuscript was prepared.

ISBN: 978-1-4798-0193-0 (hardback)
ISBN: 978-1-4798-9491-8 (paperback)

For Library of Congress Cataloging-in-Publication data, please contact
the Library of Congress.

New York University Press books are printed on acid-free paper,
and their binding materials are chosen for strength and durability.
We strive to use environmentally responsible suppliers and materials
to the greatest extent possible in publishing our books.

10 9 8 7 6 5 4 3 2 1

Also available as an ebook

To our students

CONTENTS

ACKNOWLEDGMENTS

This anthology is the culmination of almost a decade of teaching, conversation, and community building around the relationships between environment and society. We have benefited tremendously from our colleagues and students in the Department of Environmental Studies at New York University. This project was supported by Kristina Sokourenko and Allie Tsubota, who provided substantial, sustained, and diligent research assistance. Amanda Anjum delivered exceptional administrative support. Our many Environmental Studies students who have taken "Environment and Society" over the years gave us feedback and thoughtful responses to the course and readings. We also appreciated the helpful feedback from a number of anonymous referees. We thank Leo Purman for the cover photography. All royalties from the sale of this book will be donated to the NYU Environmental Studies Scholarship Fund.

*   *   *

The editors wish to express their appreciation to the following for permissions to reprint the selections in this volume:

Edward Abbey, "Freedom and Wilderness, Wilderness and Freedom," from *The Journey Home*. Copyright © 1977 by Edward Abbey. Used by permission of Dutton, an imprint of Penguin Publishing Group, a division of Penguin Random House LLC.

Jonathan H. Adler, "About Free-Market Environmentalism," from *Ecology, Liberty, and Property: A Free-Market Environmental Reader* (Washington, DC: Competitive Enterprise Institute, 2000). Reprinted by permission of Competitive Enterprise Institute.

Brad Allenby, "Earth Systems Engineering and Management," *IEEE Technology and Society Magazine* (Winter 2000–2001). Reprinted by permission of the publisher.

Amita Baviskar, "Between Violence and Desire: Space, Power, and Identity in the Making of Metropolitan Delhi," *International Social Science Journal* 55 (2003): 89–98. Reprinted by permission of the publisher.

Wendell Berry, "The Agrarian Standard," from *Citizenship Papers* (Berkeley: Counterpoint, 2003). Copyright © 2014 by Wendell Berry. Reprinted by permission of Counterpoint.

Norman Borlaug, "The Green Revolution Revisited and the Road Ahead," 2000 Anniversary Nobel Lecture, Norwegian Nobel Institute in Oslo, Norway. © The Nobel

Foundation (1970). Source Nobelprize.org. Reprinted by permission of Nobel Media AB, Publishing and Broadcast Rights.

Robert D. Bullard, "Environmentalism and Social Justice," from *Dumping in Dixie: Race, Class, and Environmental Quality*, 3rd ed. (Boulder, CO: Westview, 2000). Reprinted by permission of the publisher.

Rachel Carson, "A Fable for Tomorrow" and "The Obligation to Endure," from *Silent Spring*. Copyright © 1962 by Rachel L. Carson, renewed 1990 by Roger Christie. Reprinted by permission of Houghton Mifflin Harcourt Publishing Company. All rights reserved.

Marian R. Chertow, "The IPAT Equation and Its Variants," *Journal of Industrial Ecology* 4 (2000): 13–29. Copyright © 2000 by Marian R. Chertow. Reproduced with permission of Blackwell Publishing Ltd.

Gretchen C. Daily and Paul R. Ehrlich, "Socioeconomic Equity, Sustainability, and Earth's Carrying Capacity," *Ecological Applications* 6 (1996): 991–1001. Reproduced by permission of the publisher.

Pope Francis, *Laudato Si*, 2015. Reprinted by permission of the Vatican Press.

Robert E. Goodin, "Naturalness as a Source of Value," from *Green Political Theory* (Cambridge, UK: Polity, 1992). Reprinted by permission of the publisher.

Robert Gottlieb, "Where We Live, Work, and Play," from *Forcing the Spring*. Copyright © 2005 by Robert Gottlieb. Reproduced by permission of Island Press, Washington, DC.

Ramachandra Guha, "The Paradox of Global Environmentalism," *Current History*, October 2000, 367–370. Reprinted with permission from *Current History* magazine. Copyright © 2015 by Current History, Inc.

Garrett Hardin, "The Tragedy of the Commons," *Science* 13 (1968): 1243–1248. Reprinted by permission of the publisher.

Alan Holland, "Sustainability," from *A Companion to Environmental Philosophy*, ed. Dale Jamieson (Malden, MA: Blackwell, 2001). Reproduced by permission of the publisher.

Marion Hourdequin, "Climate, Collective Action and Individual Ethical Obligations," *Environmental Values* 19 (2010): 444–445, 452–462. Reprinted by permission of The White Horse Press.

David W. Keith, "The Earth Is Not Yet an Artifact," *IEEE Technology and Society Magazine*, Winter 2000–2001. Reprinted by permission of the publisher.

Leon Kolankiewicz, "Overpopulation versus Biodiversity," from *Life on the Brink* (Athens: University of Georgia Press, 2012), 75–90. Reprinted by permission of the publisher.

Aldo Leopold, from "The Land Ethic," from *A Sand County Almanac*, 239–40. Copyright © 1968 by Oxford University Press, USA. Reprinted by permission of the publisher.

Thomas Malthus, *An Essay on the Principle of Population* (London, 1798).

Michael Maniates, excerpt from "In Search of Consumptive Resistance: The Voluntary Simplicity Movement," from *Confronting Consumption*, ed. Thomas Princen, Michael Maniates, and Ken Conca. Copyright © 2002 by Massachusetts Institute of Technology, by permission of The MIT Press.

William McDonough and Michael Braungart, "The NEXT Industrial Revolution," *Atlantic*, October 1998. Reprinted with permission of the authors.

Bill McKibben, excerpts from *The End of Nature* (New York: Penguin Random House, 2006), 40–47, 49–56. Copyright © 1989, 2006 by William McKibben. Used by permission of Random House, an imprint and division of Penguin Random House LLC. All rights reserved.

Carolyn Merchant, "Reinventing Eden: Western Culture as a Recovery Narrative," from *Uncommon Ground*, edited by William Cronon. Copyright © 1995 by William Cronon. Used by permission of W. W. Norton & Company, Inc.

John Muir, with illustrations by Galen A. Rowell, "Hetch Hetchy Valley," from *The Yosemite: The Original John Muir Text* (San Francisco: Sierra Club Books, 1989).

Ted Nordhaus and Michael Shellenberger, excerpts from "The Death of Environmentalism," from *Break Through: From the Death of Environmentalism to the Politics of Possibility*, by Ted Nordhaus and Michael Shellenberger. Copyright © 2007 by Ted Nordhaus and Michael Shellenberger. Reprinted by permission of Houghton Mifflin Harcourt Publishing Company. All rights reserved.

Elinor Ostrom, Joanna Burger, Christopher B. Field, Richard B. Norgaard, and David Policansky, "Revisiting the Commons: Local Lessons, Global Challenges," *Science* 284 (1999): 278–282. Reprinted by permission of the publisher.

David Owen, "More like Manhattan," from *Green Metropolis*. Copyright © 2009 by David Owen. Used by permission of Riverhead, an imprint of Penguin Publishing Group, a division of Penguin Random House LLC.

John Passmore, "Conservation," from *Man's Responsibility for Nature* (London: Gerald Duckworth, 1974), 73–81, 98–100. Reprinted by permission of Gerald Duckworth & Co. Ltd.

Peggy Petrzelka and Michael M. Bell, "Rationality and Solidarities," *Human Organization Journal* 59 (2000): 343–352. Reproduced by permission of the Society for Applied Anthropology.

David Schlosberg, "Theorising Environmental Justice: The Expanding Sphere of a Discourse," *Environmental Politics* 22 (2013): 40–45, 49–51. Reprinted by permission of Taylor and Francis Group.

Gary Snyder, excerpts from "The Etiquette of Freedom," from *The Practice of the Wild*. Copyright © 1990 by Gary Snyder. Reprinted by permission of Counterpoint.

Will Steffen, Paul J. Crutzen, and John R. McNeill. "The Anthropocene: Are Humans Now Overwhelming the Great Forces of Nature?," *Ambio* 36, no. 8 (2007): 614–621. Reprinted by permission of Springer Publishing.

Peter J. Taylor and Frederick H. Buttel, "How Do We Know We Have Global Environmental Problems? Science and the Globalization of Environmental Discourse," *Geoforum* 23 (1992): 405–416. Reprinted by permission of Elsevier.

Henry David Thoreau, from "Walking," from *The Writings of Henry David Thoreau*, vol. 5, *Excursions and Poems* (Boston: Houghton Mifflin, 1906), 657–674.

Mark Van Vugt, "Averting the Tragedy of the Commons," *Current Directions in Psychological Science* 18 (2009): 169–173. Reprinted by permission of the publisher.

Alan Weisman, *The World without Us* (New York: Thomas Dunne Books / St. Martin's, 2007). Reprinted by permission of St. Martin's Press.

Introduction

In an era marked by climate change, rapid urbanization, and new geographies of resource scarcity, Environmental Studies has emerged as an important scholarly arena for engaging pressing questions in an interdisciplinary way. *Environment and Society: A Reader* traces an intellectual genealogy of the field while connecting its core themes to central issues and debates for the twenty-first century. This volume assembles canonical and contemporary texts about environment and society, constituting a systematic survey of central concepts and issues. These include questions about social mobilization on behalf of environmental objectives; the relationships among human population, economic growth, and stresses on the planet's natural resources; debates about the relative effects of collective and individual action; and unequal distribution of the social costs of environmental degradation. The parts are organized around a set of themes, which together constitute the core concerns of the field. Through the collection of central papers, each part simultaneously explores the theme and maps the genealogy of debates related to that theme. The aim of the volume is to acquaint readers with the essential history of Environmental Studies while demonstrating how its uniquely interdisciplinary intellectual arena affords meaningful engagement with the critical issues of the present.

Environment and Society: A Reader balances breadth and depth by covering a wide range of central questions for Environmental Studies while focusing on cases and specific challenges. It covers both theoretical and applied dimensions. It engages readers to fully consider how society shapes the environment, how the environment shapes society, why it matters, and what can be done in response.

All of the coeditors have taught the course "Environment and Society" at New York University. This course is one of the two gateway courses in the Department of Environmental Studies at New York University. This anthology is intended to serve as a textbook for this course and for comparable courses in other universities and colleges and is structured for this purpose. The aim of this anthology is to promote an analytical approach to understanding the environment and environmental movements. We feature both canonical and contemporary selections that are interdisciplinary within the social sciences and humanities. *Environment and Society: A Reader* is organized around six categories—ideas, movements, population, public goods, values, and controversies—and offers an intellectual genealogy of the field of Environmental Studies that connects its core themes to central issues and debates about the environment. While these themes are applicable to problem solving and resource management, they also engage the major theoretical tensions within Environmental Studies.

Environment and Society: A Reader covers a broader historical range, from the early environmental movement to today. The texts are selected to highlight the recurring questions that have arisen concerning human interactions with the environment. It offers a wider range of disciplinary and interdisciplinary texts (as a result of the breadth of expertise of the coeditors) and emphasizes major conceptual challenges for the field of Environmental Studies. The selections are essential readings engaging central questions about the environment, supplemented by coauthored introductions, discussion questions, and suggestions for further readings. *Environment and Society: A Reader* is broadly interdisciplinary, inclusive of both historical and contemporary texts, and designed to engage central questions and concepts in Environmental Studies.

We have edited some entries and have maintained spelling customs of the original texts.

PART I

Ideas of Nature

Environmentalists work both to improve environmental quality and to protect nature. It is thus not surprising that people sometimes use the word "environment" as if it were synonymous with "nature." This is incorrect, however. Everything around us is part of the environment, but not everything is part of nature. When we preserve historic buildings, we protect the environment, but we do not preserve nature. When we establish a nature reserve such as the Arctic National Wildlife Refuge, we protect nature, but we do not improve the environment (except for a very few people and a great many caribou). Buildings, people, plants, animals, and toxic-waste dumps are all part of the environment, but toxic-waste dumps and buildings are not part of nature. Still, there are important connections between the environment and nature, and it is not a coincidence that environmentalists have been interested in both.

The English word "nature" comes from the Latin word *natura*, which in turn is a translation of the Greek word *physis*. *Natura* in its oldest uses means "birth." We hear the echo of this when people speak of "Mother Nature" or nature as the source of all life. When we speak of nature, we often mean to be referring to what is essential, as when we talk of "human nature." Sometimes nature is thought to be irresistible ("it's in his nature"), other times as something to be overcome (when it takes the form of disease), and other times still it is regarded as good ("natural ingredients"). While environmentalists do not always endorse what they take to be natural, there is a tendency in environmental thought to think of nature as providing a norm or ideal that is at least presumptively good. While there is more to environmentalism than respecting nature, investigating the idea of nature is a good place to begin our understanding of environment and society.

This part begins with selections from the writer and activist Bill McKibben's *The End of Nature*. In 1989, when climate change was much less well understood than it is today, McKibben went to what he saw as the heart of the problem. Yes, climate change will damage people and property, and that is very important. But the most profound consequence of climate change is that it will extinguish an important source of human meaning. Throughout human history, temperature and rainfall have always been the work of "some separate, uncivilizable force." Now they are in part "a product of our habits, our economics, our ways of life." Nature, as we have always understood it, is coming to an end. And the end of nature—the extinction of the idea that we live in a world larger than ourselves and beyond our control—brings sorrow, nostalgia, and a loss of meaning.

This new era in which humanity dominates nature is increasingly being called the "Anthropocene." In our second selection, the distinguished scientists Will Steffen and Paul J. Crutzen and the environmental historian John R. McNeill describe this era and chart its history. The Anthropocene began, according to our authors, around 1800 with the onset of industrialization. Since about 1945, we have been in "the great acceleration." Since "humankind will remain a major geological force for many millennia," the challenge is to become responsible stewards of the Earth system.

The journalist Alan Weisman cautions us not to mistake the ubiquity of our impact with its permanence. In a delightful thought experiment, he asks us to imagine a world without us. Very quickly, it turns out, the world without humanity is quickly reclaimed by nature. A few signs of humanity's impact may persist for awhile (e.g., feral cats feeding on starlings); but soon enough the glaciers will return, and all that will be left is a layer of the Earth's crust that incorporates "an unnatural concentration of a reddish metal, which briefly had assumed the form of wiring and plumbing."

When seen from a long-enough distance, everything of significance seems to disappear. The historian Carolyn Merchant returns us to the idea of nature in Western culture, which she sees as encompassing a recovery narrative. In the beginning was Eden, which was defiled by sin. According to this story, "aided by the Christian doctrine of redemption and the inventions of science, technology, and capitalism," the Earth will be transformed into a "vast cultivated garden." Merchant relates this recovery narrative to gendered conceptions of nature and technology and our fascination with biotechnology.

Pope Francis sees the Christian teachings in a different light. They have the potential to help us achieve the proper balance between the recognition of human uniqueness and the intrinsic value of all forms of life. According to Francis, "all of us are linked by unseen bonds and together form a kind of universal family."

In the essay "The Etiquette of Freedom," the Buddhist poet Gary Snyder reveals "the lessons of the wild." He relates notions such as nature and wild to the Chinese *tao* and the Buddhist *dharma*. He finds the wild within as well as without: "The body is . . . in the mind. They are both wild."

The part ends with a short passage from Aldo Leopold, the twentieth-century American scientist who is regarded as one of the foundational figures in American environmentalism. Here Leopold explains the "the land ethic," which expands the moral community to include "soils, waters, plants, and animals."

These ideas of nature, sometimes complementary and at times competing, form the background of the way that we think about the environment and its relation to society.

1

Excerpts from *The End of Nature*

BILL MCKIBBEN

Almost every day, I hike up the hill out my back door. Within a hundred yards the woods swallows me up, and there is nothing to remind me of human society—no trash, no stumps, no fence, not even a real path. Looking out from the high places, you can't see road or house; it is a world apart from man. But once in a while someone will be cutting wood farther down the valley, and the snarl of a chain saw will fill the woods. It is harder on those days to get caught up in the timeless meaning of the forest, for man is nearby. The sound of the chain saw doesn't blot out all the noises of the forest or drive the animals away, but it does drive away the feeling that you are in another, separate, timeless, wild sphere.

Now that we have changed the most basic forces around us, the noise of that chain saw will always be in the woods. We have changed the atmosphere, and that will change the weather. The temperature and rainfall are no longer to be entirely the work of some separate, uncivilizable force, but instead in part a product of our habits, our economies, our ways of life. Even in the most remote wilderness, where the strictest laws forbid the felling of a single tree, the sound of that saw will be clear, and a walk in the woods will be changed—tainted—by its whine. The world outdoors will mean much the same thing as the world indoors, the hill the same thing as the house.

An idea, a relationship, can go extinct, just like an animal or a plant. The idea in this case is "nature," the separate and wild province, the world apart from man to which he adapted, under whose rules he was born and died. In the past, we spoiled and polluted parts of that nature, inflicted environmental "damage." But that was like stabbing a man with toothpicks: though it hurt, annoyed, degraded, it did not touch vital organs, block the path of the lymph or blood. We never thought that we had wrecked nature. Deep down, we never really thought we could: it was too big and too old; its forces—the wind, the rain, the sun—were too strong, too elemental.

But, quite by accident, it turned out that the carbon dioxide and other gases we were producing in our pursuit of a better life—in pursuit of warm houses and eternal economic growth and of agriculture so productive it would free most of us from farming—could alter the power of the sun, could increase its heat. And that increase could change the patterns of moisture and dryness, breed storms in new places, breed deserts. Those things may or may not have yet begun to happen, but it is too late to altogether prevent them from happening. We have produced the carbon dioxide—we are ending nature. . . .

* * *

The argument that nature is ended is complex; profound objections to it are possible, and I will try to answer them. But to understand what's ending requires some attention to the past. Not the ancient past, not the big bang or the primal stew. The European exploration of this continent is far enough back, for it is man's idea of nature that is important here, and it was in response to this wild new world that much of our modern notion of nature developed. North America, of course, was not entirely unaltered by man when the colonists arrived, but it's previous occupants had treated it fairly well. In many places, it was wilderness.

And most of it was wilderness still on the eve of Revolution, when William Bartram, one of America's first professional naturalists, set out from his native Philadelphia to tour the South. His report on that trip through "North and South Carolina, Georgia, East and West Florida, the Cherokee Country, the Extensive Territories of the Muscogulges, or Creek Confederacy, and the Country of the Choctaws" is a classic; it gives the sharpest early picture of the fresh continent. Though some of the land he traveled had been settled (he spent a number of his nights with gentlemen farmers on their plantations), the settlement was sparse, and the fields of indigo and rice gave way quickly to the wilderness. And not the dark and forbidding wilderness of European fairy tales but a blooming, humming, fertile paradise. Every page of Bartram's long journal shouts of the fecundity, the profligacy, of this fresh land. "I continued several miles [reaching] verdant swelling knolls, profusely productive of flowers and fragrant strawberries, their rich juice dyeing my horse's feet and ankles." When he stops for dinner, he catches a trout, picks a wild orange, and stews the first in the juices of the second over his fire.

Whatever direction he struck off in, Bartram found vigorous beauty. He could not even stumble in this New World without discovering something: near the Broad River, while ascending a "steep, rocky hill," he slips and reaching for a shrub to steady himself he tears up several plants of a new species of Caryopbyliata (*Geum odoratissimum*). Fittingly, their roots "filled the air with animating scents of cloves and spicy perfumes." His diary brims over with the grand Latin binomials of a thousand plants and animals—*Kalmia latifolia*, "snowy mantled" *Philadelphia modonts, Pinus sylvestris, Populus tremula, Dionea muscipula* ("admirable are the properties" of these "sportive vegetables"!), *Rheum rhubarbamm, Magnolia grandiflora*—and also with the warm common names: the bank martin, the water wagtail, the mountain cock, the chattering plover, the bumblebee. But the roll call of his adjectives is even more indicative of his mood. On one page, in the account of a single afternoon, he musters fruitful, fragrant, sylvan (twice), moderately warm, exceedingly pleasant, charming, fine, joyful, most beautiful, pale gold, golden, russet, silver (twice), blue green, velvet black, orange, prodigious, gilded, delicious, harmonious, soothing, tuneful, sprightly, elevated, cheerful (twice), high and airy, brisk and cool, clear, moonlit, sweet, and healthy. Even where he can't see, he imagines marvels: the fish disappearing into subterranean streams, "where, probably, they are separated from each other by innumerable paths, or secret rocky avenues, and after encountering various obstacles, and beholding new and unthought-of scenes of pleasure and disgust, after many days' absence from the surface of the world emerge again from the dreary vaults, and appear exulting in gladness and sporting in the transparent waters of some far distant lake." But he is no Disney—this is no Fantasia. He is a scientist recording his observations, and words like "cheerful" and "sweet" seem to have been technical descriptions of the untouched world in which he wandered.

This sort of joy in the natural was not a literary convention, a given; as Paul Brooks points out in *Speaking for Nature*, much of the literature had regarded wilderness as ugly and crude until the Romantic movement of the late eighteenth century. Andrew Marvell, for instance, referred to mountains as "ill-designed excrescences." This silliness changed into a new silliness with the Romantics; Chateaubriand's immensely popular *Atala*, for instance, describes the American wilderness as filled with bears "drunk with grapes, and reeling on the branches of the elm trees." But the rapturous fever took on a healthier aspect in this country. If most of the pioneers, to be sure, saw a buffalo as something to hunt, a forest as something to cut down, a flock of passenger pigeons as a call to heavy artillery (farmers would bring their hogs to feed on the carcasses of pigeons raining down in the slaughter), there were always a good many (even, or especially, among the hunters and loggers) who recognized and described the beauty and order of this early time. . . .

* * *

Such visions of the world as it existed outside human history became scarcer with each year that passed, of course. By the 1930s, when Bob Marshall, the founder of the Wilderness Society, set off to explore Alaska's Brooks Range, all the lower forty-eight states had been visited, mapped, and named. "Often, as when visiting Yosemite or Glacier Park or the Grand Canyon or Avalanche Lake or some other natural scene of surpassing beauty, I had wished selfishly enough that I might have had the joy of being the first person to discover it," he wrote. "I had been thrilled reading Captain Lewis's glowing account of the great falls of the Missouri. I yearned for adventures comparable to those of Lewis and Clark." And he found them, on the upper reaches of the Koyukuk River, where no one, not even an Alaskan Eskimo, seems ever to have been before. Each day brought eight, ten, a dozen ridges and streams and peaks under his eye and hence into human history. One morning he came around a corner to discover that "the Clear River emerged from none of three gorges we had imagined, but from a hidden valley which turned almost at right angles to the west. I cannot convey in words my feeling in finding this broad valley lying there, just as fresh and untrammeled as at the dawn of geological eras hundreds of millions of years ago. Nor is there any adequate way of describing the scenery. . . . I could make mention of thousand-foot sheer precipices; I could liken the valley to a Yosemite without waterfalls, but with rock domes beside which the world-renowned Half Dome would be trivial—yet with all that I would not have conveyed the sense of the continuous, exulting feeling of immensity. . . . Best of all it was fresh—gloriously fresh. At every step there was the exhilarating feeling of breaking new ground. There were no musty signs of human occupation. This, beyond a doubt, was an unbeaten path." . . .

We are rarely reminded anymore of the continent's newness. That era of discovery is as firmly closed to us as the age of knights and dragons. Katahdin, though preserved as a park, is so popular that the authorities must strictly limit the number of campers—some days hundreds are at the summit simultaneously. The trail up Mt. Marcy on a holiday weekend is like the Macy's escalators with a heavy balsam scent. I once interviewed a man who was rowing to Antarctica from Tierra del Fuego because, he explained, "you can't be the first to explore the blank spots on the map or to climb the mountains anymore. It has a lot more to do with style now." (He had previously skied around Mt. Everest.) Not even the moon to conquer!

Over time, though, we've reconciled ourselves to the idea that we'll not be the first up any hill, and, indeed, we've come to appreciate the history of a spot as a source of added pleasure and interest. On the prairies we search for the rutted tracks left by the wagon trains; at Walden Pond, where Thoreau sought to escape man, we dutifully trek around the shore to see the site of the cabin. In something of the same fashion, we have come to accept, and enjoy, the intrusion of scientific explanation—to know that we can marvel with undiminished awe at the south wall of the Grand Canyon even while understanding the geologic forces that carved it. The Grand Canyon is so . . . grand that we can cope with not being the first people to see it. The wonder of nature does not depend on its freshness. . . .

But still we feel the need for pristine places, places substantially unaltered by man. Even if we do not visit them, they matter to us. We need to know that though we are surrounded by buildings there are vast places where the world goes on as it always had. The Arctic National Wildlife Refuge, on Alaska's northern shore, is reached by just a few hundred people a year, but it has a vivid life in the minds of many more, who are upset that oil companies want to drill there. And upset not only because it may or may not harm the caribou but because there is a vast space free of roads and buildings and antennas, a blank spot if not on the map then on the surface. It sickens us to hear that "improper waste disposal practices" at the American Antarctic research station in McMurdo Sound have likely spread toxic waste on that remote continent, or that an Exxon tanker has foundered off the port of Valdez, tarring the beaches with petroleum.

One proof of the deep-rooted desire for pristine places is the decision that Americans and others have made to legislate "wilderness"—to set aside vast tracts of land where, in the words of the federal statute, "the earth and its community of life are untrammeled by man, where man himself is a visitor who does not remain." Pristine nature, we recognize, has been overwhelmed in many places, even in many of our national parks. But in these few spots it makes a stand. If we can't have places where no man has ever been, we can at least have spots where no man is at the moment.

The idea of wildness, in other words, can survive most of the "normal" destruction of nature. Wildness can survive in our minds once the land has been discovered and mapped and even chewed up. It can survive all sorts of pollution, even the ceaseless munching of a million cows. If the ground is dusty and trodden, we look at the sky; if the sky is smoggy, we travel someplace where it's clear; if we can't travel to someplace where it's clear, we imagine ourselves in Alaska or Australia or some place where it is, and that works nearly as well. Nature, while often fragile in reality, is durable in our imaginations. Wildness, the idea of wildness, has outlasted the exploration of the entire globe. It has endured the pesticides and the pollution. When the nature around us is degraded, we picture it fresh and untainted elsewhere. When elsewhere, too, it rains acid or DDT, we can still imagine that someday soon it will be better, that we will stop polluting and despoiling and instead "restore" nature. (And, indeed, people have begun to do just this sort of work: here in the Adirondacks, helicopters drop huge quantities of lime into lakes in order to reduce their acidity.) In our minds, nature suffers from a terrible case of acne, or even skin cancer—but our faith in its essential strength remains, for the damage always seems local. . . . If you travel by plane and dog team and snowshoe to the farthest corner of the Arctic and it is a mild summer day, you will not know whether the temperature is what it is "supposed" to be, or whether, thanks to the extra carbon dioxide, you are standing in the equivalent of a heated room. . . .

It is also true that this is not the first huge rupture in the globe's history. Perhaps thirty times since the earth formed, planetesimals up to ten miles in diameter and traveling at sixty times the speed of sound have crashed into the earth, releasing, according to James Lovelock, perhaps a thousand times as much energy as would be liberated by the explosion of all present stocks of nuclear weapons. Such events, some scientists say, may have destroyed 90 percent of all living organisms. On an even larger scale, the sun has steadily increased its brightness; it has grown nearly 30 percent more luminous since life on earth began, forcing that life to keep forever scrambling to stay ahead—a race it will eventually lose, though perhaps not for some billions of years. Or consider an example more closely resembling the sharp divide we have now crossed. About two billion years ago, the microbiologist Lynn Margulis writes, the spread of certain sorts of bacteria caused, in short order, an increase in atmospheric oxygen from one part in a million to one part in just five—from 0.0001 percent to 21 percent. Compared to that, the increase in carbon dioxide from 280 to 560 parts per million is as the hill behind my house to Annapurna. "This was by far the greatest pollution crisis the earth has ever endured," Margulis writes. Oxygen poisoned most microbial life, which "had no defense against this cataclysm except the standard way of DNA replication and duplication, gene transfer and mutation." And, indeed, these produced the successful oxygen-synthesizing life forms that now dominate the earth.

But each of these examples is different from what we now experience, for they were "natural," as opposed to man made. A pint-sized planet cracks into the earth; the ice advances; the sun, by the immutable laws of stars, burns brighter till its inevitable explosion; genetic mutation sets certain bacteria to spewing out oxygen and soon they dominate the planet, a "strictly natural" pollution.

One can, of course, argue that the current crisis, too, is "natural," because man is part of nature. This echoes the views of the earliest Greek philosophers, who saw no difference between matter and consciousness—nature included everything. James Lovelock wrote some years ago that "our species with its technology is simply an inevitable part of the natural scene," nothing more than mechanically advanced beavers. In this view, to say that we "ended" nature, or even damaged nature, makes no sense, since we are nature, and nothing we can do is "unnatural." This view can be, and is, carried to even greater lengths; Lynn Margulis, for instance, ponders the question of whether robots can be said to be living creatures, since any "invention of human beings is ultimately based on a variety of processes including that of DNA replication, no matter the separation in space or time of that replication from the invention."

But one can argue this forever and still not really feel it. It is a debater's point, a semantic argument. When I say that we have ended nature, I don't mean, obviously, that natural processes have ceased—there is still sunshine and still wind, still growth, still decay. Photosynthesis continues, as does respiration. But we have ended the thing that has, at least in modern times, defined nature for us—its separation from human society.

That separation is quite real. It is fine to argue, as certain poets and biologists have, that we must learn to fit in with nature, to recognize that we are but one species among many, and so on. But none of us, on the inside, quite believe it. The Sophists contrasted the "natural" with the "conventional"—what exists originally with what it becomes as the result of human intervention. And their distinction, filtered through Plato and Christianity and a dozen other screens, survives, because it agrees with our instinctive sense of the world. I sit writing here in my office. On the wall facing me there is a shelf of reference books—dictionaries, the Guinness Book of Records, a set of encyclopedias—and a typewriter and a computer. There's another shelf of books, all dealing with American history, on my left, and, on my right, pictures of my family, a stack of mail-order catalogs for Christmas shopping, and a radio broadcasting a Cleveland performance of Ravel's Piano Concerto in D for the left hand. Visible through the window is a steep mountain with nearly a mile of bare ridge and a pond almost at the peak.

The mountain and the office are separate parts of my life; I do not really think of them as connected. At night it's dark out there; save for the streetlamp by the lake there's not a light for twenty miles to the west and thirty to the south. But in here the light shines. Its beams stretch a few yards into the night and then falter, turn to shadow, then back. In the winter it's cold out there, but in here the fire warms us until near dawn, and when it dwindles the oil burner kicks in.

What happens in here I control; what happens out there has always been the work of some independent force. That is not to say that the outside world isn't vitally important; I moved here so I could get to the mountains easily, and I think nature means a good deal even to the most inured city dweller. But it is enough for now to say that in our modern minds nature and human society are separate things. It is this separate nature I am talking about when I use the word—"nature," if you like.

One could also argue that we destroyed this independent nature long ago, that there's no present need for particular distress. That the day man made his first tool he irrevocably altered nature, or the day he planted his first crop. Walter Truett Anderson, in his recent book *To Govern Evolution*, makes the case that everything people do—including our attempts to set aside wilderness or protect endangered species—is "one way or another human intervention." California, his home, was permanently changed by the 1870s, he contends, when early agribusiness followed gold miners and shepherds. Technically, of course, he is correct. Any action alters its environment—even a bird building a nest—and it is true that we cannot, as he puts it, "return to a natural order untouched by human society." But Anderson's argument, and others like it that have often been employed as a rationale for further altering the environment, is too broad. Independent nature was not dead in California in 1870; in 1870, John Muir was just beginning his sojourn in Yosemite that would yield some of the greatest hymns to and insights into that world beyond man. As long as some places remained free and wild, the idea of the free and wild could live.

2

The Anthropocene

Are Humans Now Overwhelming the Great Forces of Nature?

WILL STEFFEN, PAUL J. CRUTZEN, AND JOHN R. MCNEILL

Introduction

Global warming and many other human-driven changes to the environment are raising concerns about the future of Earth's environment and its ability to provide the services required to maintain viable human civilizations. The consequences of this unintended experiment of humankind on its own life support system are hotly debated, but worst-case scenarios paint a gloomy picture for the future of contemporary societies.

Underlying global change (Box 2.1) are human-driven alterations of *i)* the biological fabric of the Earth; *ii)* the stocks and flows of major elements in the planetary machinery such as nitrogen, carbon, phosphorus, and silicon; and *iii)* the energy balance at the Earth's surface.[1] The term *Anthropocene* (Box 2.2) suggests that the Earth has now left its natural geological epoch, the present interglacial state called the Holocene. Human activities have become so pervasive and profound that they rival the great forces of Nature and are pushing the Earth into planetary *terra incognita*. The Earth is rapidly moving into a less biologically diverse, less forested, much warmer, and probably wetter and stormier state.

The phenomenon of global change represents a profound shift in the relationship between humans and the rest of nature. Interest in this fundamental issue has escalated rapidly in the international research community, leading to innovative new research projects like Integrated History and future of People on Earth (IHOPE).[2] The objective of this paper is to explore one aspect of the IHOPE research agenda—the evolution of humans and our societies from hunter-gatherers to a global geophysical force.

To address this objective, we examine the trajectory of the human enterprise through time, from the arrival of humans on Earth through the present and into the next centuries. Our analysis is based on a few critical questions:

- Is the imprint of human activity on the environment discernible at the global scale? How has this imprint evolved through time?
- How does the magnitude and rate of human impact compare with the natural variability of the Earth's environment? Are human effects similar to or greater than the great forces of nature in terms of their influence on Earth System functioning?

- What are the socioeconomic, cultural, political, and technological developments that change the relationship between human societies and the rest of nature and lead to accelerating impacts on the Earth System?

Pre-Anthropocene Events

Before the advent of agriculture about 10,000–12,000 years ago, humans lived in small groups as hunter-gatherers. In recent centuries, under the influence of noble savage myths, it was often thought that preagricultural humans lived in idyllic harmony with their environment. Recent research has painted a rather

---

BOX 2.1. GLOBAL CHANGE AND THE EARTH SYSTEM

The term *Earth System* refers to the suite of interacting physical, chemical and biological global-scale cycles and energy fluxes that provide the life-support system for life at the surface of the planet.[a] This definition of the Earth System goes well beyond the notion that the geophysical processes encompassing the Earth's two great fluids—the ocean and the atmosphere—generate the planetary life-support system on their own. In our definition biological/ecological processes are an integral part of the functioning of the Earth System and not merely the recipient of changes in the coupled ocean-atmosphere part of the system. A second critical feature is that forcings and feedbacks within the Earth System are as important as external drivers of change, such as the flux of energy from the sun. Finally, the Earth System includes humans, our societies, and our activities; thus, humans are not an outside force perturbing an otherwise natural system but rather an integral and interacting part of the Earth System itself.

We use the term *global change* to mean both the biophysical and the socio-economic changes that are altering the structure and the functioning of the Earth System. Global change includes alterations in a wide range of global-scale phenomena: land use and land cover, urbanization, globalization, coastal ecosystems, atmospheric composition, riverine flow, nitrogen cycle, carbon cycle, physical climate, marine food chains, biological diversity, population, economy, resource use, energy, transport, communication, and so on. Interactions and linkages between the various changes listed above are also part of global change and are just as important as the individual changes themselves. Many components of global change do not occur in linear fashion but rather show strong nonlinearities.

NOTES

[a] F. Oldfield and W. Steffen, "The Earth System," in *Global Change and the Earth System: A Planet under Pressure*, ed. W. Steffen, A. Sanderson, P. Tyson, J. Jäger, P. Matson, B. Moore III, F. Oldfield, K. Richardson, et al., IGBP Global Change Series (Berlin: Springer-Verlag; New York: Heidelberg, 2004), 7.

BOX 2.2. THE ANTHROPOCENE

Holocene ("Recent Whole") is the name given to the postglacial geological epoch of the past ten to twelve thousand years as agreed upon by the International Geological Congress in Bologna in 1885.[a] During the Holocene, accelerating in the industrial period, humankind's activities became a growing geological and morphological force, as recognized early by a number of scientists. Thus, in 1864, Marsh published a book with the title "Man and Nature," more recently reprinted as "The Earth as Modified by Human Action."[b] Stoppani in 1873 rated human activities as a "new telluric force which in power and universality may be compared to the greater forces of earth" (quoted from Clark).[c] Stoppani already spoke of the anthropozoic era. Humankind has now inhabited or visited all places on Earth; he has even set foot on the moon. The great Russian geologist and biologist Vernadsky in 1926 recognized the increasing power of humankind in the environment with the following excerpt: ". . . the direction in which the processes of evolution must proceed, namely towards increasing consciousness and thought, and forms having greater and greater influence on their surroundings."[d] He, the French Jesuit priest P. Teilhard de Chardin and E. Le Roy in 1924 coined the term "noösphere," the world of thought, knowledge society, to mark the growing role played by humankind's brainpower and technological talents in shaping its own future and environment. A few years ago the term "Anthropocene" has been introduced by one of the authors (P.J.C.)[e] for the current geological epoch to emphasize the central role of humankind in geology and ecology. The impact of current human activities is projected to last over very long periods. For example, because of past and future anthropogenic emissions of CO_2, climate may depart significantly from natural behavior over the next 50,000 years.

NOTES

[a] "Holocene," Encyclopædia Britannica: Micropædia, vol. 9 (London: Encyclopædia Britannica, 1976).
[b] G. P. Marsh, The Earth as Modified by Human Action (Cambridge, MA: Belknap Press of Harvard University, 1965), 504.
[c] W. C. Clark, chapter 1 in Sustainable Development of the Biosphere, ed. W. C. Clark and R. E. Munn (Cambridge: Cambridge University Press, 1986), 491.
[d] V. I. Vernadski, The Biosphere (New York: Copernicus, 1998), 192.
[e] P. J. Crutzen, "Geology of Mankind: The Anthropocene," Nature 415 (2002): 23.

different picture, producing evidence of widespread human impact on the environment through predation and the modification of landscapes, often through use of fire.[3] However, as the examples below show, the human imprint on environment may have been discernible at local, regional, and even continental scales, but preindustrial humans did not have the technological or organizational capability to match or dominate the great forces of nature.

The mastery of fire by our ancestors provided humankind with a powerful monopolistic tool unavailable to other species, that put us firmly on the long path towards the Anthropocene. Remnants of charcoal from human hearths indicate that the first use of fire by our bipedal ancestors, belonging to the genus *Homo erectus*, occurred a couple of million years ago. Use of fire followed the earlier development of stone tool and weapon making, another major step in the trajectory of the human enterprise.

Early humans used the considerable power of fire to their advantage.[4] Fire kept dangerous animals at a respectful distance, especially during the night, and helped in hunting protein-rich, more easily digestible food. The diet of our ancestors changed from mainly vegetarian to omnivorous, a shift that led to enhanced physical and mental capabilities. Hominid brain size nearly tripled up to an average volume of about 1300 cm^3, and gave humans the largest ratio between brain and body size of any species.[5] As a consequence, spoken and then, about 10,000 years ago, written language could begin to develop, promoting communication and transfer of knowledge within and between generations of humans, efficient accumulation of knowledge, and social learning over many thousands of years in an impressive catalytic process, involving many human brains and their discoveries and innovations. This power is minimal in other species.

Among the earliest impacts of humans on the Earth's biota are the late Pleistocene megafauna extinctions, a wave of extinctions during the last ice age extending from the woolly mammoth in northern Eurasia to giant wombats in Australia.[6] A similar wave of extinctions was observed later in the Americas. Although there has been vigorous debate about the relative roles of climate variability and human predation in driving these extinctions, there is little doubt that humans played a significant role, given the strong correlation between the extinction events and human migration patterns. A later but even more profound impact of humans on fauna was the domestication of animals, beginning with the dog up to 100,000 years ago[7] and continuing into the Holocene with horses, sheep, cattle, goats, and the other familiar farm animals. The concomitant domestication of plants during the early to mid-Holocene led to agriculture, which initially also developed through the use of fire for forest clearing and, somewhat later, irrigation.[8]

According to one hypothesis, early agricultural development, around the mid-Holocene, affected Earth System functioning so fundamentally that it prevented the onset of the next ice age.[9] The argument proposes that clearing of forests for agriculture about 8000 years ago and irrigation of rice about 5000 years ago led to increases in atmospheric carbon dioxide (CO_2) and methane (CH_4) concentrations, reversing trends of concentration decreases established in the early Holocene. These rates of forest clearing, however, were small compared with the massive amount of land transformation that has taken place in the last 300 years.[10] Nevertheless, deforestation and agricultural development in the 8000 to 5000 BP period may have led to small increases in CO_2 and CH_4 concentrations (maybe about 5–10 parts per million for CO_2) but increases that were perhaps large enough to stop the onset of glaciation in northeast Canada thousands of years ago. However, recent analyses of solar forcing in the late Quaternary[11] and of natural carbon cycle dynamics[12] argue that natural processes can explain the observed pattern of atmospheric CO_2 variation through the Holocene. Thus, the hypothesis that the advent of agriculture thousands of years ago changed the course of glacial-interglacial dynamics remains an intriguing but unproven beginning of the Anthropocene.

The first significant use of fossil fuels in human history came in China during the Song Dynasty (960–1279 AD).[13] Coal mines in the north, notably Shanxi province, provided abundant coal for use in China's growing iron industry. At its height, in the late 11th century, China's coal production reached levels equal to all of Europe (not including Russia) in 1700. But China suffered many setbacks, such as epidemics and invasions, and the coal industry apparently went into a long decline. Meanwhile in England coal mines provided fuel for home heating, notably in London, from at least the 13th century.[14] The first commission charged to investigate the evils of coal smoke began work in 1285.[15] But as a concentrated fuel, coal had its advantages, especially when wood and charcoal grew dear, so by the late 1600s London depended heavily upon it and burned some 360,000 tons annually. The iron forges of Song China and the furnaces of medieval London were regional exceptions, however; most of the world burned wood or charcoal rather than resorting to fuel subsidies from the Carboniferous.

Preindustrial human societies indeed influenced their environment in many ways, from local to continental scales. Most of the changes they wrought were based on knowledge, probably gained from observation and trial-and-error, of natural ecosystem dynamics and its modification to ease the tasks of hunting, gathering, and eventually of farming. Preindustrial societies could and did modify coastal and terrestrial ecosystems but they did not have the numbers, social and economic organization, or technologies needed to equal or dominate the great forces of Nature in magnitude or rate. Their impacts remained largely local and transitory, well within the bounds of the natural variability of the environment.

The Industrial Era (ca. 1800–1945): Stage 1 of the Anthropocene

One of the three or four most decisive transitions in the history of humankind, potentially of similar importance in the history of the Earth itself, was the onset of industrialization. In the footsteps of the Enlightenment, the transition began in the 1700s in England and the Low Countries for reasons that remain in dispute among historians.[16] Some emphasize material factors such as wood shortages and abundant water power and coal in England, while others point to social and political structures that rewarded risk-taking and innovation, matters connected to legal regimes, a nascent banking system, and a market culture. Whatever its origins, the transition took off quickly and by 1850 had transformed England and was beginning to transform much of the rest of the world.

What made industrialization central for the Earth System was the enormous expansion in the use of fossil fuels, first coal and then oil and gas as well. Hitherto humankind had relied on energy captured from ongoing flows in the form of wind, water, plants, and animals, and from the 100- or 200-year stocks held in trees. Fossil fuel use offered access to carbon stored from millions of years of photosynthesis: a massive energy subsidy from the deep past to modern society, upon which a great deal of our modern wealth depends.

Industrial societies as a rule use four or five times as much energy as did agrarian ones, which in turn used three or four times as much as did hunting and gathering societies.[17] Without this transition to a high-energy society it is inconceivable that global population could have risen from a billion around 1820 to more than six billion today, or that perhaps one billion of the more fortunate among us could lead lives of comfort unknown to any but kings and courtiers in centuries past.

Prior to the widespread use of fossil fuels, the energy harvest available to humankind was tightly constrained. Water and wind power were available only in favored locations, and only in societies where the relevant technologies of watermills, sailing ships, and windmills had been developed or imported. Muscular energy derived from animals, and through them from plants, was limited by the area of suitable land for crops and forage, in many places by shortages of water, and everywhere by inescapable biological inefficiencies: plants photosynthesize less than a percent of the solar energy that falls on the Earth, and animals eating those plants retain only a tenth of the chemical energy stored in plants. All this amounted to a bottleneck upon human numbers, the global economy, and the ability of humankind to shape the rest of the biosphere and to influence the functioning of the Earth System.

The invention (some would say refinement) of the steam engine by James Watt in the 1770s and 1780s and the turn to fossil fuels shattered this bottleneck, opening an era of far looser constraints upon energy supply, upon human numbers, and upon the global economy. Between 1800 and 2000, population grew more than six-fold, the global economy about 50-fold, and energy use about 40-fold.[18] It also opened an era of intensified and ever-mounting human influence upon the Earth System.

Fossil fuels and their associated technologies—steam engines, internal combustion engines—made many new activities possible and old ones more efficient. For example, with abundant energy it proved possible to synthesize ammonia from atmospheric nitrogen, in effect to make fertilizer out of air, a process pioneered by the German chemist Fritz Haber early in the 20th century. The Haber-Bosch synthesis, as it would become known (Carl Bosch was an industrialist), revolutionized agriculture and sharply increased crop yields all over the world, which, together with vastly improved medical provisions, made possible the surge in human population growth.

The imprint on the global environment of the industrial era was, in retrospect, clearly evident by the early to mid-20th century.[19] Deforestation and conversion to agriculture were extensive in the midlatitudes, particularly in the northern hemisphere. Only about 10% of the global terrestrial surface had been "domesticated" at the beginning of the industrial era around 1800, but this figure rose significantly to about 25–30% by 1950.[20] Human transformation of the hydrological cycle was also evident in the accelerating number of large dams, particularly in Europe and North America.[21] The flux of nitrogen compounds through the coastal zone had increased over 10-fold since 1800.[22]

The global-scale transformation of the environment by industrialization was, however, nowhere more evident than in the atmosphere. The concentrations of CH_4 and nitrous oxide (N_2O) had risen by 1950 to about 1250 and 288 ppbv, respectively, noticeably above their preindustrial values of about 850 and 272 ppbv.[23] By 1950 the atmospheric CO_2 concentration had pushed above 300 ppmv, above its preindustrial value of 270–275 ppmv, and was beginning to accelerate sharply.[24]

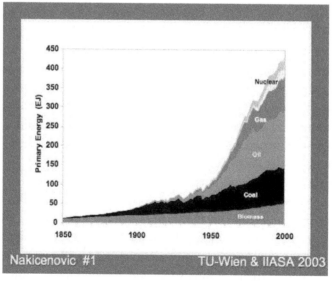

Figure 2.1. The mix of fuels in energy systems at the global scale from 1850 to 2000. Note the rapid relative decrease in traditional renewable energy sources and the sharp rise in fossil-fuel-based energy systems since the beginning of the Industrial Revolution, and particularly after 1950. By 2000 fossil-fuel-based energy systems generated about 80% of the total energy used to power the global economy.

TABLE 2.1. Atmospheric CO_2 Concentration during the Existence of Fully Modern Humans on Earth

Year/Period	Atmospheric CO_2 concentration (ppmv)[a]
250,000–12,000 years BP:[b]	
Range during interglacial periods:	262–287
Minimum during glacial periods:	182
12,000–2,000 years BP:	260–285
Holocene (current interglacial)	
1000	279
1500	282
1600	276
1700	277
1750	277
1775	279
1800 (Anthropocene Stage I begins)	283
1825	284
1850	285
1875	289
1900	296
1925	305
1950 (Anthropocene Stage II begins)	311
1975	331
2000	369
2005	379

[a] The CO_2 concentration data were obtained from: (1) http:/cdiac.ornl.gov/trends/trends.htm for the 250,000–12,000 BP period and for the 1000 AD–2005 AD period. More specifically, data were obtained from R. Costanza, L. Graumlich, and W. Steffen, eds., *Integrated History and Future of People on Earth*, Dahlem Workshop Report 96 (Cambridge, MA: MIT Press, 2006), 495 (25,000–12,000 BP), A. Indermuhle, T. F. Stocker, H. Fischer, H. J. Smith, F. Joos, M. Wahlen, B. Deck, D. Mastroianni, et al., "High-Resolution Holocene CO_2 Record from the Taylor Dame Ice Core (Antarctica)," *Nature* 398 (1999): 121–126 (12,000–2,000 BP), J.-M. Barnola, D. Raynaud, C. Lorius, and N. I. Barkov, "Historical CO2 Record from the Vostok Ice Core," in *Trends: A Compendium of Data on Global Change* (Oak Ridge, TN: Carbon Dioxide Information Analysis Center, Oak Ridge National Laboratory, U.S. Department of Energy, 2003) (1000–1950 AD), and D. M. Etheridge, L. P. Steele, R. I. Langenfelds, R. J. Francey, J.-M. Barnola, and V.I. Morgan, "Historical CO2 Records from the Law Dome DE08, DE08-2, and DSS Ice Cores," in *Trends: A Compendium of Data on Global Change* (Oak Ridge, TN: Carbon Dioxide Information Analysis Center, Oak Ridge National Laboratory, U.S. Department of Energy, 1998) (1975–2000 AD). (2) CO_2 concentrations for the 12,000–2,000 BP period (the Holocene) were obtained from C. D. Keeling and T. P. Whorf, "Atmospheric CO2 Records from Sites in the SIO Air Sampling Network," in *Trends: A Compendium of Data on Global Climate Change* (Oak Ridge, TN: Carbon Dioxide Information Analysis Center, Oak Ridge National Laboratory, U.S. Department of Energy, 2005).
[b] The period 250,000–12,000 years BP encompasses two interglacial periods prior to the current interglacial (the Holocene) and two glacial periods. The values listed in the table are the maximum and minimum CO_2 concentrations recorded during the two interglacial periods and the minimum CO_2 concentration recorded over the two glacial periods. According to mtDNA evidence, the first appearance of fully modern humans was approximately 250,000 years BP.

Quantification of the human imprint on the Earth System can be most directly related to the advent and spread of fossil-fuel-based energy systems (Fig. 2.1), the signature of which is the accumulation of CO_2 in the atmosphere roughly in proportion to the amount of fossil fuels that have been consumed. We propose that atmospheric CO_2 concentration can be used as a single, simple indicator to track the progression of the Anthropocene, to define its stages quantitatively, and to compare the human imprint on the Earth System with natural variability (Table 2.1).

Around 1850, near the beginning of Anthropocene Stage 1, the atmospheric CO_2 concentration was 285 ppm, within the range of natural variability for inter-glacial periods during the late Quaternary period. During the course of Stage 1 from 1800/50 to 1945, the CO_2 concentration rose by about 25 ppm, enough to surpass the upper limit of natural variation through the Holocene and thus provide the first indisputable evidence that human activities were affecting the environment at the global scale. We therefore assign the beginning of the Anthropocene to coincide with the beginning of the industrial era, in the 1800–1850 period. This first stage of the Anthropocene ended abruptly around 1945, when the most rapid and pervasive shift in the human-environment relationship began.

The Great Acceleration (1945–ca. 2015): Stage 2 of the Anthropocene

The human enterprise suddenly accelerated after the end of the Second World War[25] (Fig. 2.2). Population doubled in just 50 years, to over 6 billion by the end of the 20th century, but the global economy increased by more than 15-fold. Petroleum consumption has grown by a factor of 3.5 since 1960, and the number of motor vehicles increased dramatically from about 40 million at the end of the War to nearly 700 million by 1996. From 1950 to 2000 the percentage of the world's population living in urban areas grew from 30% to 50% and continues to grow strongly. The interconnectedness of cultures is increasing rapidly with the explosion in electronic communication, international travel and the globalization of economies.

The pressure on the global environment from this burgeoning human enterprise is intensifying sharply. Over the past 50 years, humans have changed the world's ecosystems more rapidly and extensively than in any other comparable period in human history.[26] The Earth is in its sixth great extinction event, with rates of species loss growing rapidly for both terrestrial and marine ecosystems.[27] The atmospheric concentrations of several important greenhouse gases have increased substantially, and the Earth is warming rapidly.[28] More nitrogen is now converted from the atmosphere into reactive forms by fertilizer production and fossil fuel combustion than by all of the natural processes in terrestrial ecosystems put together[29] (Fig. 2.3).

The remarkable explosion of the human enterprise from the mid-20th century, and the associated global-scale impacts on many aspects of Earth System functioning, mark the second stage of the Anthropocene—the Great Acceleration.[30] In many respects the stage had been set for the Great Acceleration by 1890 or 1910. Population growth was proceeding faster than at any previous time in human history, as well as economic growth. Industrialization had gathered irresistible momentum, and was spreading quickly in North America, Europe, Russia, and Japan. Automobiles and airplanes had appeared, and soon rapidly

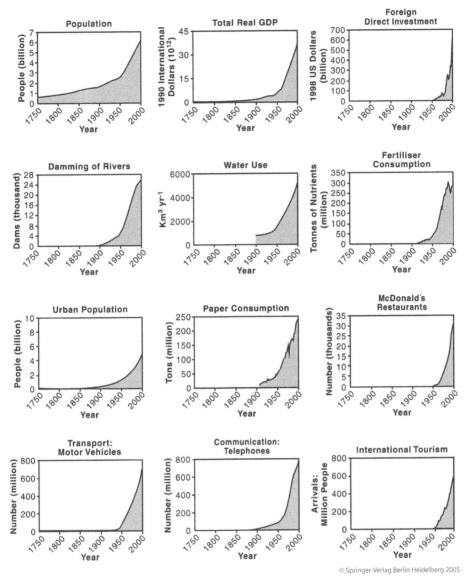

Figure 2.2. The change in the human enterprise from 1750 to 2000. The Great Acceleration is clearly shown in every component of the human enterprise included in the figure. Either the component was not present before 1950 (e.g., foreign direct investment) or its rate of change increased sharply after 1950 (e.g., population). Source: W. Steffen, A. Sanderson, P. D. Tyson, J. Jager, P. Matson, B. Moore III, F. Oldfield, K. Richardson, et al., *Global Change and the Earth System: A Planet Under Pressure*, IGBP Global Change Series (Berlin: Springer-Verlag; New York: Heidelberg, 2004).

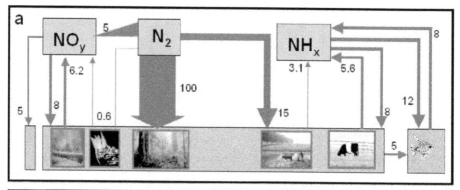

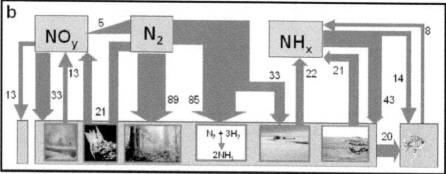

Figure 2.3. Global terrestrial nitrogen budget for (a) 1890 and (b) 1990 in Tg N yr⁻¹. The emissions to the NOy box from the coal reflect fossil fuel combustion. Those from the vegetation include agricultural and natural soil emissions and combustion of biofuel, biomass (savanna and forest) and agricultural waste. The NHx emissions from the cow and feedlot reflect emissions from animal wastes. The transfers to the fish box represent the lateral flow of dissolved inorganic nitrogen from terrestrial systems to the coastal seas. Note the enormous amount of N_2 converted to NH_3 in the 1990 panel compared to the 1980. This represents human fixation of nitrogen through the Haber-Bosch process, made possible by the development of fossil-fuel-based energy systems. Source: K. A. Hibbard, P. J. Crutzen, E. F. Lambin, D. Liverman, N. J. Mantua, J. R. McNeill, B. Messerli, and W. Steffen, "Decadal Interactions of Humans and the Environment," in *Integrated History and Future of People on Earth*, Dahlem Workshop Report 96, ed. R. Costanza, L. Graumlich, and W. Steffen (Cambridge, MA: MIT Press, 2006), 341–375.

transformed mobility. The world economy was growing ever more tightly linked by mounting flows of migration, trade, and capital. The years 1870 to 1914 were, in fact, an age of globalization in the world economy. Mines and plantations in diverse lands such as Australia, South Africa, and Chile were opening or expanding in response to the emergence of growing markets for their products, especially in the cities of the industrialized world.

At the same time, cities burgeoned as public health efforts, such as checking waterborne disease through sanitation measures, for the first time in world history made it feasible for births consistently to outnumber deaths in urban environments. A major transition was underway in which the characteristic habitat of the human species, which for several millennia had been the village, now was becoming the city. (In 1890 perhaps 200 million people lived in cities worldwide, but by 2000 the figure had leapt to three billion, half of the human population.) Cities had long been the seats of managerial and technological innovation and engines of economic growth, and in the Great Acceleration played that role with even greater effect.

However, the Great Acceleration truly began only after 1945. In the decades between 1914 and 1945 the Great Acceleration was stalled by changes in politics and the world economy. Three great wrenching events lay behind this: World War I, the Great Depression, and World War II. Taken together, they slowed population growth, checked—indeed temporarily reversed—the integration and growth of the world economy. They also briefly checked urbanization, as city populations led the way in reducing their birth rates. Some European cities in the 1930s in effect went on reproduction strikes, so that (had they maintained this reluctance) they would have disappeared within decades. Paradoxically, however, these events also helped to initiate the Great Acceleration.

The lessons absorbed about the disasters of world wars and depression inspired a new regime of international institutions after 1945 that helped create conditions for resumed economic growth. The United States in particular championed more open trade and capital flows, reintegrating much of the world economy and helping growth rates reach their highest ever levels in the period from 1950 to 1973. At the same time, the pace of technological change surged. Out of World War II came a number of new technologies—many of which represented new applications for fossil fuels—and a commitment to subsidized research and development, often in the form of alliances among government, industry, and universities. This proved enormously effective and, in a climate of renewed prosperity, ensured unprecedented funding for science and technology, unprecedented recruitment into these fields, and unprecedented advances as well.

The Great Acceleration took place in an intellectual, cultural, political, and legal context in which the growing impacts upon the Earth System counted for very little in the calculations and decisions made in the world's ministries, boardrooms, laboratories, farmhouses, village huts, and, for that matter, bedrooms. This context was not new, but it too was a necessary condition for the Great Acceleration.

The exponential character of the Great Acceleration is obvious from our quantification of the human imprint on the Earth System, using atmospheric CO_2 concentration as the indicator (Table 2.1). Although by the Second World War the CO_2 concentration had clearly risen above the upper limit of the Holocene, its growth rate hit a take-off point around 1950. Nearly three-quarters of the anthropogenically driven rise in CO_2 concentration has occurred since 1950 (from about 310 to 380 ppm), and about half of the total rise (48 ppm) has occurred in just the last 30 years.

Stewards of the Earth System? (ca. 2015–?): Stage 3 of the Anthropocene

Humankind will remain a major geological force for many millennia, maybe millions of years, to come. To develop a universally accepted strategy to ensure the sustainability of Earth's life support system against human-induced stresses is one of the greatest research and policy challenges ever to confront humanity. Can humanity meet this challenge?

Signs abound to suggest that the intellectual, cultural, political and legal context that permitted the Great Acceleration after 1945 has shifted in ways that could curtail it.[31] Not surprisingly, some reflective people noted human impact upon the environment centuries and even millennia ago. However, as a major societal concern it dates from the 1960s with the rise of modern environmentalism. Observations showed incontrovertibly that the concentration of CO_2 in the atmosphere was rising markedly.[32] In the 1980s temperature measurements showed global warming was a reality, a fact that encountered political opposition because of its implications, but within 20 years was no longer in serious doubt.[33] Scientific observations showing the erosion of the earth's stratospheric ozone layer led to international agreements reducing the production and use of CFCs (chlorofluorocarbons).[34] On numerous ecological issues local, national, and international environmental policies were devised, and the environment routinely became a consideration, although rarely a dominant one, in political and economic calculations.

This process represents the beginning of the third stage of the Anthropocene, in which the recognition that human activities are indeed affecting the structure and functioning of the Earth System as a whole (as opposed to local- and regional-scale environmental issues) is filtering through to decision-making at many levels. The growing awareness of human influence on the Earth System has been aided by *i)* rapid advances in research and understanding, the most innovative of which is interdisciplinary work on human-environment systems; *ii)* the enormous power of the internet as a global, self-organizing information system; *iii)* the spread of more free and open societies, supporting independent media; and *iv)* the growth of democratic political systems, narrowing the scope for the exercise of arbitrary state power and strengthening the role of civil society. Humanity is, in one way or another, becoming a self-conscious, active agent in the operation of its own life support system.[35]

This process is still in train, and where it may lead remains quite uncertain. However, three broad philosophical approaches can be discerned in the growing debate about dealing with the changing global environment.[36]

Business-as-Usual

In this conceptualization of the next stage of the Anthropocene, the institutions and economic system that have driven the Great Acceleration continue to dominate human affairs. This approach is based on several assumptions. First, global change will not be severe or rapid enough to cause major disruptions to the global economic system or to other important aspects of societies, such as human health. Second, the existing market-oriented economic system can deal autonomously with any adaptations that are required. This assumption is based on the fact that as societies have become wealthier, they have dealt effectively with some local and regional pollution problems.[37] Examples include the clean-up of major European rivers and the amelioration of the acid rain problem in western Europe and eastern North America. Third, resources required to mitigate global change proactively would be better spent on more pressing human needs.

The business-as-usual approach appears, on the surface, to be a safe and conservative way forward. However, it entails considerable risks. As the Earth System changes in response to human activities, it operates at a time scale that is mismatched with human decision-making or with the workings of the economic system. The long-term momentum built into the Earth System means that by the time humans realize that a business-as-usual approach may not work, the world will be committed to further decades or even centuries of environmental change. Collapse of modern, globalized society under uncontrollable environmental change is one possible outcome.

An example of this mismatch in time scales is the stability of the cryosphere, the ice on land and ocean and in the soil. Depending on the scenario and the model, the Intergovernmental Panel on Climate Change (IPCC)[38] projected a global average warming of 1.1–6.4°C for 2094–2099 relative to 1980–1999, accompanied by a projected sea-level rise of 0.18–0.59 m (excluding contributions from the dynamics of the large polar ice sheets). However, warming is projected to be more than twice as large as the global average in the polar regions, enhancing ice sheet instability and glacier melting. Recent observations of glacial dynamics suggest a higher degree of instability than estimated by current cryospheric models, which would lead to higher sea-level rise through this century than estimated by the IPCC in 2001.[39] It is now conceivable that an irreversible threshold could be crossed in the next several decades, eventually (over centuries or a millennium) leading to the loss of the Greenland ice sheet and consequent sea-level rise of about 5 m.

Mitigation

An alternative pathway into the future is based on the recognition that the threat of further global change is serious enough that it must be dealt with proactively. The mitigation pathway attempts to take the human pressure off of the Earth System by vastly improved technology and management, wise use of Earth's resources, control of human and domestic animal population, and overall careful use and restoration of the natural environment. The ultimate goal is to reduce the human modification of the global environment to avoid dangerous or difficult-to-control levels and rates of change,[40] and ultimately to allow the Earth System to function in a pre-Anthropocene way.

Technology must play a strong role in reducing the pressure on the Earth System.[41] Over the past several decades rapid advances in transport, energy, agriculture, and other sectors have led to a trend of dematerialization in several advanced economies. The amount and value of economic activity continue to grow but the amount of physical material flowing through the economy does not.

There are further technological opportunities. Worldwide energy use is equivalent to only 0.05% of the solar radiation reaching the continents. Only 0.4% of the incoming solar radiation, 1 W m^{-2}, is converted to chemical energy by photosynthesis on land. Human appropriation of net primary production is about 10%, including agriculture, fiber, and fisheries.[42] In addition to the many opportunities for energy conservation, numerous technologies—from solar thermal and photovoltaic through nuclear fission and fusion to wind power and biofuels from forests and crops—are available now or under development to replace fossil fuels.

Although improved technology is essential for mitigating global change, it may not be enough on its own. Changes in societal values and individual behavior will likely be necessary.[43] Some signs of these changes are now evident, but the Great Acceleration has considerable momentum and appears to be intensifying.[44] The critical question is whether the trends of dematerialization and shifting societal values become strong enough to trigger a transition of our globalizing society towards a much more sustainable one.

Geo-engineering Options

The severity of global change, particularly changes to the climate system, may force societies to consider more drastic options. For example, the anthropogenic emission of aerosol particles (e.g., smoke, sulphate, dust, etc.) into the atmosphere leads to a net cooling effect because these particles and their influence on cloud properties enhance backscattering of incoming solar radiation. Thus, aerosols act in opposition to the greenhouse effect, masking some of the warming we would otherwise see now.[45] Paradoxically, a clean-up of air pollution can thus increase greenhouse warming, perhaps leading to an additional 1°C of warming and bringing the Earth closer to "dangerous" levels of climate change. This and other amplifying effects, such as feedbacks from the carbon cycle as the Earth warms,[46] could render mitigation efforts largely ineffectual. Just to stabilize the atmospheric concentration of CO_2, without taking into account these amplifying effects, requires a reduction in anthropogenic emissions by more than 60%—a herculean task considering that most people on Earth, in order to increase their standard of living, are in need of much additional energy. One engineering approach to reducing the amount of CO_2 in the atmosphere is its sequestration in underground reservoirs.[47] This "geo-sequestration" would not only alleviate the pressures on climate, but would also lessen the expected acidification of the ocean surface waters, which leads to dissolution of calcareous marine organisms.[48]

In this situation some argue for geo-engineering solutions, a highly controversial topic. Geo-engineering involves purposeful manipulation by humans of global-scale Earth System processes with the intention of counteracting anthropogenically driven environmental change such as greenhouse warming.[49] One proposal is based on the cooling effect of aerosols noted in the previous paragraph.[50] The idea is to artificially enhance the Earth's albedo by releasing sunlight-reflective material, such as sulphate particles, in the stratosphere, where they remain for 1–2 years before settling in the troposphere. The sulphate particles would be produced by the oxidation of SO_2, just as happens during volcanic eruptions. In order to compensate for a doubling of CO_2, if this were to happen, the input of sulphur would have to be about 1–2 Tg S y^{-1} (compared to an input of about 10 Tg S by Mount Pinatubo in 1991). The sulphur injections would have to occur for as long as CO_2 levels remain high.

Looking more deeply into the evolution of the Anthropocene, future generations of *H. sapiens* will likely do all they can to prevent a new ice age by adding powerful artificial greenhouse gases to the atmosphere. Similarly, any drop in CO_2 levels to low concentrations, causing strong reductions in photosynthesis and agricultural productivity, might be combated by artificial releases of CO_2, maybe from earlier CO_2 sequestration. And likewise, far into the future, *H. sapiens* will deflect meteorites and asteroids before they could hit the Earth.

For the present, however, just the suggestion of geo-engineering options can raise serious ethical questions and intense debate. In addition to fundamental ethical concerns, a critical issue is the possibility for unintended and unanticipated side effects that could have severe consequences. The cure could be worse than the disease. For the sulphate injection example described above, the residence time of the sulphate particles in the atmosphere is only a few years, so if serious side effects occurred, the injections could be discontinued and the climate would relax to its former high CO_2 state within a decade.

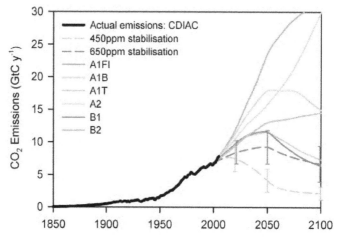

Figure 2.4. The observed trajectory from 1850 to 2005 of carbon emissions due to fossil fuel combustion. Note the acceleration in emissions since 2000. The gap between current emission rates and those required to stabilize atmospheric CO_2 concentration at various levels (450, 650, and 1000 ppm) is growing rapidly. Source: M. R. Raupach, G. Marland, P. Ciais, C. Le Quere, J. G. Canadell, G. Klepper, and C. B. Field, "Global and Regional Drivers of Accelerating CO_2 Emissions," *Proc. Nat. Acad. Sci. USA* 104, no. 24 (2007): 10288–10293.

The Great Acceleration is reaching criticality (Fig. 2.4). Enormous, immediate challenges confront humanity over the next few decades as it attempts to pass through a bottleneck of continued population growth, excessive resource use and environmental deterioration. In most parts of the world the demand for fossil fuels overwhelms the desire to significantly reduce greenhouse gas emissions. About 60% of ecosystem services are already degraded and will continue to degrade further unless significant societal changes in values and management occur.[51] There is also evidence for radically different directions built around innovative, knowledge-based solutions. Whatever unfolds, the next few decades will surely be a tipping point in the evolution of the Anthropocene.

NOTES

1. J. Hansen, L. Nazarenko, R. Ruedy, M. Sato, J. Willis, A. Del Genio, D. Koch, A. Lacis, et al., "Earth's Energy Imbalance: Confirmation and Implications," *Science* 308 (2005): 1431–1435.

2. R. Costanza, L. Graumlich, and W. Steffen, eds., *Integrated History and Future of People on Earth*, Dahlem Workshop Report 96 (Cambridge, MA: MIT Press, 2006).

3. S. Pyne, *World Fire: The Culture of Fire on Earth* (Seattle: University of Washington Press, 1997).

4. Ibid.

5. P. V. Tobias, "The Brain in Hominid Evolution," in *Encyclopædia Britannica: Macropædia*, vol. 5 (London: Encyclopædia Britannica, 1976), 1032.

6. P. S. Martin and R. G. Klein, *Quaternary Extinctions: A Prehistoric Revolution* (Tucson: University of Arizona Press, 1984), 892; J. Alroy, "A Multispecies Overkill Simulation of the End-Pleistocene Megafaunal Mass Extinction," *Science* 292 (2001): 1893–1896; R. G. Roberts, T. F. Flannery, L. K. Ayliffe, H. Yoshida, J. M. Olley, G. J. Prideaux, G. M. Laslett, A. Baynes, et al., "New Ages for the Last Australian Megafauna: Continent-Wide Extinction About 46,000 Years Ago," *Science* 292 (2001): 1888–1892.

7. H. M. Leach, "Human Domestication Reconsidered," *Curr. Anthropol.* 44 (2003): 349–368.

8. B. D. Smith, *The Emergence of Agriculture*, Scientific American Library (New York and Oxford, UK: W. H. Freeman, 1995), 231.

9. W. F. Ruddiman, "The Anthropogenic Greenhouse Era Began Thousands of Years Ago," *Climate Change* 61 (2003): 261–293.

10. E. F. Lambin and H. J. Geist, eds., *Land-Use and Land-Cover Change: Local Processes and Global Impacts*, IGBP Global Change Series (Berlin: Springer-Verlag; New York: Heidelberg, 2006).

11. EPICA Community Members, "Eight Glacial Cycles from an Antarctic Ice Core," *Nature* 429 (2004): 623–628.

12. W. C. Broecker and T. F. Stocker, "The Holocene CO_2 Rise: Anthropogenic or Natural?," *Eos* 87, no. 3 (2006): 27–29; F. Joos, S. Gerber, I. C. Prentice, B. L. Otto-Bliesner, and P. J. Valdes, "Transient Simulations of Holocene Atmospheric Carbon Dioxide and Terrestrial Carbon since the Last Glacial Maximum," *Global Biogeochemical Cycles* 18 (2004): GB2002.

13. R. Hartwell, "A Revolution in the Iron and Coal Industries during the Northern Sung," *J. Asian Stud.* 21 (1962): 153–162; R. Hartwell, "A Cycle of Economic Change in Imperial China: Coal and Iron in Northeast China, 750–1350," *J. Soc. and Econ. Hist. Orient* 10 (1967): 102–159.

14. W. H. TeBrake, "Air Pollution and Fuel Crisis in Preindustrial London, 1250–1650," *Technol. Culture* 16 (1975): 337–359.

15. P. Brimblecombe, *The Big Smoke: A History of Air Pollution in London since Medieval Times* (London: Methuen, 1987), 185.

16. J. Mokyr, ed., *The British Industrial Revolution: An Economic Perspective* (Boulder, CO: Westview, 1999).

17. R. P. Sieferle and H. Breuninger, *Der Europäische Sonderweg: Ursachen und Factoren* (Stuttgart: Breuninger-Stiftung, 2001), 53 (in German).

18. J. R. McNeill, *Something New under the Sun* (New York and London: Norton, 2001), 416.

19. W. Steffen, A. Sanderson, P. D. Tyson, J. Jäger, P. Matson, B. Moore III, F. Oldfield, K. Richardson, et al., *Global Change and the Earth System: A Planet under Pressure*, IGBP Global Change Series (Berlin: Springer-Verlag; New York: Heidelberg, 2004), 336.

20. E. F. Lambin, and H. J. Geist, *Land-Use and Land-Cover Change: Local Processes and Global Impacts*, IGBP Global Change Series (Berlin: Springer-Verlag; New York: Heidelberg, 2006), 222.

21. C. J. Vörösmarty, K. Sharma, B. Fekete, A. H. Copeland, J. Holden, J. Marble, and J. A. Lough, "The Storage and Aging of Continental Runoff in Large Reservoir Systems of the World," *Ambio* 26 (1997): 210–219.

22. F. T. Mackenzie, L. M. Ver, and A. Lerman, "Century-Scale Nitrogen and Phosphorus Controls of the Carbon Cycle," *Chem. Geol.* 190 (2002): 13–32.

23. T. Blunier, J. Chappellaz, J. Schwander, J.-M. Barnola, T. Desperts, B. Stauffer, and D. Raynaud, "Atmospheric Methane Record from a Greenland Ice Core over the Last 1000 Years," *J. Geophys. Res.* 20 (1993): 2219–2222; T. Machida, T. Nakazawa, Y. Fujii, S. Aoki, and O. Watanabe, "Increase in the Atmospheric Nitrous Oxide Concentration during the Last 250 Years," *Geophys. Res. Lett.* 22 (1995): 2921–2924.

24. D. M. Etheridge, L. P. Steele, R. L. Langenfelds, R. J. Francey, J-M. Barnola, and V. I. Morgan, "Natural and Anthropogenic Changes in Atmospheric CO_2 over the Last 1000 Years from Air in Antarctic Ice and Firn," *J. Geophys. Res.* 101 (1996): 4115–4128.

25. J. R. McNeill, *Something New under the Sun* (New York: Norton, 2001), 416.

26. Millennium Ecosystem Assessment, *Ecosystems & Human Wellbeing: Synthesis* (Washington, DC: Island, 2005).

27. S. L. Pimm, G. J. Russell, J. L. Gittleman, and T. M. Brooks, "The Future of Biodiversity," *Science* 269 (1995): 347–350.

28. Intergovernmental Panel on Climate Change (IPCC), *Climate Change 2007: The Physical Science Basis*, Summary for Policymakers (Geneva, Switzerland: IPCC Secretariat, World Meteorological Organization, 2007), 18.

29. J. N. Galloway and E. B. Cowling, "Reactive Nitrogen and the World: Two Hundred Years of Change," *Ambio* 31 (2002): 64–71.

30. K. A. Hibbard, P. J. Crutzen, E. F. Lambin, D. Liverman, N. J. Mantua, J. R. McNeill, B. Messerli, and W. Steffen, "Decadal Interactions of Humans and the Environment," in *Integrated History and Future of People on Earth*, Dahlem Workshop Report 96, ed. R. Costanza, L. Graumlich, and W. Steffen (Cambridge, MA: MIT Press, 2006), 341–375.

31. Ibid.

32. C. D. Keeling, and T. P. Whorf, "Atmospheric CO_2 Records from Sites in the SIO Air Sampling Network," in *Trends: A Compendium of Data on Global Change* (Oak Ridge, TN: Carbon Dioxide Information Analysis Center, Oak Ridge National Laboratory, U.S. Department of Energy, 2005).

33. Intergovernmental Panel on Climate Change (IPCC), *Climate Change 2007: The Physical Science Basis*, Summary for Policymakers (Geneva, Switzerland: IPCC Secretariat, World Meteorological Organization, 2007), 18.

34. P. Crutzen, "My Life with O_3, NOx and Other YZOxs," in *Les Prix Nobel (The Nobel Prizes) 1995* (Stockholm: Almqvist and Wiksell International, 1995), 123–157.

35. H.-J. Schellnhuber, "Discourse: Earth System Analysis: The Scope of the Challenge," in *Earth System Analysis*, ed. H.-J. Schellnhuber and V. Wetzel (Berlin: Springer-Verlag; New York: Heidelberg, 2004), 3–195.

36. W. Steffen, A. Sanderson, P. D. Tyson, J. Jäger, P. Matson, B. Moore III, F. Oldfield, K. Richardson, et al., "Global Change and the Earth System: A Planet under Pressure," IGBP Global Change Series (Berlin: Springer-Verlag; New York: Heidelberg, 2004), 336; H.-J. Schellnhuber, "Discourse: Earth System Analysis: The Scope of the Challenge," in *Earth System Analysis*, ed. H.-J. Schellnhuber and V. Wetzel (Berlin: Springer-Verlag; New York: Heidelberg, 1998), 3–195.

37. B. Lomborg, *The Skeptical Environmentalist: Measuring the Real State of the World* (Cambridge: Cambridge University Press, 2001), 548.

38. Intergovernmental Panel on Climate Change (IPCC), *Climate Change 2007: The Physical Science Basis*, Summary for Policymakers (Geneva, Switzerland: IPCC Secretariat, World Meteorological Organization, 2007),18.

39. S. Rahmstorf, "A Semi-empirical Approach to Projecting Future Sea-Level Rise," *Science* 315 (2007): 368–370.

40. H.-J. Schellnhuber, W. Cramer, N. Nakicenovic, T. Wigley, and G. Yohe, eds., *Avoiding Dangerous Climate Change* (Cambridge: Cambridge University Press, 2006).

41. W. Steffen, "Will Technology Spare the Planet?," in *Challenges of a Changing Earth: Proceedings of the Global Change Open Science Conference: Amsterdam, The Netherlands, 10–13 July 2001*, ed. W. Steffen, J. Jäger, D. Carson, and C. Bradshaw, IGBP Global Change Series (Berlin: Springer-Verlag; New York: Heidelberg, 2002), 189–191.

42. H. Haberl, "The Energetic Metabolism of the European Union and the United States, Decadal Energy Inputs with an Emphasis on Biomass," *J. Ind. Ecol.* 10 (2006): 151–171.

43. J. Fischer, A. D. Manning, W. Steffen, D. B. Rose, K. Danielle, A. Felton, S. Garnett, B. Gilna, et al., "Mind the Sustainability Gap," *Trends Ecol. Evol.* 22, no. 12 (2007): 621–624.

44. S. Rahmstorf, A. Cazenave, J. A. Church, J. E. Hansen, R. F. Keeling, D. E. Parker, R. C. J. Somerville, et al., "Recent Climate Observations Compared to Projections," *Science* 316 (2007): 709.

45. M. O. Andreae, C. D. Jones, and P. M. Cox, "Strong Present-Day Aerosol Cooling Implies a Hot Future," *Nature* 435 (2005): 1187–1190.

46. P. Friedlingstein, P. Cox, R. Betts, L. Bopp, W. von Bloh, V. Brovkin, V. S. Doney, M. I. Eby, et al., "Climate-Carbon Cycle Feedback Analysis: Results from the C⁴MIP Model Intercomparison," *J. Clim.* 19 (2006): 3337–3353.

47. Intergovernmental Panel on Climate Change (IPCC), *Carbon Dioxide Capture and Storage,* A Special Report of Working Group III (Geneva, Switzerland: Intergovernmental Panel on Climate Change, 2005), 430.

48. The Royal Society, *Ocean Acidification Due to Increasing Atmospheric Carbon Dioxide,* policy document 12/05 (London: The Royal Society, 2005), 68.

49. S. H. Schneider, "Earth Systems Engineering and Management," *Nature* 409 (2001): 417–421.

50. P. J. Crutzen, "Albedo Enhancement by Stratospheric Sulfur Injections: A Contribution to Resolve a Policy Dilemma," *Clim. Chang.* 77 (2006): 211–220.

51. Millennium Ecosystem Assessment, *Ecosystems & Human Well-Being: Synthesis* (Washington, DC: Island, 2005).

3

Excerpts from *The World without Us*

ALAN WEISMAN

The notion that someday nature could swallow whole something so colossal and concrete as a modern city doesn't slide easily into our imaginations. The sheer titanic presence of a New York City resists efforts to picture it wasting away. The events of September 2001 showed only what human beings with explosive hardware can do, not crude processes like erosion or rot. The breathtaking, swift collapse of the World Trade Center towers suggested more to us about their attackers than about mortal vulnerabilities that could doom our entire infrastructure. And even that once-inconceivable calamity was confined to just a few buildings. Nevertheless, the time it would take nature to rid itself of what urbanity has wrought may be less than we might suspect. . . .

When Dr. Eric Sanderson leads a tour through the park, he and his flock usually pass Jagiello without pausing, because they are lost in another century altogether—the 17th. Bespectacled under his wide-brimmed felt hat, a trim beard graying around his chin and a laptop jammed in his backpack, Sanderson is a landscape ecologist with the Wildlife Conservation Society, a global squadron of researchers trying to save an imperiled world from itself. At its Bronx Zoo headquarters, Sanderson directs the Mannahatta Project, an attempt to re-create, virtually, Manhattan Island as it was when Henry Hudson's crew first saw it in 1609: a pre-urban vision that tempts speculation about how a posthuman figure might look. . . .

Eric Sanderson sees water flowing everywhere in town, much of it bubbling from underground ("which is how Spring Street got its name"). He's identified more than 40 brooks and streams that traversed what was once a hilly, rocky island: in the Algonquin tongue of its first human occupants, the Lenni Lenape, *Mannahatta* referred to those now-vanished hills. When New York's 19th-century planners imposed a grid on everything north of Greenwich Village— the jumble of original streets to the south being impossible to unsnarl—they behaved as if topography were irrelevant. Except for some massive, unmovable schist outcrops in Central Park and at the island's northern tip, Manhattan's textured terrain was squashed and dumped into streambeds, then planed and leveled to receive the advancing city.

Later, new contours arose, this time routed through rectilinear forms and hard angles, much as the water that once sculpted the island's land was now forced underground through a lattice of pipes. Eric Sanderson's Mannahatta Project has plotted how closely the modern sewer system follows the old water-courses, although man-made sewer lines can't wick away runoff as efficiently as nature. In a city that buried its rivers, he observes, "rain still falls. It has to go somewhere."

As it happens, that will be the key to breaching Manhattan's hard shell if nature sets about dismantling it. It would begin very quickly, with the first strike at the city's most vulnerable spot: its underbelly.

*

New York Transit's Paul Schuber and Peter Briffa, superintendent of Hydraulics and level one maintenance supervisor of Hydraulics Emergency Response, respectively, understand perfectly how this would work. Every day, they must keep 13 million gallons of water from overpowering New York's subway tunnels.

"That's just the water that's already underground," notes Schuber.

"When it rains, the amount is . . ." Briffa shows his palms, surrendering. "It's incalculable."

Maybe not actually incalculable, but it doesn't rain any less now than before the city was built. Once, Manhattan was 27 square miles of porous ground interlaced with living roots that siphoned the 47.2 inches of average annual rainfall up trees and into meadow grasses, which drank their fill and exhaled the rest back into the atmosphere. Whatever the roots didn't take settled into the island's water table. In places, it surfaced in lakes and marshes, with the excess draining off to the ocean via those 40 streams—which now lie trapped beneath concrete and asphalt.

Today, because there's little soil to absorb rainfall or vegetation to transpire it, and because buildings block sunlight from evaporating it, rain collects in puddles or follows gravity down sewers—or it flows into subway vents, adding to the water already down there. Below 131st Street and Lenox Avenue, for example, a rising underground river is corroding the bottom of the A, B, C, and D subway lines. Constantly, men in reflective vests and denim rough-outs like Schuber's and Briffa's are clambering around beneath the city to deal with the fact that under New York, groundwater is always rising.

Whenever it rains hard, sewers clog with storm debris—the number of plastic garbage bags adrift in the world's cities may truly exceed calculation—and the water, needing to go somewhere, plops down the nearest subway stairs. Add a nor'easter, and the surging Atlantic Ocean bangs against New York's water table until, in places like Water Street in lower Manhattan or Yankee Stadium in the Bronx, it backs up right into the tunnels, shutting everything down until it subsides. Should the ocean continue to warm and rise even faster than the current inch per decade, at some point it simply won't subside. Schuber and Briffa have no idea what will happen then.

Add to all that the 1930s-vintage water mains that frequently burst, and the only thing that has kept New York from flooding already is the incessant vigilance of its subway crews and 753 pumps. Think about those pumps: New York's subway system, an engineering marvel in 1903, was laid underneath an already-existing burgeoning city. As that city already had sewer lines, the only place for subways to go was below them. "So," explains Schuber, "we have to pump uphill." In this, New York is not alone: cities like London, Moscow, and Washington built their subways far deeper, often to double as bomb shelters. Therein lies much potential disaster.

Shading his eyes with his white hard hat, Schuber peers down into a square pit beneath the Van Siclen Avenue station in Brooklyn, where each minute 650 gallons of natural groundwater gush from the bedrock. Gesturing over the roaring cascade, he indicates four submersible cast-iron pumps that take turns laboring against gravity to stay ahead. Such pumps run on electricity. When the power fails, things can get difficult very fast. Following the World Trade Center attack, an emergency pump train bearing a jumbo portable diesel generator pumped out 27 times the volume of Shea Stadium. Had the Hudson River actually burst through the PATH train tunnels that connect New York's subways to New Jersey, as was greatly feared, the pump train—and possibly much of the city—would simply have been overwhelmed.

In an abandoned city, there would be no one like Paul Schuber and Peter Briffa to race from station to flooded station whenever more than two inches of rain falls—as happens lately with disturbing frequency—sometimes snaking hoses up stairways to pump to a sewer down the streets, sometimes navigating these tunnels in inflatable boats. With no people, there would also be no power. The pumps will go off, and stay off. "When this pump facility shuts down," says Schuber, "in half an hour water reaches a level where trains can't pass anymore."

Briffa removes his safety goggles and rubs their eyes. "A flood in one zone would push water into the others. Within 36 hours, the whole thing could fill."

Even if it weren't raining, with subway pumps stilled, that would take no more than a couple of days, they estimate. At that point, water would start sluicing away soil under the pavement. Before long, streets start to crater. With no one unclogging sewers, some new watercourses form on the surface. Others appear suddenly as waterlogged subway ceilings collapse. Within 20 years, the water-soaked steel columns that support the street above the East Side's 4, 5, and 6 trains corrode and buckle. As Lexington Avenue caves in, it becomes a river.

*

Well before then, however, pavement all over town would have already been in trouble. According to Dr. Jameel Ahman, chairman of the civil engineering department at New York's Cooper Union, things will begin to fall apart during the first month of March after humans vacate Manhattan. Each March, temperatures normally flutter back and forth around 32°F as many as 40 times (presumably, climate change could push this back to February). Whenever it is, the repeated freezing and thawing make asphalt and cement split. When snow thaws, water seeps into these fresh cracks. When it freezes, the water expands, and cracks widen.

Call it water's retaliation for being squished under all that cityscape. Almost every other compound in nature contracts when frozen, but H_2O molecules do the opposite, organizing themselves into elegant hexagonal crystals that take up about 9 percent more space than they did when sloshing around in a liquid state. Pretty six-sided crystals suggest snowflakes so gossamer it's hard to conceive of them pushing apart slabs of sidewalk. It's even more difficult to imagine carbon steel water pipes built to withstand 7,500 pounds of pressure per square inch exploding when they freeze. Yet that's exactly what happens.

As pavement separates, weeds like mustard, shamrock, and goosegrass blow in from Central Park and work their way down the new cracks, which widen further. In the current world, before they get too far, city maintenance usually shows up, kills the weeds, and fills the fissures. But in the post-people world, there's no one left to continually patch up New York. The weeds are followed by the city's most prolific exotic species, the Chinese ailanthus tree. Even with 8 million people around, ailanthus—otherwise innocently known as the tree-of-heaven—are implacable invaders capable of rooting up in tiny chinks in subway tunnels, unnoticed until their spreading leaf canopies start poking from sidewalk grates. With no one to yank their seedlings, within five years powerful ailanthus roots are heaving up sidewalks and wreaking havoc in sewers—which are already stressed by all the plastic bags and old newspaper mush that no one is clearing away. As soil long trapped beneath pavement gets exposed to sun and rain, other species jump in, and soon leaf litter adds to the rising piles of debris clogging the sewer grates.

The early pioneer plants won't even have to wait for the pavement to fall apart. Starting from the mulch collecting in gutters, a layer of soil will start forming atop New York's sterile hard shell, and seedlings will sprout. With far less organic material available to it—just windblown dust and urban soot—precisely that has happened in an abandoned elevated iron bed of the New York Central Railroad on Manhattan's West Side. Since trains stopped running there in 1980, the inevitable ailanthus trees have been joined by a thickening ground cover of onion grass and fuzzy lamb's ear, accented by stands of goldenrod. In some places, the track emerges from the second stories of warehouses it once serviced into elevated lanes of wild crocuses, irises, evening primrose, asters, and Queen Anne's lace. So many New Yorkers, glancing down from windows in Chelsea's art district, were moved by the sight of this untended, flowering green ribbon, prophetically and swiftly laying claim to a dead slice of their city, that it was dubbed the High Line and officially designated a park.

*

In the first few years with no heat, pipes burst all over town, the freeze-thaw cycle moves indoors, and things start to seriously deteriorate. Buildings groan as their innards expand and contract; joints between walls and rooflines separate. Where they do, rain leaks in, bolts rust, and facing pops off, exposing insulation. If the city hasn't burned yet, it will now. Collectively, New York architecture isn't as combustible as, say, San Francisco's incendiary rows of clapboard Victorians. But with no firemen to answer the call, a dry lightning strike that ignites a decade of dead branches and leaves piling up in Central Park will spread flames through the streets. Within two decades, lightning rods have begun to rust and snap, and roof fires leap among buildings, entering paneled offices filled with paper fuel. Rain and snow blow in, and soon even poured concrete floors are freezing, thawing, and starting to buckle. Burnt insulation and charred wood add nutrients to Manhattan's growing soil cap. Native Virginia creeper and poison ivy claw at walls covered with lichens, which thrive in the absence of air pollution. Red-tailed hawks and peregrine falcons nest in increasingly skeletal high-rise structures.

Within two centuries, estimates Brooklyn Botanical Garden vice president Steven Clemants, colonizing trees will have substantially replaced pioneer weeds. Gutters buried under tons of leaf litter provide new, fertile ground for native oaks and maples from city parks. Arriving black locust and autumn olive shrubs fix nitrogen, allowing sunflowers, bluestem, and white snakeroot to move in along with apple trees, their seeds expelled by proliferating birds.

Biodiversity will increase even more, predicts Cooper Union civil engineering chair Jameel Ahman, as buildings tumble and smash into each other, and lime from crushed concrete raises soil pH, inviting in trees, such as buckthorn and birch, that need less acidic environments. Ahmad, a hearty silver-haired man whose hands talk in descriptive circles, believes that process will begin faster than people might think. A native of Lahore, Pakistan, a city of ancient mosaic-encrusted mosques, he now teaches how to design and retrofit buildings to withstand terrorist attacks, and has accrued a keen understanding of structural weakness.

"Even buildings anchored into hard Manhattan schist, like most New York skyscrapers," he notes, "weren't intended to have their steel foundations waterlogged." Plugged sewers, deluged tunnels, and streets reverting to rivers, he says, will conspire to undermine subbasements and destabilize their huge loads. In a future that portends stronger and more-frequent hurricanes striking North America's Atlantic coast, ferocious winds will pummel tall, unsteady structures.

Some will topple, knocking down others. Like a gap in the forest when a giant tree falls, new growth will rush in. Gradually, the asphalt jungle will give way to a real one.

*   *   *

The New York Botanical Garden, located on 250 acres across from the Bronx Zoo, possesses the largest herbarium anywhere outside of Europe. Among its treasures are wildflower specimens gathered on Captain Cook's 1769 Pacific wanderings, and a shred of moss from Tierra del Fuego, with accompanying notes written in watery black ink and signed by its collector, C. Darwin. Most remarkable, though, is the NYBG's 40-acre tract of original, old-growth, virgin New York forest, never logged.

Never cut, but mightily changed. Until only recently, it was known as the Hemlock Forest for its shady stands of that graceful conifer, but almost every hemlock here is now dead, slain by a Japanese insect smaller than the period at the end of this sentence, which arrived in New York in the mid-1980s. The oldest and biggest oaks, dating back to when this forest was British, are also crashing down, their vigor sapped by acid rain and heavy metals such as lead from automobile and factory fumes, which have soaked into the soil. It's unlikely that they'll come back, because most canopy trees here long ago stopped regenerating. Every resident native species now harbors its own pathogen: some fungus, insect, or disease that seizes the opportunity to ravish trees weakened by chemical onslaught. As if that weren't enough, as the NYBG forest became an island of greenery surrounded by hundreds of square miles of gray urbanity, it became the primary refuge for Bronx squirrels. With natural predators gone and no hunting permitted, there's nothing to stop them from devouring every acorn or hickory nut before it can germinate. Which they do. . . .

As human beings learned to transport themselves all over the world, they took living things with them and brought back others. Plants from the Americas changed not only ecosystems in European countries but also their very identities: think of Ireland before potatoes, or Italy before tomatoes. In the opposite direction, Old World invaders not only forced themselves on hapless women of vanquished new lands, but broadcast other kinds of seed, beginning with wheat, barley, and rye. In a phrase coined by the American geographer Alfred Crosby, this ecological imperialism helped European conquerors to permanently stamp their image on their colonies. . . .

*

The other invasion that has accosted natives—metals such as lead, mercury, and cadmium—will not wash quickly from the soil, because these are literally heavy molecules. One thing is certain: when cars have stopped for good, and factories go dark and stay that way, no more such metals will be deposited. For the first 100 years or so, however, corrosion will periodically set off time bombs left in petroleum tanks, chemical and power plants, and hundreds of dry cleaners. Gradually, bacteria will feed on residues of fuel, laundry solvents, and lubricants, reducing them to more-benign organic hydrocarbons—although a whole spectrum of man-made novelties, ranging from certain pesticides to plasticizers to insulators, will linger for many millennia until microbes evolve to process them.

Yet with each new acid-free rainfall, trees that still endure will have fewer contaminants to resist as chemicals are gradually flushed from the system. Over centuries, vegetation will take up decreasing levels of heavy metals, and will recycle, redeposit, and dilute them further. As plants die, decay, and lay down more soil cover, the industrial toxins will be buried deeper, and each succeeding crop of native seedlings will do better.

And although many of New York's heirloom trees are endangered if not actually dying, few if any are already extinct. Even the deeply mourned American chestnut, devastated everywhere after a fungal blight entered New York around 1900 in a shipment of Asian nursery plants, still hangs on in the New York Botanical Garden's old forest—literally by its roots. It sprouts, sends up skinny shoots two feet high, gets knocked back by blight, and does it again. One day, perhaps, with no human stresses sapping its vigor, a resistant strain will finally emerge. Once the tallest hardwood in American eastern forests, the resurrected chestnut trees will have to coexist with robust non-natives that are probably here to stay—Japanese barberry, Oriental bittersweet, and surely ailanthus. The ecosystem here will be a human artifact that will persist in our absence, a cosmopolitan botanical mixture that would never have occurred without us. . . .

*   *   *

In the millennial year 2000, a harbinger of a future that might revive the past appeared in the form of a coyote that managed to reach Central Park. Subsequently, two more made it into town, as well as a wild turkey. The rewilding of New York City may not wait until people leave.

That first advance coyote scout arrived via the George Washington Bridge, which Jerry Del Tufo managed for the Port Authority of New York and New Jersey. Later, he took over the bridges that link Staten Island to the mainland and Long Island. A structural engineer in his forties, he considers bridges among the loveliest ideas humans ever conceived, gracefully spanning chasms to bring people together. . . .

On a February afternoon he heads through snow flurries to the Bayonne Bridge, chatting with his crew over his radio. The underside of the approach on the Staten Island side is a powerful steel matrix that converges in a huge concrete block anchored to the bedrock, an abutment that bears half the load of the Bayonne's main span. To stare up directly into its labyrinthine load-bearing I-beams and bracing members, interlocked with half-inch-thick steel plates, flanges, and several million half-inch rivets and bolts, recalls the crushing awe that humbles pilgrims gaping at the soaring Vatican dome of St. Peter's Cathedral: something this mighty is here forever. Yet Jerry Del Tufo knows exactly how these bridges, without humans to defend them, would come down.

It wouldn't happen immediately, because the most immediate threat will disappear with us. It's not, says Del Tufo, the incessant pounding traffic.

"These bridges are so overbuilt, traffic's like an ant on an elephant." In the 1930s, with no computers to precisely calculate tolerances of construction materials, cautious engineers simply heaped on excess mass and redundancy. "We're living off the overcapacity of our forefathers. The GW alone has enough galvanized steel wire in its three-inch main cables to wrap the Earth four times. Even if every other suspender rope deteriorated, the bridge wouldn't fall down."

Enemy number one is the salt that highway departments spread on the roadways each winter—ravenous stuff that keeps eating steel once it's done with the ice. Oil, antifreeze, and snowmelt dripping from cars wash salt into catch basins and crevices where maintenance crews must find it and flush it. With no more people, there won't be salt. There will, however, be rust, and quite a bit of it, when no one is painting the bridges.

At first, oxidation forms a coating on steel plate, twice as thick or more as the metal itself, which slows the pace of chemical attack. For steel to completely rust through and fall apart might take centuries, but it won't be necessary to wait that long for New York's bridges to start dropping. The reason is a metallic version of the freeze-thaw drama. Rather than crack like concrete, steel expands when it warms and contracts when it cools. So that steel bridges can actually get longer in summer, they need expansion joints.

In winter, when they shrink, the space inside expansion joints open wider, and stuff blows in. Wherever it does, there's less room for the bridge to expand when things warm up. With no one painting bridges, joints fill not only with debris but also with rust, which swells to occupy far more space than the original metal.

"Come summer," says Del Tufo, "the bridge is going to get bigger whether you like it or not. If the expansion joint is clogged, it expands toward the weakest link—like where two different materials connect." He points to where four lanes of steel meet the concrete abutment. "There, for example. The concrete could crack where the beam is bolted to the pier. Or, after a few seasons, that bolt could shear off. Eventually, the beam could walk itself right off and fall."

Every connection is vulnerable. Rust that forms between two steel plates bolted together exerts forces so extreme that either the plates bend or rivets pop, says Del Tufo. Arch bridges like the Bayonne—or the Hell Gate over the East River, made to hold railroads—are the most overbuilt of all. They might hold for the next 1,000 years, although earthquakes rippling through one of several faults under the coastal plain could shorten that period. (They would probably do better than the 14 steel-lined, concrete subway tubes beneath the East River— one of which, leading to Brooklyn, dates back to horses and buggies. Should any of their sections separate, the Atlantic Ocean would rush in.) The suspension and truss bridges that carry automobiles, however, will last only two or three centuries before their rivets and bolts fail and entire sections fall into the waiting waters.

*

Until then, more coyotes follow the footsteps of the intrepid ones that managed to reach Central Park. Deer, bear, and finally wolves, which have reentered New England from Canada, arrive in turn. By the time most of its bridges are gone, Manhattan's newer buildings have also been ravaged, as wherever leaks reach their embedded steel reinforcing bars, they rust, expand, and burst the concrete that sheaths them. Older stone buildings such as Grand Central—especially with no more acid rain to pock their marble—will outlast every shiny modern box.

Ruins of high-rises echo the love songs of frogs breeding in Manhattan's reconstituted streams, now stocked with alewives and mussels dropped by seagulls. Herring and shad have returned to the Hudson, though they spent some generations adjusting to radioactivity trickling out of Indian Point Nuclear Power Plant, 35 miles north of Times Square, after its reinforced concrete succumbed. Missing, however, are nearly all fauna adapted to us. The seemingly invincible cockroach, a tropical import, long ago froze in unheated apartment buildings. Without garbage, rats starved or became lunch for the raptors nesting in burnt-out skyscrapers.

Rising water, tides, and salt corrosion have replaced the engineered shoreline, circling New York's five boroughs with estuaries and small beaches. With no dredging, Central Park's ponds and reservoir have been reincarnated as marshes. Without natural grazers—unless horses used by hansom cabs and by park policemen managed to go feral and breed—Central Park's grass is gone. A maturing forest is in its place, radiating down former streets and invading empty foundations. Coyotes, wolves, red foxes, and bobcats have brought squirrels back into balance with oak trees tough enough to outlast the lead we deposited, and after 500 years, even in a warming climate the oaks, beeches, and moisture-loving species such as ash dominate.

Long before, the wild predators finished off the last descendants of pet dogs, but a wily population of feral house cats persists, feeding on starlings. With bridges finally down, tunnels flooded, and Manhattan truly an island again, moose and bears swim a widened Harlem river to feast on the berries that the Lenape once picked. . . .

. . . Three times in the past 100,000 years, glaciers have scraped New York clean. Unless humankind's Faustian affair with carbon fuels ends up tipping the atmosphere past the point of no return, and runaway global warming transfigures Earth into Venus, at some unknown date glaciers will do so again. The mature beech-oak-ash-ailanthus forest will be mowed down. The four giant mounts of entombed garbage at the Fresh Kills landfill on Staten Island will be flattened, their vast accumulation of stubborn PVC plastic and of one of the most durable human creations of all—glass—ground to powder.

After the ice recedes, buried in the moraine and eventually in geologic layers below will be an unnatural concentration of a reddish metal, which briefly had assumed the form of wiring and plumbing. Then it was hauled to the dump and returned to the Earth. The next toolmaker to arrive or evolve on this planet might discover and use it, but by then there would be nothing to indicate that it was us who put it there.

4

Excerpts from "Reinventing Eden: Western Culture as a Recovery Narrative"

CAROLYN MERCHANT

A Penobscot Indian story from northern New England explains the origin of maize. A great famine had deprived people of food and water. A beautiful Indian maiden appeared and married one of the young men of the tribe, but soon succumbed to another lover, a snake. On discovery she promised to alleviate her husband's sorrow if he would plant a blade of green grass clinging to her ankle. First he must kill her with his ax, then drag her body through the forest clearing until all her flesh had been stripped, and finally bury her bones in the center of the clearing. She then appeared to him in a dream and taught him how to tend, harvest, and cook corn and smoke tobacco.[1]

This agricultural origin story taught the Indians not only how to plant their corn in forest clearings but also that the earth would continue to regenerate the human body through the corn plant. It features a woman, the maiden, and a male lover as central actors. It begins with the state of nature as drought and famine. Nature is a desert, a poor place for human existence The plot features a woman as savior. Through a willing sacrifice in which her body brings forth new life, she introduces agriculture to her husband and to the women who subsequently plant the corn, beans, and squash that sustain the life of the tribe. The result is an ecological system based on planting of interdependent polycultures in forest gardens. The story type is ascensionist and progressive. Women transform nature from a desert into a garden. From a tragic situation of despair and death, a comic, happy, and optimistic situation of continued life results. In this story the valence of women as corn mothers is good; they bring bountiful gifts. The valence of nature ends as a good. The earth is an agent of regeneration. Death is transformed into life through a reunification of the corn mother's body with the earth. Even death therefore results in a higher good.[2]

Into this bountiful world of corn mothers, enter the Puritan fathers bringing their own agricultural origin story of Adam and Eve. The biblical myth begins where the Indian story ends, with an ecological system of polycultures in the Garden of Eden. A woman, Eve, shows "the man," Adam, how to pick fruit from the Tree of the Knowledge of Good and Evil and to harvest the fruits of the garden. Instead of attaining a resultant good, the couple is cast out of the garden into a desert. Instead of moving from desert to garden, as in the Indian story, the biblical story moves from garden to desert. The Fall from paradise is caused by a woman. Men must labor in the earth by the sweat of their brow to produce food. Here a woman is also the central actress and, like the Indian story, the biblical story contains violence toward women. But the plot is declensionist and tragic, not progressive and comic as in the Indian story. The end result is a poorer state of nature than in the beginning. The valence of woman is bad. The end valence of nature is bad. Here men become the agents of transformation. They become saviors, who through their own agricultural labor have the capacity to re-create the lost garden on earth.[3]

According to Benjamin Franklin, Indians quickly perceived the difference between the two accounts. Franklin satirically writes that when the Indians were apprised of the "historical facts on which our [own] religion is founded, such as the fall of our first parents by eating an apple, . . . an Indian orator stood up" to thank the Europeans for their information. "What you have told us . . . is all very good. It is, indeed, bad to eat apples. It is much better to make them all into cider. We are much obliged by your kindness in coming so far to tell us these things which you have heard from your mothers; in return I will tell you some of those which we have heard from ours."[4]

Historical events reversed the plots of the European and the Indian origin stories. The Indians' comic happy ending changed to a story of decline and conquest, while Euramericans were largely successful in creating a New World garden. Indeed, the story of Western civilization since the seventeenth century and its advent on the American continent can be conceptualized as a grand narrative of fall and recovery. The concept of recovery, as it emerged in the seventeenth century, not only meant a recovery from the Fall but also entailed restoration of health, reclamation of land, and recovery of property.[5] The recovery plot is the long, slow process of returning humans to the Garden of Eden through labor in the earth. Three subplots organize its argument: Christian religion, modern science, and capitalism. The Genesis story of the Fall provides the beginning; science and capitalism, the middle; recovery of the garden, the end. The initial lapsarian moment (i.e., the lapse from innocence) is the decline from garden to desert as the first couple is cast from the light of an ordered paradise into a dark, disorderly wasteland.

The Bible, however, offered two versions of the origin story that led to the Fall. In the Genesis 1 version, God created the land, sea, grass, herbs, and fruit; the stars, sun, and moon; and the birds, whales, cattle, and beasts—after which he made "man in his own image . . . ; male and female created he them." Adam and Eve were instructed, "Be fruitful and multiply, and replenish the earth, and subdue it," and were given "dominion over the fish of the sea, and over the fowl of the air, and over every living thing that moveth upon the earth." In the Genesis 2 version, thought to have derived from a different tradition, God first created the plants and herbs, next "man" from dust, and then the garden of Eden with its trees for food (including the Tree of Life and the Tree of the Knowledge of Good and Evil in the center) and four rivers flowing out of it. He then put "the man" in the garden "to dress and keep it," formed the beasts and fowls from dust, and brought them to Adam to name. Only then did he create Eve from Adam's rib. Genesis 3 narrates the Fall from the garden, beginning with Eve's temptation by the serpent, the consumption of the fruit from the Tree of the Knowledge of Good and Evil (which in the Renaissance becomes an apple), the expulsion of Adam and Eve from the garden "to till the ground from which he was taken," and finally God's placement of the cherubims and flaming sword at the entrance of the garden to guard the Tree of Life.[6]

During the Renaissance, artists illustrated the Garden of Eden story through woodcuts and paintings, one of the most famous of which is Lucas Cranach's 1526 painting of Eve offering the apple to Adam, after having been enticed by the snake coiled around the Tree of the Knowledge of Good and Evil (Figure 4.1).

Figure 4.1. Lucas Cranach, Adam and Eve, 1526. (Courtesy Courtauld Institute Galleries)

Writers from Dante to Milton depicted the Fall and subsequent quest for paradise, while explorers searched for the garden first in the Old World and then in the New. Although settlers endowed new lands and peoples with Eden-like qualities, a major effort to re-create the Garden of Eden on earth ultimately ensued. Seventeenth-century botanical gardens and zoos marked early efforts to reassemble the parts of the garden dispersed throughout the world after the Fall and the Flood.[7]

But beginning in the seventeenth century and proceeding to the present, New World colonists have undertaken a massive effort to reinvent the whole earth in the image of the Garden of Eden. Aided by the Christian doctrine of redemption and the inventions of science, technology, and capitalism ("arte and industria"), the long-term goal of the recovery project has been to turn the earth itself into a vast cultivated garden. The strong interventionist version in Genesis 1 legitimates recovery through domination, while the softer Genesis 2 version advocates dressing and keeping the garden through human management (stewardship). Human labor would redeem the souls of men and women, while cultivation and domestication would redeem the earthly wilderness. The End Drama envisions a reunification of the earth with God (the Parousia), in which the redeemed earthly garden merges into a higher heavenly paradise. The Second Coming of Christ was to occur either at the outset of the thousand-year period of his reign on earth (the millennium) or at the Last Judgment, when the faithful were reunited with God at the resurrection.[8]

Greek philosophy offered the intellectual framework for the modern version of the recovery project. Parmenidean oneness represents the unchanging natural law that has lapsed into the appearances of the Platonic world. This fallen phenomenal world is incomplete, corrupt, and inconstant. Only by recollection of the pure, unchanging forms can the fallen partake of the original unity. Recovered and Christianized in the Renaissance, Platonism provided paradigmatic ideals (such as that of the Garden of Eden) through which to interpret the earthly signs and signatures leading to the recovery.[9]

Modern Europeans added two components to the Christian recovery project—mechanistic science and laissez-faire capitalism—to create a grand master narrative of Enlightenment. Mechanistic science supplies the instrumental knowledge for reinventing the garden on earth. The Baconian-Cartesian-Newtonian project is premised on the power of technology to subdue and dominate nature, on the certainty of mathematical law, and on the unification of natural laws into a single framework of explanation. Just as the alchemists tried to speed up nature's labor through human intervention in the transformation of base metals into gold, so science and technology hastened the recovery project by inventing the tools and knowledge that could be used to dominate nature. Francis Bacon saw science and technology as the way to control nature and hence recover the right to the garden given to the first parents. "Man by the fall, fell at the same time from his state of innocency and from his dominion over creation. Both of these losses can in this life be in some part repaired; the former by religion and faith; the latter by arts and science." Humans, he asserted, could "recover that right over nature which belongs to it by divine bequest," and should endeavor "to establish and extend the power and dominion of the human race itself over the [entire] universe."[10]

The origin story of capitalism is a movement from desert back to garden through the transformation of undeveloped nature into a state of civility and order.[11] Natural resources—"the ore in the mine, the stone unquarried [and] the timber unfelled"—are converted by human labor into commodities to be exchanged on the market. The good state makes capitalist production possible by imposing order on the fallen worlds of nature and human nature. Thomas Hobbes's nation-state was the end result of a social contract created for the purpose of controlling people in the violent and unruly state of nature. John Locke's political theory rested on the improvement of undeveloped nature by mixing human labor with the soil and subduing the earth through human dominion. Simultaneously, Protestantism helped to speed the recovery by sanctioning increased human labor just as science and technology accelerated nature's labor.[12]

Crucial to the structure of the recovery narrative is the role of gender encoded into the story. In the Judeo-Christian tradition the original oneness is male and the Fall is caused by a female, Eve, with Adam, the innocent bystander, being forced to pay the consequences as his sons are pushed into developing both pastoralism and farming.[13] While fallen Adam becomes the inventor of the tools and technologies that will restore the garden, fallen Eve becomes the nature that must be tamed into submission. In the Western tradition, fallen nature is opposed by male science and technology. The good state that keeps unruly nature in check is invented, engineered, and operated by men. The good economy that organizes the labor needed to restore the garden is likewise a male-directed project.

Nature, in the Edenic recovery story, appears in three forms. As original Eve, nature is virgin, pure, and light—land that is pristine or barren, but that has the potential for development. As fallen Eve, nature is disorderly and chaotic; a wilderness, wasteland, or desert requiring improvement; dark and witch-like, the victim and mouthpiece of Satan as serpent. As mother Eve, nature is an improved garden, a nurturing earth bearing fruit, a ripened ovary, maturity. Original Adam is the image of God as creator, initial agent, activity. Fallen Adam appears as the agent of earthly transformation, the hero who redeems the fallen land. Father Adam is the image of God as patriarch, law, and rule, the model for the kingdom and state. These meanings of nature as female and agency as male are encoded as symbols and myths into American lands as having the potential for development, but needing the male hero, Adam. Such symbols are not essences because they do not represent characteristics necessary or essential to being female or male. Rather, they are historically constructed meanings deriving from the origin stories of European settlers and European cultural and economic practices transported to and developed in the American New World. That they may appear to be essences is a result of their historical construction in Western history, not their immutable characteristics.

The Enlightenment idea of progress is rooted in the recovery of the garden lost in the Fall—the bringing of light to the dark world of inchoate nature. The lapsarian origin story is thus reversed by the grand narrative of Enlightenment that lies at the very heart of modernism. The controlling image of Enlightenment is the transformation from desert wilderness to cultivated garden. This complex of Christian, Greco-Roman, and Enlightenment components touched and reinforced each other at critical nodal points. As a powerful narrative, the idea of recovery functioned as ideology and legitimation for settlement of the New World, while capitalism, science, and technology provided the means of transforming the material world. . . .

Female Nature in the Recovery Narrative

An account of the history of American settlement as a lapsarian and recovery narrative must also consider the crucial role of nature conceptualized as female in the very structure of the plot. The rhetoric of American settlement is filled with language that casts nature as female object to be transformed and men as the agents of change. Allusions to Eve as virgin land to be subdued, as fallen nature to be redeemed through reclamation, and as fruitful garden to be harvested and enjoyed are central to the particular ways in which American lands were developed. The extraction of resources from "nature's bosom," the penetration of "her womb" by science and technology, and the "seduction" of female land by male agriculture reinforced capitalist expansion.[14]

Images of nature as female are deeply encoded into the texts of American history, art, and literature and function as ideologies for settlement. Thus Thomas Morton in praising New England as a new Canaan likened its potential for development by "art and industry" to a "faire virgin longing to be sped and meete her lover in a Nuptiall bed." Now, however, "her fruitful wombe, not being enjoyed is like a glorious tombe."[15] Male agriculturalists saw in plow technology a way to compel female nature to produce. Calling Bacon "the grand master of philosophy" in 1833, the Massachusetts agricultural improver Henry Colman promoted Bacon's approach to recovering the garden through agriculture. "The effort to extend the dominion of man over nature," he wrote, "is the most healthy and most noble of all ambitions." He characterized the earth as a female whose productivity could help to advance the progress of the human race. "Here man exercises dominion over nature; . . . commands the earth on which he treads to waken her mysterious energies . . . compels the inanimate earth to teem with life; and to impart sustenance and power, health and happiness to the countless multitudes who hang on her breast and are dependent on her bounty."[16] . . .

The narrative of frontier expansion is a story of male energy subduing female nature, taming the wild, plowing the land, re-creating the garden lost by Eve. American males lived the frontier myth in their everyday lives, making the land safe for capitalism and commodity production. Once tamed by men, the land was safe for women. To civilize was to bring the land out of a state of savagery and barbarism into a state of refinement and enlightenment. This state of domestication, of civility, is symbolized by woman and "womanlike" man. "The man of training, the civilizee," reported *Scribner's Monthly* in 1880, "is less manly than the rough, the pioneer."[17]

But the taming of external nature was intimately linked to the taming of internal nature, the exploitation of nonhuman nature to the exploitation of human nature. The civilizing process not only removed wild beasts from the pastoral lands of the garden; it suppressed the wild animal in men. Crèvecoeur in 1782 noted that on the frontier "men appear to be no better than carnivorous animals . . . living on the flesh of wild animals." Those who farmed the middle settlements, on the other hand, were "like plants," purified by the "simple cultivation of the earth," becoming civilized through reading and political discourse.[18] Or as Richard Burton put it in 1861, "The civilizee shudders at the idea of eating wolf."[19] Just as the earth is female to the farmer who subdues it with the plow, so wilderness is female to the male explorer, frontiersman, and pioneer who tame it with the brute strength of the ax, the trap, and the gun. Its valence, however, varies from the negative satanic forest of William Bradford and the untamed wilderness of the pioneer (fallen Eve) to the positive pristine Eden and mother earth of John Muir (original and Mother Eve) and the parks of Frederick Law Olmsted. A wilderness vanishes before advancing civilization, its remnants must be preserved as test zones for men (epitomized by Theodore Roosevelt) to hone male strength and skills.[20]

Civilization is the final end, the telos, toward which "wild" nature is destined. The progressive narrative undoes the declension of the Fall. The "end of nature" is civilization. Civilization is thus nature natured, *Natura naturata*—the natural order, or nature ordered and tamed. It is no longer nature naturing, *Natura naturans*—nature as creative force. Nature passes from inchoate matter endowed with a formative power to a reflection of the civilized natural order designed by God. The unruly energy of wild female nature is suppressed and pacified. The final, happy state of nature natured is female and civilized—the restored garden of the world.[21] . . .

The City in the Garden . . .

. . . Nature, wilderness, and civilization are socially constructed concepts that change over time and serve as stage settings in the progressive narrative. So too are the concepts of male and female and the roles that men and women play on the stage of history. The authors of such powerful narratives as laissez-faire capitalism, mechanistic science, manifest density, and the frontier story are usually privileged elites with access to power and patronage. Their words are read by persons of power who add the new stories to the older biblical story. As such the books become the library of Western culture. The library, in turn, functions as ideology when ordinary people read, listen to, internalize, and act out the stories told by their elders—the ministers, entrepreneurs, newspaper editors, and professors who teach and socialize the young.

The most recent chapter of the book of the recovery narrative is the transformation of nature through biotechnology. From genetically engineered apples to Flavr-Savr tomatoes, the fruits of the original (evolved) garden are being redesigned so that the salinated irrigated desert can continue to blossom as the rose. In the recovered Garden of Eden, fruits will ripen faster, have fewer seeds, need less water, require fewer pesticides, contain less saturated fat, and have longer shelf lives. The human temptation to engineer nature is reaching too close to the powers of God, warn the Jeremiahs who depict the snake coiled around the Tree of Knowledge of Good and Evil as the DNA spiral. But the progressive engineers who design the technologies that allow the recovery to accelerate see only hope in the new fabrications.

The twentieth-century Garden of Eden is the enclosed shopping mall decorated with trees, flowers, and fountains in which people can shop for nature at the Nature Company, purchase "natural" clothing at Esprit, sample organic foods and Rainforest Crunch in kitchen gardens, buy twenty-first-century products at Sharper Image, and play virtual reality games in which SimEve is reinvented in Cyberspace. This garden in the city re-creates the pleasures and temptations of the original garden and the golden age where people can peacefully harvest the fruits of earth with gold grown by the market. The mall, enclosed by the desert of the parking lots surrounding it, is covered by glass domes reaching to heaven, accessed by spiral staircases and escalators affording a vista over the whole garden of shops. The "river that went out of Eden to water the garden" is reclaimed in meandering streams lined with palm trees and filled with bright orange carp. Today's malls feature stone grottoes, trellises decorated with flowers, life-sized trees, statues, birds, animals, and even indoor beaches that simulate paradigmatic nature as a cultivated, benign garden. With their engineered spaces and commodity fetishes, they epitomize consumer capitalism's vision of the recovery from the Fall.[22]

Critiques of the Recovery Narrative

The modern version of the recovery narrative, however, has been subjected to scathing criticism. Postmodern thinkers contest its Enlightenment assumptions, while cultural feminists and environmentalists reverse its plot, depicting a slow decline from a prior golden age, not a progressive ascent to a new garden on earth. The critics' plot does not move from the tragedy of the Fall to the comedy of an earthly paradise but descends from an original state of oneness with nature to the tragedy of nature's destruction. Nevertheless, they too hope for a recovery, one rapid enough to save the earth and society by the mid-twenty-first century. The metanarrative of recovery does not change, but the declensionist plot, into which they have cast prior history, must be radically reversed. The postmodern critique of modernism is both a deconstruction of Enlightenment thought and a set of reconstructive proposals for the creation of a better world.

The identification of modernism as a problem rather than as progress was sharply formulated by Max Horkheimer and Theodor Adorno in the opening sentences of their 1944 *Dialectic of Enlightenment*: "The fully enlightened earth radiates disaster triumphant. The program of the enlightenment was the disenchantment of the world; the dissolution of myths and the substitution of knowledge for fancy." They criticized both Francis Bacon's concept of the domination of nature and Karl Marx and Friedrich Engels's optimism that the control of nature would lead to advancement. They faulted the reduction of nature to mere number by mechanistic science and capitalism: "Number becomes the canon of the Enlightenment. The same equations dominate bourgeois justice and commodity exchange. . . . Myth turns into enlightenment and nature into mere objectivity."[23]

Among the critics of modernism are many feminists and environmentalists who propose a reversal that will initiate a new millennium in the twenty-first century. Cultural feminists and ecofeminists see the original oneness as female, the *terra mater* of the neolithic era, from which emerged the consciousness of differences between humans and animals, male and female, people and nature, leading to dominance and submission. The advent of patriarchy initiates a long decline in the status of women and nature. Men's plow agriculture took over women's gathering and horticultural activities, horse-mounted warriors injected violence into a largely peaceful Old European culture, and male gods replaced female deities in origin stories. In the proposed recovery, Eve is revisioned as the first scientist, Sophia as ultimate wisdom, and the goddess as symbol of female power and creativity. Feminist religious history redirects inquiry into the gendered nature of the original oneness as both male and female. The recovery would therefore be a feminist or an egalitarian world.[24]

Feminist science sees the original mind as having no sex, and hence accessible to male and female minds alike. It has been men, many feminists would argue, who have invented the science and technology and organized the market economies that have made nature victim in the ascent of "man." For such feminists the new narrative would entail reclaiming women's roles in the history of science and asserting female power in contemporary science and technology. Hence both sexes can participate in the recovery.[25] . . .

Like feminists, environmentalists want to rewrite the modern progressive story. Having seen the plot as declensionist rather than progressive, they nevertheless opt for a recovery that must be put in place by the mid-twenty-first century. "Sustainability" is a new vision of the recovered garden, one in which humanity will live in a relationship of balance and harmony with the natural world. Environmentalists who press for sustainable development see the recovery as achievable through the spread of nondegrading forms of agriculture and industry. Preservationists and deep ecologists strive to save pristine nature as wilderness before it can be destroyed by development. Restoration ecologists wish to marshal human labor to restore an already degraded nature to an earlier, pristine state. Social ecologists and green parties devise new economic and political structures that overcome the domination of human beings and nonhuman nature. Women and nature, minorities and nature, other animals and nature will be fully included in the recovery. The regeneration of nature and people will be achieved through social and environmental justice. The End Drama envisions a postpatriarchal, socially just ecotopia for the postmillennial world of the twenty-first century.[26]

NOTES

1. Ronald Nelson, Penobscot, as recorded by Frank Speck, "Penobscot Tales and Religious Beliefs," *Journal of American Folklore* 48 (Jan.–Mar. 1935): 1–107, on 75. This corn mother origin story is a variant on a number of eastern U.S. and Canadian transformative accounts, recorded from oral traditions, that attribute the origins of corn to a mythical corn mother who produces corn from her body, grows old, and then instructs her lover or son how to plant and tend corn. The killing of the corn mother in most of the origin stories may symbolize a transition from gathering-hunting to active corn cultivation. The snake lover may be an influence from the Christian tradition or a more universal symbol of the renewal of life (snakes shed their skins) and/or the male sexual organ. On corn mother origin stories, see John Witthoft, *Green Corn Ceremonialism in the Eastern Woodlands* (Ann Arbor: Univ. of Michigan Press, 1949), 77–85; Joe Nicholas, Malechite, Tobique Point, Canada, Aug. 1910, as recorded by W. H. Mechling, *Malenchite Tales* (Ottawa: Government Printing Bureau, 1914), 87–88; for the Passamaquoddy variant, see *Journal of American Folklore* 3 (1890): 214; for Creek and Natchez variants, see J. R. Swanton, "Myths and Tales of the Southeastern Indians," *Bulletin of the Bureau of American Ethnology*, no. 88 (1929): 9–17; on Iroquois variants, see Jesse Cornplanter, *Legends of the Longhouse* (Philadelphia: J. B. Lippincott, 1938), and Arthur Parker, "Iroquois Use of Maize and Other Food Plants," *New York State Museum Bulletin*, no. 144 (1910): 36–39; Gudmund Hatt, "The Corn Mother in America and Indonesia," *Anthropos* 46 (1951): 835–914. Examples of corn mother origin stories from the Southwest include the Pueblo emergence from the dark interior of the earth into the light of the fourth world, where corn mother plants thought woman's gift of corn. See Ramón

Gutiérrez, *When Jesus Came the Corn Mothers Went Away* (Stanford, CA: Stanford Univ. Press, 1991). For a discussion of the relationship of the corn mother to mother earth, see Sam Gill, *Mother Earth: An American Story* (Chicago: Univ. of Chicago Press, 1987), 4, 125.

2. On Great Plains environmental histories as progressive and declensionist plots, see William Cronon, "A Place for Stories: Nature, History, and Narrative," *Journal of American History* 78 (1992): 1347–76. The Indian and European origin stories can be interpreted from a variety of standpoints other than the declensionist and progressive narrative formats I have empha- sized here (such as romance and satire). Additionally, the concepts of desert, wilderness, and garden are nuanced and elaborate motifs that change valences over time in ways I have not tried to deal with here.

3. Genesis, chap. 1. On the comic and tragic visions of the human animal, vegetable, mineral and unformed worlds, see Northup Frye, *Fables of Identity* (New York: Harcourt, Brace, 1963), 19–20. In the comic state, or vision, the human world is a community, the animal world consists of domesticated flocks and birds of peace, the vegetable world is a garden or park with trees, the mineral world is a city or temple with precious stones and starlit domes, and the unformed world is a river. In the tragic state or vision, the human world is an anarchy of individuals, the animal world is filled with beasts and birds of prey (such as wolves, vultures, and serpents), the vegetable world is a wilderness, desert, or sinister forest, the mineral world is filled with rocks and ruins, and the unformed world is a sea or flood. The plot of the tragedy moves from a better or comic state to a worse or tragic state; the comedy from an initial tragic state, to a comic or happy outcome. I think Hayden White for this reference. On history as narrative, see Hayden White, *Metahistory: The Historical Imagination in Nineteenth-Century Europe* (Baltimore: Johns Hopkins Univ. Press, 1973); idem, *Tropics of Discourse: Essays in Cultural Criticism* (Baltimore: Johns Hopkins Univ. Press, 1978); idem, *The Content of the Form: Narrative Discourse and Historical Representation* (Baltimore: Johns Hopkins Univ. Press, 1987).

4. Benjamin Franklin, "Remarks concerning the Savages of North America," in Richard E. Amacher, ed., *Franklin's Wit and Folly: The Bagatelles* (New Brunswick, NJ: Rutgers Univ. Press, 1953), 89–98. Franklin's story is probably satirical rather than literal.

5. The concept of a recovery from the original Fall appears in the early modern period. See the *Oxford English Dictionary*, s.v. "recovery": "The act of recovering oneself from a mishap, mistake, fall, etc." See Bishop Edward Stillingfleet, *Origines Sacrae* (London, 1662), II, I, sec. 1: "The conditions on which fallen man may expect a recovery." William Cowper, *Retirement* (1781), 138: "To . . . search the themes, important above all, Ourselves, and our recovery from our fall." See also Richard Eden, *The Decades of the Newe Worlde or West India* (1555), 168: "The recouerie of the kyngedome of Granata." The term "recovery" also embraced the idea of regaining a "natural position" after falling and a return to health after sickness. It acquired a legal meaning in the sense of gaining possession of property by a verdict or judgment of the court. In common recovery, an estate was transferred from one party to another. John Cowell, *The Interpreter* (1607), s.v. "recouerie": "A true recouerie is an actuall or reall recouerie of anything, or the value thereof by Judgement." Another meaning was the restoration of a person or thing to a healthy or normal condition, or a return from a lapsed state to a higher or better state, including the reclamation of land and of resources such as soil. Anonymous, *Captives bound in Chains . . . the misery of graceless Sinners, and the hope of their recovery by Christ* (1674); Bishop Joseph Butler, *The Anthology of Religion Natural and Revealed* (1736), II, 295: "Indeed neither Reason nor Analogy would lead us to think . . . that the Interposition of Christ . . . would be of that Efficacy for Recovery of the World, which Scripture teaches us it was." Joseph Gilbert, *The Christian Atonement* (1836), I, 24: "A modified system, which shall include the provision of means for recovery from a lapsed

state." James Martineau, *Essays, Reviews, and Addresses* (1890–91), II, 310: "He is fitted to be among the prophets of recovery, who may prepare for us a more wholesome future." John Henry Newman, *Historical Sketches* (1872–73) II, 1, iii, 121: "The special work of his reign was the recovery of the soil."

6. On the Genesis 1, or priestly, version (Genesis P), composed in the fifth century BC, versus the Genesis 2, or Yahwist, version (Genesis J), composed in the ninth or tenth century BC, and their relationships to the environmental movement, see J. Baird Callicott, "Genesis Revisited: Muirian Musings on the Lynn White, Jr. Debate," *Environmental Review* 14, nos. 1–2 (Spring–Summer 1990): 65–92. Callicott argues that Lynn White, Jr., mixed the two versions in his famous article "The Historical Roots of Our Ecologic Crisis," *Science* 155 (1967): 1203–7. On the historical traditions behind the Genesis stories, see Arthur Weiser, *The Old Testament: Its Formation and Development*, trans. Dorothea M. Barton (New York: Association Press, 1961).

7. John Prest, *The Garden of Eden: The Botanic Garden and the Re-creation of Paradise* (New Haven, CT: Yale Univ. Press, 1981), 1–37; J. A. Phillips, *Eve: The History of an Idea* (San Francisco: Harper and Row, 1984).

8. "Paradise" derives from the old Persian word for "enclosure" and in Greek and Latin takes on a meaning of garden. Its meanings include heaven, a state of bliss, an enclosed garden or park, and the Garden of Eden. "Parousia" derives from the Latin *parere*, meaning to produce or bring forth. The Parousia is the idea of the End of the World, expressed as the hope set forth in the New Testament that "he shall come again to judge both the quick and the dead." See A. L. Moore, *The Parousia in the New Testament* (Leiden: E. J. Brill, 1966). I thank Anthony Chennells for bringing this concept to my attention. Capitalism and Protestantism were initially mutually reinforcing in their common hope of a future golden age. But as capitalism became more materialistic and worldly it began to undercut the church's Parousia hope. Communism retained the idea of a future golden age in its concern for community and future direction (ibid., 2–3). The Parousia hope was a driving force behind the church's missionary work in its early development and in the New World (5). The age of glory was a gift of God, an acknowledgment of the future inbreaking of God (JHWH) into history (16, 17). "The scene of the future consummation is a radically transformed earth. The coming of this kingdom was conceptualized as a sudden catastrophic moment, or as preceded by the Messianic kingdom, during which it was anticipated that progressive work would take place" (20). "Concerning the central figure in the awaited End-drama there is considerable variation. In some visions the figure of Messiah is entirely absent. In such cases 'the kingdom was always represented as under the immediate sovereignty of God'" (21). "The divine intervention in history was the manifestation of the Kingdom of God.... [T]his would involve a total transformation of the present situation, hence the picture of world renewal enhanced sometimes by the idea of an entirely supernatural realm" (25–26). "The fourth Eclogue of Virgil presents the hope of a 'golden age' but in fundamental contrast to apocalyptic expectation; although it is on a cosmic scale, it is the hope of revolution from within rather than of intervention from without" (28).

9. Max Oelschlaeger, *The Idea of Wilderness: From Prehistory to the Age of Ecology* (New Haven, CT: Yale Univ. Press, 1991), 49–60.

10. Francis Bacon, *Novum Organum*, in *Works*, ed. James Spedding, Robert Leslie Ellis, and Douglas Devon Heath, 14 vols. (London: Longmans Green, 1870), 4:247–48, 114–15. See also Bacon's statement "I mean (according to the practice in civil causes) in this great plea or suit granted by the divine favor and providence (whereby the human race seeks to recover its right over nature) to examine nature herself and the arts upon interrogatories." Bacon, "Preparative towards a Natural and Experimental History," in *Works*, 4:263. William Leiss,

The Domination of Nature (New York: George Braziller, 1972), 48–52; Carolyn Merchant, *The Death of Nature: Women, Ecology and the Scientific Revolution* (San Francisco: Harper and Row, 1980), 185–86; Charles Whitney, *Francis Bacon and Modernity* (New Haven, CT: Yale Univ. Press, 1986), 25.

11. Marshall Sahlins, *Culture and Practical Reason* (Chicago: Univ. of Chicago Press, 1976), 53: "The development from a Hobbesian state of nature is the origin myth of Western capitalism."

12. On the definition of natural resources, see John Yeats, *Natural History of Commerce* (London, 1870), 2. Thomas Hobbes, *Leviathan* (1651), in *English Works*, 11 vols. (reprint ed., Aalen, W. Germany: Scientia, 1966), 3:145, 158. John Locke, *Two Treatises of Government* (1690), ed. Peter Laslett (Cambridge: Cambridge Univ. Press, 1960), Second Treatise, chap. 5, secs. 28, 32, 35, 37, 46, 48.

13. The Fall from Eden may be interpreted (as can the corn mother origin story; see note 1) as representing a transition from gathering-hunting to agriculture. In the Garden of Eden, Adam and Eve pick the fruits of the trees without having to labor in the earth (Genesis 1:29–30; Genesis 2:9). After the Fall they had to till the ground "in the sweat of thy face" and eat "the herb of the field" (Genesis 3:18, 19, 23). In Genesis 4, Abel, "keeper of sheep," is the pastoralist, while Cain, "tiller of the ground," is the farmer. Although God accepted Abel's lamb as a firstfruit, he did not accept Cain's offering. Cain's killing of Abel may represent the ascendancy of farming over pastoralism. Agriculture requires more intensive labor than either pastoralism or gathering. See Oelschlaeger, *Idea of Wilderness*; Callicott, "Genesis Revisited," 81.

14. On images and metaphors of nature as female in American History, see Annette Kolodny, *The Lay of the Land: Metaphor as Experience and History in American Life and Letters* (Chapel Hill: Univ. of North Carolina Press, 1975); idem, *The Land before Her: Fantasy and Experience of the American Frontier, 1630–1860* (Chapel Hill: Univ. of North Carolina Press, 1984); Vera Norwood and Janice Monk, eds., *The Desert Is No Lady: Southwestern Landscapes in Women's Writing and Art* (New Haven: Yale Univ. Press, 1987); Vera Norwood, *Made from This Earth: American Women and Nature* (Chapel Hill: Univ. of North Carolina Press, 1993); Sam Gill, *Mother Earth* (Chicago: Univ. of Chicago Press, 1987).

15. Thomas Morton, *New English Canaan*, in Peter Force, ed., *Tracts and Other Papers . . .* (Washington, D.C., 1838), 2:10.

16. Henry Colman, "Address before the Hampshire, Franklin and Hampden Agricultural Society Delivered in Greenfield, Oct. 23, 1833" (Greenfield, MA: Phelps and Ingersoll, 1833), 5–6, 15, 27.

17. *Scribner's Monthly*, Nov. 1880, 61. On the association of women with civilization and culture in nineteenth-century America, see Christopher Lasch, *The New Radicalism in America, 1889–1963* (New York: Norton, 1965), 65; Nancy Woloch, *Women and the American Experience* (New York: Knopf, 1984), chap. 6; Merchant, *Ecological Revolutions*, chap. 7.

18. J. Hector St. John de Crèvecoeur, "What Is an American?," *Letters from an American Farmer* (1782) (New York: E. P. Dutton, 1957), 39–43.

19. Richard F. Burton, *The City of the Saints and across the Rocky Mountains to California* (1861) (New York: Knopf, 1963), 72.

20. See also Roderick Nash, *Wilderness and the American Mind*, 3rd ed. (New Haven, CT: Yale Univ. Press, 1982); Richard Slotkin, *Regeneration through Violence: The Mythology of the Frontier* (Middletown, CT: Wesleyan Univ. Press, 1973); idem, *Gunfighter Nation: The Myth of the Frontier in Twentieth-Century America* (New York: Atheneum, 1992). For future discussions of these themes, see the chapters in this book by Kenneth Olwig, Anne Spirn, and William Cronon.

21. On the Renaissance distinction between *Natura naturans* and *Natura naturata*, see Eustace M. W. Tillyard, *The Elizabethan World Picture* (New York: Vintage, 1959), 46: "This giving a soul to nature—nature, that is, in the sense of *natura naturans*, the creative force, not of *natura naturata*, the natural creation—was a mildly unorthodox addition to the spiritual or intellectual beings. . . . Hooker, orthodox as usual, is explicit on this matter. [Nature] cannot be allowed a will of her own. . . . She is not even an agent . . . [but] is the direct and involuntary tool of God himself." See also Whitney, *Bacon and Modernity*, 123: "[T]he extreme dehumanization of [nature by] the Baconian scientist . . . is linked not simply to a complementary dehumanization of the feminine object of study, but to a somewhat anachronistic return to a more robust feminine image of nature as *natura naturans*." Spinoza likewise used the two terms, but with meanings rather different from those implied here. See *Spinoza Selections*, ed. John Wild (New York: Charles Scribner's Sons, 1930), 80–82; Harry A. Wolfson, *The Philosophy of Spinoza*, 2 vols. (1934) (New York: Meridian, 1958), 1:253–55.

22. Philip Elmer-Dewitt, "Fried Gene Tomatoes," *Time*, May 30, 1994, 54–55; Richard Keller Simon, "The Formal Garden in the Age of Consumer Culture: A Reading of the Twentieth-Century Shopping Mall," in Wayne Franklin and Michael Steiner, eds., *Mapping American Culture* (Iowa City: Univ. of Iowa Press, 1992), 231–50. . . .

23. Max Horkheimer and Theodor Adorno, *Dialectic of Enlightenment* (1944) (New York: Continuum, 1993), quotations on 3, 7, 9.

24. Maria Gimbutas, *The Goddesses and Gods of Old Europe, 6500–3500 B.C.* (Berkeley: Univ. of California Press, 1982); Merlin Stone, *When God Was a Woman* (New York: Harcourt Brace Jovanovich, 1976); Riane Eisler, *The Chalice and the Blade* (San Francisco: Harper and Row, 1988); Elinor Gadon, *The Once and Future Goddess* (San Francisco: Harper and Row, 1989); Monica Sjöö and Barbara Mor, *The Great Cosmic Mother: Rediscovering the Religion of the Earth* (San Francisco: Harper and Row, 1987); Pamela Berger, *The Goddess Obscured: The Transformation of the Grain Protectress from Goddess to Saint* (Boston: Beacon, 1985). On cultural ecofeminism see some of the essays in Irene Diamond and Gloria Orenstein, eds., *Reweaving the World: The Emergence of Ecofeminism* (San Francisco: Sierra Club Books, 1990).

25. Londa Schienbinger, *The Mind Has No Sex? Women in the Origins of Modern Science* (Cambridge, MA: Harvard Univ. Press, 1989); Evelyn Fox Keller, *Reflections on Gender and Science* (New Haven, CT: Yale Univ. Press, 1985).

26. Carolyn Merchant, *Radical Ecology: The Search for a Livable World* (New York: Routledge, 1992).

5

Excerpts from *Laudato Si*

POPE FRANCIS

A Universal Communion

89. The created things of this world are not free of ownership: "For they are yours, O Lord, who love the living" (*Wis* 11:26). This is the basis of our conviction that, as part of the universe, called into being by one Father, all of us are linked by unseen bonds and together form a kind of universal family, a sublime communion which fills us with a sacred, affectionate and humble respect. Here I would reiterate that "God has joined us so closely to the world around us that we can feel the desertification of the soil almost as a physical ailment, and the extinction of a species as a painful disfigurement."[1]

90. This is not to put all living beings on the same level nor to deprive human beings of their unique worth and the tremendous responsibility it entails. Nor does it imply a divinization of the earth which would prevent us from working on it and protecting it in its fragility. Such notions would end up creating new imbalances which would deflect us from the reality which challenges us.[2] At times we see an obsession with denying any pre-eminence to the human person; more zeal is shown in protecting other species than in defending the dignity which all human beings share in equal measure. Certainly, we should be concerned lest other living beings be treated irresponsibly. But we should be particularly indignant at the enormous inequalities in our midst, whereby we continue to tolerate some considering themselves more worthy than others. We fail to see that some are mired in desperate and degrading poverty, with no way out, while others have not the faintest idea of what to do with their possessions, vainly showing off their supposed superiority and leaving behind them so much waste which, if it were the case everywhere, would destroy the planet. In practice, we continue to tolerate that some consider themselves more human than others, as if they had been born with greater rights.

91. A sense of deep communion with the rest of nature cannot be real if our hearts lack tenderness, compassion and concern for our fellow human beings. It is clearly inconsistent to combat trafficking in endangered species while remaining completely indifferent to human trafficking, unconcerned about the poor, or undertaking to destroy another human being deemed unwanted. This compromises the very meaning of our struggle for the sake of the environment. It is no coincidence that, in the canticle in which Saint Francis praises God for his creatures, he goes on to say: "Praised be you my Lord, through those who give

pardon for your love." Everything is connected. Concern for the environment thus needs to be joined to a sincere love for our fellow human beings and an unwavering commitment to resolving the problems of society.

92. Moreover, when our hearts are authentically open to universal communion, this sense of fraternity excludes nothing and no one. It follows that our indifference or cruelty towards fellow creatures of this world sooner or later affects the treatment we mete out to other human beings. We have only one heart, and the same wretchedness which leads us to mistreat an animal will not be long in showing itself in our relationships with other people. Every act of cruelty towards any creature is "contrary to human dignity."[3] We can hardly consider ourselves to be fully loving if we disregard any aspect of reality: "Peace, justice and the preservation of creation are three absolutely interconnected themes, which cannot be separated and treated individually without once again falling into reductionism."[4] Everything is related, and we human beings are united as brothers and sisters on a wonderful pilgrimage, woven together by the love God has for each of his creatures and which also unites us in fond affection with brother sun, sister moon, brother river and mother earth. . . .

The Crisis and Effects of Modern Anthropocentrism

115. Modern anthropocentrism has paradoxically ended up prizing technical thought over reality, since "the technological mind sees nature as an insensate order, as a cold body of facts, as a mere 'given,' as an object of utility, as raw material to be hammered into useful shape; it views the cosmos similarly as a mere 'space' into which objects can be thrown with complete indifference."[5] The intrinsic dignity of the world is thus compromised. When human beings fail to find their true place in this world, they misunderstand themselves and end up acting against themselves: "Not only has God given the earth to man, who must use it with respect for the original good purpose for which it was given, but, man too is God's gift to man. He must therefore respect the natural and moral structure with which he has been endowed."[6]

116. Modernity has been marked by an excessive anthropocentrism which today, under another guise, continues to stand in the way of shared understanding and of any effort to strengthen social bonds. The time has come to pay renewed attention to reality and the limits it imposes; this in turn is the condition for a more sound and fruitful development of individuals and society. An inadequate presentation of Christian anthropology gave rise to a wrong understanding of the relationship between human beings and the world. Often, what was handed on was a Promethean vision of mastery over the world, which gave the impression that the protection of nature was something that only the faint-hearted cared about. Instead, our "dominion" over the universe should be understood more properly in the sense of responsible stewardship.[7]

117. Neglecting to monitor the harm done to nature and the environmental impact of our decisions is only the most striking sign of a disregard for the message contained in the structures of nature itself. When we fail to acknowledge as part of reality the worth of a poor person, a human embryo, a person with disabilities—to offer just a few examples—it becomes difficult to hear the cry of nature itself; everything is connected. Once the human being declares independence from reality and behaves with absolute dominion, the very foundations of our life begin to crumble, for "instead of carrying out his role as a cooperator with God in the work of creation, man sets himself up in place of God and thus ends up provoking a rebellion on the part of nature."[8]

118. This situation has led to a constant schizophrenia, wherein a technocracy which sees no intrinsic value in lesser beings coexists with the other extreme, which sees no special value in human beings. But one cannot prescind from humanity. There can be no renewal of our relationship with nature without a renewal of humanity itself. There can be no ecology without an adequate anthropology. When the human person is considered as simply one being among others, the product of chance or physical determinism, then "our overall sense of responsibility wanes."[9] A misguided anthropocentrism need not necessarily yield to "biocentrism," for that would entail adding yet another imbalance, failing to solve present problems and adding new ones. Human beings cannot be expected to feel responsibility for the world unless, at the same time, their unique capacities of knowledge, will, freedom and responsibility are recognized and valued.

119. Nor must the critique of a misguided anthropocentrism underestimate the importance of interpersonal relations. If the present ecological crisis is one small sign of the ethical, cultural and spiritual crisis of modernity, we cannot presume to heal our relationship with nature and the environment without healing all fundamental human relationships. Christian thought sees human beings as possessing a particular dignity above other creatures; it thus inculcates esteem for each person and respect for others. Our openness to others, each of whom is a "thou" capable of knowing, loving and entering into dialogue, remains the source of our nobility as human persons. A correct relationship with the created world demands that we not weaken this social dimension of openness to others, much less the transcendent dimension of our openness to the "Thou" of God. Our relationship with the environment can never be isolated from our relationship with others and with God. Otherwise, it would be nothing more than romantic individualism dressed up in ecological garb, locking us into a stifling immanence.

NOTES

1. Pope Francis, *Evangelii Gaudium: Apostolic Exhortation* (The Holy See: Vatican Publishing House, 2013), 215: *Acta Apostolicae Sedis* 105 (The Holy See: Vatican Publishing House, 2013), 1109.

2. Cf. Benedict XVI, "Encyclical Letter," in *Caritas in Veritate* (The Holy See: Vatican Publishing House, 2009), 14: *Acta Apostolicae Sedis* 101 (2009): 650.

3. Pope John Paul II, *Catechism of the Catholic Church* (The Holy See: Vatican Publishing House, 1992), 2418.

4. Conference of Dominican Bishops, *Sobre la relación del hombre con la naturaleza*, Pastoral Letter (21 January 1987).

5. Romano Guardini, *Das Ende der Neuzeit* (Schussenried, Germany: Hess Verlag, 1950), 63.

6. Pope John Paul II, "Encyclical Letter," in *Centesimus Annus* (The Holy See: Vatican Publishing House, 1991), 38: *Acta Apostolicae Sedis* 83 (1991): 841.

7. Cf. *Love for Creation: An Asian Response to the Ecological Crisis*, Declaration of the Colloquium sponsored by the Federation of Asian Bishops' Conferences (Tagatay, 31 January–5 February 1993), 3.3.2.

8. Pope John Paul II, "Encyclical Letter," in *Centesimus Annus* (The Holy See: Vatican Publishing House, 1991), 37: *Acta Apostolicae Sedis* 83 (1991): 840.

9. Benedict XVI, *Message for the 2010 World Day of Peace* (The Holy See: Vatican Publishing House, 2010), 2: *Acta Apostolicae Sedis* 102 (2010): 41.

6

Excerpts from "The Etiquette of Freedom"

GARY SNYDER

Coyote and Ground Squirrel do not break the compact they have with each other that one must play predator and the other play game. In the wild a baby Black-tailed Hare gets maybe one free chance to run across a meadow without looking up. There won't be a second. The sharper the knife, the cleaner the line of the carving. We can appreciate the elegance of the forces that shape life and the world, that have shaped every line of our bodies—teeth and nails, nipples and eyebrows. We also see that we must try to live without causing unnecessary harm, not just to fellow humans but to all beings. We must try not to be stingy, or to exploit others. There will be enough pain in the world as it is.

Such are the lessons of the wild. The school where these lessons can be learned, the realms of caribou and elk, elephant and rhinoceros, orca and walrus, are shrinking day by day. Creatures who have traveled with us through the ages are now apparently doomed, as their habitat—and the old, old habitat of humans—falls before the slow-motion explosion of expanding world economies. If the lad or lass is among us who knows where the secret heart of this Growth-Monster is hidden, let them please tell us where to shoot the arrow that will slow it down. And if the secret heart stays secret and our work is made no easier, I for one will keep working for wildness day by day.

*   *   *

"Wild and free." An American dream-phase loosing images: a long-maned stallion racing across the grasslands, a V of Canada Geese high and honking, a squirrel chattering and leaping limb to limb overhead in an oak. It also sounds like an ad for a Harley-Davidson. Both words, profoundly political and sensitive as they are, have become consumer baubles. I hope to investigate the meaning of *wild* and how it connects with *free* and what one would want to do with these meanings. To be truly free one must take on the basic conditions as they are—painful, impermanent, open, imperfect—and then be grateful for impermanence and the freedom it grants us. For in a fixed universe there would be no freedom. With that freedom we improve the campsite, teach children, oust tyrants. The world is nature, and in the long run inevitably wild, because the wild, as the process and essence of nature, is also an ordering of impermanence.

Although *nature* is a term that is not of itself threatening, the idea of the "wild" in civilized societies—both European and Asian—is often associated with unruliness, disorder, and violence. The Chinese word for nature, *zi-ran* (Japanese *shizen*) means "self-thus." It is a bland and general word. The word for wild in Chinese, *ye* (Japanese *ya*), which basically means "open country," has a wide set of meanings: in various combinations the term becomes illicit connection, desert country, an illegitimate child (open-country child), prostitute (open-country flower), and such. In an interesting case, *ye-man zi-yu* ("open-country southern-tribal-person-freedom") means "wild license." In another context "open-country story" becomes "fiction and fictitious romance." Other associations are usually with the rustic and uncouth. In a way *ye* is taken to mean "nature at its worst." Although the Chinese and Japanese have long given lip service to nature, only the early Daoists might have thought that wisdom could come of wildness.

Thoreau says "give me a wildness no civilization can endure." That's clearly not difficult to find. It is harder to imagine a civilization that wildness can endure, yet this is just what we must try to do. Wildness is not just the "preservation of the world," it *is* the world. Civilizations east and west have long been on a collision course with wild nature, and now the developed nations in particular have the witless power to destroy not only individual creatures but whole species, whole processes, of the earth. We need a civilization that can live fully and creatively together with wildness. We must start growing it right here, in the New World.

When we think of wilderness in America today, we think of remote and perhaps designated regions that are commonly alpine, desert, or swamp. Just a few centuries ago, when virtually *all* was wild in North America, wilderness was not something exceptionally severe. Pronghorn and bison trailed through the grasslands, creeks ran full of salmon, there were acres of clams, and grizzlies, cougar, and bighorn sheep were common in the lowlands. There were human beings, too: North America was *all populated*. One might say yes, but thinly—which raises the question of according to whom. The fact is, people were everywhere. When the Spanish foot soldier Alvar Núñez Cabeza de Vaca and his two companions (one of whom was African) were wrecked on the beach of what is now Galveston, and walked to the Rio Grande valley and then south back into Mexico between 1528 and 1536, there were few times in the whole eight years that they were not staying at a native settlement or camp. They were always on trails.

It has always been part of basic human experience to live in a culture of wilderness. There has been no wilderness without some kind of human presence for several hundred thousand years. Nature is not a place to visit, it is *home*—and within that home territory there are more familiar and less familiar places. Often there are areas that are difficult and remote, but all are *known* and even named. One August I was at a pass in the Brooks Range of northern Alaska at the headwaters of the Koyukuk River, a green three-thousand-foot tundra pass between the broad ranges, open and gentle, dividing the waters that flow to the Arctic Sea from the Yukon. It is as remote a place as you could be in North America, no roads, and the trails are those made by migrating caribou. Yet this pass has been steadily used by Inupiaq people of the north slope and Athapaskan people of the Yukon as a regular north-south trade route for at least seven thousand years.

All of the hills and lakes of Alaska have been named in one or another of the dozen or so languages spoken by the native people, as the researches of Jim Kari (*Denaina Elnena, Tanaina Country* [Fairbanks: University of Alaska Native Languages Center, 1982]; *Native Place Names in Alaska: Trends in Policy and Research* [Montreal: McGill University Symposium on Indigenous Names in the North, 1985]) and others have shown. Euro-American mapmakers name these places after transient exploiters, or their own girlfriends, or home towns in the Lower 48. The point is: it's all in the native story, yet only the tiniest trace of human presence through all that time shows. The place-based stories the people tell, and the naming they've done, is their archaeology, architecture, and *title* to the land. Talk about living lightly.

Cultures of wilderness live by the life and death lessons of subsistence economies. But what can we now mean by the words *wild* and for that matter *nature*? Languages meander like great rivers leaving oxbow traces over forgotten beds, to be seen only from the air or by scholars. Language is like some kind of infinitely interfertile family of species spreading or mysteriously declining over time, shamelessly and endlessly hybridizing, changing its own rules as it goes. Words are used as signs, as stand-ins, arbitrary and temporary, even as language reflects (and informs) the shifting values of the peoples whose minds it inhabits and glides through. We have faith in "meaning" the way we might believe in wolverines—putting trust in the occasional reports of others or on the authority of once seeing a pelt. But it is sometimes worth tracking these tricksters back.

The Words *Nature, Wild,* and *Wilderness*

Take *nature* first. The word *nature* is from Latin *natura*, "birth, constitution, character, course of things"—ultimately from *nasci*, to be born. So we have *nation, natal, native, pregnant*. The probable Indo-European root (via Greek *gna*—hence cognate, agnate) is *gen* (Sanskrit *jan*), which provides *generate* and *genus*, as well as *kin* and *kind*.

The word gets two slightly different meanings. One is "the outdoors"—the physical world, including all living things. Nature by this definition is a norm of the world that is apart from the features or products of civilization and human will. The machine, the artifact, the devised, or the extraordinary (like a two-headed calf) is spoken of as "unnatural." The other meaning, which is broader, is "the material world or its collective objects and phenomena," including the products of human action and intention. As an agency nature is defined as "the creative and regulative physical power which is conceived of as operating in the material world and as the immediate cause of all its phenomena." Science and some sorts of mysticism rightly propose that *everything* is natural. By these lights there is nothing unnatural about New York City, or toxic wastes, or atomic energy, and nothing—by definition—that we do or experience in life is "unnatural."

(The "supernatural"? One way to deal with it is to say that "the supernatural" is a name for phenomena which are reported by so few people as to leave their reality in doubt. Nonetheless these events—ghosts, gods, magical transformations, and such—are described often enough to make them continue to be intriguing and, for some, credible.)

The physical universe and all its properties—I would prefer to use the word *nature* in this sense. But it will come up meaning "the outdoors" or "other-than-human" sometimes even here.

*   *   *

The world *wild* is like a gray fox trotting off through the forest, ducking behind bushes, going in and out of sight. Up close, first glance, it is "wild"—then farther into the woods next glance it's "wyld" and it recedes via Old Norse *villr* and Old Teutonic *wilthijaz* into a faint pre-Teutonic *ghweltijos* which means, still, and wild and maybe wooded (*wald*) and lurks back there with possible connections to *will*, to Latin *silva* (forest, sauvage), and to the Indo-European root *ghwer*, base of Latin *ferus* (feral, fierce), which swings us around to Thoreau's "awful ferity" shared by virtuous people and lovers. The Oxford English Dictionary has it this way:

Of animals—not tame, undomesticated, unruly.

Of plants—not cultivated.

Of land—uninhabited, uncultivated.

Of foodcrops—produced or yielded without cultivation.

Of societies—uncivilized, rude, resisting constituted government.

Of individuals—unrestrained, insubordinate, licentious, dissolute, loose. "Wild and wanton widowes"—1614

Of behavior—violent, destructive, cruel, unruly.

Of behavior—artless, free, spontaneous. "Warble this native wood-notes wild"—John Milton.

Wild is largely defined in our dictionaries by what—from a human standpoint—it is not. It cannot be seen by this approach for what it *is*. Turn it the other way:

Of animals—free agents, each with its own endowments, living within natural systems.

Of plants—self-propagating, self-maintaining, flourishing in accord with innate qualities.

Of land—a place where the original and potential vegetation and fauna are intact and in full interaction and the landforms are entirely the result of nonhuman forces. Pristine.

Of foodcrops—food supplies made available and sustainable by the natural excess and exuberance of wild plants in their growth and in the production of quantities of fruit or seeds.

Of societies—societies whose order has grown from within and is maintained by the force of consensus and custom rather than explicit legislation. Primary cultures, which consider themselves the original and eternal inhabitants of their territory. Societies which resist economic and political domination by civilization. Societies whose economic system is in a close and sustainable relation to the local eco-system.

Of individuals—following local custom, style, and etiquette without concern for the standards of the metropolis or nearest trading post. Unintimidated, self-reliant, independent. "Proud and free."

Of behavior—fiercely resisting any oppression, confinement, or exploitation. Far-out, outrageous, "bad," admirable.

Of behavior—artless, free, spontaneous, unconditioned. Expressive, physical, openly sexual, ecstatic.

Most of the senses in this second set of definitions come very close to being how the Chinese define the term *Dao*, the *way* of Great Nature: eluding analysis, beyond categories, self-organizing, self-informing, playful, surprising, impermanent, insubstantial, independent, complete, orderly, unmediated, freely manifesting, self-authenticating, self-willed, complex, quite simple. Both empty and real at the same time. In some cases we might call it sacred. It is not far from the Buddhist term *Dharma* with its original senses of forming and firming.

* * *

The word *wilderness*, earlier *wyldernesse*, Old English *wildeornes*, possibly from "wild-deer-ness" (*deor*, deer and other forest animals) but more likely "wildernness," has the meanings:

A large area of wild land, with original vegetation and wildlife, ranging from dense jungle or rainforest to arctic or alpine "white wilderness."

A wasteland, as an area unused or useless for agriculture or pasture.

A space of sea or air, as in Shakespeare, "I stand as one upon a Rock, environ'd with a Wilderness of Sea" (*Titus Andronicus*). The oceans.

A place of danger and difficulty: where you take your own chances, depend on your own skills, and do not count on rescue.

This world as contrasted with heaven. "I walked through the wildernesse of this world"
   (*Pilgrim's Progress*)
A place of abundance, as in John Milton, "a wildernesse of sweets."

Milton's usage of wilderness catches the very real condition of energy and richness that is so often found in wild systems. "A wildernesse of sweets" is like the billions of herring or mackerel babies in the ocean, the cubic miles of krill, wild prairie grass seed (leading to the bread of this day, made from the germs of grasses)—all the incredible fecundity of small animals and plants, feeding the web. But from another side, wilderness has implied chaos, eros, the unknown, realms of taboo, the habitat of both the ecstatic and the demonic. In both senses it is a place of archetypal power, teaching, and challenge....

*   *   *

Wilderness is now—for much of North America—places that are formally set aside on public lands—Forest Service or Bureau of Land Management holdings or state and federal parks. Some tiny but critical tracts are held by private non-profit groups like The Nature Conservancy or the Trust for Public Land. These are the shrines saved from all the land that was once known and lived on by the original people, the little bits left as they were, the last little places where intrinsic nature totally wails, blooms, nests, glints away. They make up only 2 percent of the land of the United States.

*   *   *

But wilderness is not limited to the 2 percent formal wilderness areas. Shifting scales, it is everywhere: ineradicable populations of fungi, moss, mold, yeasts, and such that surround and inhabit us. Deer mice on the back porch, deer bounding across the freeway, pigeons in the park, spiders in the corners. There were crickets in the paint locker of the *Sappa Creek* oil tanker, as I worked as a wiper in the engine room out in mid-Pacific, cleaning brushes. Exquisite complex beings in their energy webs inhabiting the fertile corners of the urban world in accord with the rules of wild systems, the visible hardy stalks and stems of vacant lots and railroads, the persistent raccoon squads, bacteria in the loam and in our yogurt. The term *culture*, in its meaning of "a deliberately maintained aesthetic and intellectual life" and in its other meanings of "the totality of socially transmitted behavior patterns," is never far from a biological root meaning as in "yogurt culture"—a nourishing habitat. Civilization is permeable, and could be as inhabited as the wild is.

Wilderness may temporarily dwindle, but wildness won't go away. A ghost wilderness hovers around the entire planet: the millions of tiny seeds of the original vegetation are hiding in the mud on the foot of an arctic tern, in the dry desert sands, or in the wind. These seeds are each uniquely adapted to a specific soil or circumstance, each with its own little form and fluff, ready to float, freeze, or be swallowed, always preserving the germ. Wilderness will inevitably return, but it will not be as fine a world as the one that was glistening in the early morning of the Holocene. Much life will be lost in the wake of human agency on earth, that of the twentieth and twenty-first centuries. Much is already lost—the soils and waters unravel:

> "What's that dark thing in the water?
> Is it not an oil-soaked otter?"

Where do we start to resolve the dichotomy of the civilized and the wild?

*   *   *

Do you really believe you are an animal? We are now taught this in school. It is a wonderful piece of information: I have been enjoying it all my life and I come back to it over and over again, as something to investigate and test. I grew up on a small farm with cows and chickens, and with a second-growth forest right at the back fence, so I had the good fortune of seeing the human and animal as in the same realm. But many people who have been hearing this since childhood have not absorbed the implications of it, perhaps feel removed from the nonhuman world, are not sure they are animals. That's understandable: other animals might feel they are something different than "just animals" too. But we must contemplate the shared ground of our common biological being before emphasizing the differences.

Our bodies are wild. The involuntary quick turn of a head at a shout, the vertigo at looking off a precipice, the heart-in-the-throat in a moment of danger, the catch of the breath, the quiet moments relaxing, staring, reflecting—all universal responses of this mammal body. They can be seen throughout the class. The body does not require the intercession of some conscious intellect to make it breathe, to keep the heart beating. It is to a great extent self-regulating, it is a life of its own. Sensation and perception do not exactly come from outside, and the unremitting thought and image-flow are not exactly inside. The world is our consciousness, and it surrounds us. There are more things in mind, in the imagination, than "you" can keep track of—thoughts, memories, images, angers, delights, rise unbidden. The depths of mind, the unconscious, are our inner wilderness areas, and that is where a bobcat is *right now*. I do not mean personal bobcats in personal psyches, but the bobcat that roams from dream to dream. The conscious agenda-planning ego occupies a very tiny territory, a little cubicle somewhere near the gate, keeping track of some of what goes in and out (and sometimes making expansionistic plots), and the rest takes care of itself. The body is, so to speak, in the mind. They are both wild. . . .

*   *   *

The lessons we learn from the wild become the etiquette of freedom. We can enjoy our humanity with its flashy brains and sexual buzz, its social cravings and stubborn tantrums, and take ourselves as no more and no less than another being in the Big Watershed. We can accept each other all as barefoot equals sleeping on the same ground. We can give up hoping to be eternal and quit fighting dirt. We can chase off mosquitoes and fence out varmints without hating them. No expectations, alert and sufficient, grateful and careful, generous and direct. A calm and clarity attend us in the moment we are wiping the grease off our hands between tasks and glancing up at the passing clouds. Another joy is finally sitting down to have coffee with a friend. The wild requires that we learn the terrain, nod to all the plants and animals and birds, ford the streams and cross the ridges, and tell a good story when we get back home.

And when the children are safe in bed, at one of the great holidays like the Fourth of July, New Year's, or Halloween, we can bring out some spirits and turn on the music, and the men and the women who are still among the living can get loose and really wild. So that's the final meaning of "wild"—the esoteric meaning, the deepest and most scary. Those who are ready for it will come to it. Please do not repeat this to the uninitiated.

7

Excerpts from "The Land Ethic"

Aldo Leopold

All ethics so far evolved rest upon a single premise: that the individual is a member of a community of interdependent parts. His instincts prompt him to compete for his place in the community, but his ethics prompt him also to co-operate (perhaps in order that there may be a place to compete for).

The land ethic simply enlarges the boundaries of the community to include soils, waters, plants, and animals, or collectively: the land.

This sounds simple: do we not already sing our love for and obligation to the land of the free and the home of the brave? Yes, but just what and whom do we love? Certainly not the soil, which we are sending helter-skelter downriver. Certainly not the waters, which we assume to have no function except to turn turbines, float barges, and carry off sewage. Certainly not the plants, of which we exterminate whole communities without batting an eye. Certainly not the animals, of which we have already extirpated many of the largest and most beautiful species. A land ethic of course cannot prevent this alteration, management, and use of these "resources," but it does affirm their right to continued existence, and, at least in spots, their continued existence in a natural state.

In short, a land ethic changes the role of *Homo sapiens* from the conqueror of the land-community to plain member and citizen of it. It implies respect for his fellow-members, and also respect for the community as such.

Reading Questions and Further Readings

Reading Questions

1. What is the difference between nature and the environment?
2. What is the relation between nature and wildness?
3. What does Merchant mean by "recovery narrative"?
4. What is meant by "the Anthropocene"?
5. How would someone live who embraces the land ethic?

Further Readings

Callicott, J. Baird. *Thinking like a Planet: The Land Ethic and the Earth Ethic*. New York: Oxford University Press, 2014.

Daston, Lorraine, and Fernando Vidal, eds. *The Moral Authority of Nature*. Chicago: University of Chicago Press, 2003.

Gruen, Lori, Dale Jamieson, and Christopher Schlottmann, eds. *Reflecting on Nature: Readings in Environmental Ethics and Philosophy*. New York: Oxford University Press, 2012.

Jamieson, Dale, ed. *A Companion to Environmental Philosophy*. Hoboken, NJ: Wiley-Blackwell, 2003.

Quinn, Daniel. *Ishmael: An Adventure of the Mind and Spirit*. New York: Bantam Books, 1995.

Worster, Donald. *Nature's Economy: A History of Ecological Ideas*. Studies in Environment and History. New York: Cambridge University Press, 1994.

PART II

Environmentalism and Environmental Movements

Environmental Studies and the environmental movement are interconnected, both striving to understand how people interact with the environment and the conditions under which people organize to protect it. This part looks at the historical progression and change of the environmental movement from early American environmentalism, based on preservation of nature for spiritual value, to addressing contemporary problems like urban industrial pollution in developing countries. It also covers a variety of critiques of environmentalism, including that it has historically neglected issues of fairness and cultural difference and that it is better at bemoaning problems than inspiring solutions.

As the industrial revolution took off in the United States in the mid- to late 1800s, a number of thinkers and writers viewed the loss of natural landscapes in the face of urbanization as a tragedy—not "progress." The philosopher Henry David Thoreau, whose famous 1854 book *Walden* documented his two-year experiment living in a simple cottage in the woods outside of Boston, decried the city as the source of modern society's social ills and wrote that living close to the land revitalized both body and soul. The most well-known figure of early efforts to protect the countryside and wilderness was John Muir, an avid outdoorsman who succeeded in convincing the federal government to create America's first national park—Yosemite—in 1890. Despite this victory, he lost the battle to prevent a dam from being built in the park's Hetch Hetchy Valley to supply water and hydroelectric power to San Francisco. His impassioned plea to save Hetch Hetchy, excerpted in this part, captures the preservationist philosophy that animated the first wave of American environmentalism: the notion that wild places have intrinsic, even sacred, value and ought to be maintained in as pristine a condition as possible.

In the decades after World War II, environmentalists came to view the pace and scale of ecological destruction as not just a lamentable loss of wilderness but also a grave threat to humankind. Perhaps no one sounded the alarm bells as eloquently as Rachel Carson in her 1962 book *Silent Spring*, which forecast a future in which birds no longer sang because they had all been wiped out by DDT and other pesticides, which she likened to nuclear fallout. "Can anyone believe," Carson implored in a legendary passage reprinted here, "it is possible to lay down such a barrage of poisons on the surface of the earth without making it unfit for all life?" *Silent Spring* led directly to the ban of the use of DDT in the United States and is often credited with inspiring the "modern" environmental movement. Augmented by growing public outrage over a number of high-profile environmental catastrophes that culminated in the first Earth Day in 1970, the movement scored its most significant political victories in the following decade, including cajoling President Richard Nixon into creating the Environmental Protection Agency (EPA) and signing the Clean Water Act into law.

With the popular environmental movement focused on land preservation and federal regulations, many minorities and poor people—who were concentrated in cities—felt that environmentalists neglected the problems that disproportionately impacted them. Their views were supported by an explosive 1987 report titled *Toxic Waste and Race in the United States*, which found that race was the most significant predictor of living close to a hazardous waste facility (e.g., garbage incinerator, sewage treatment plant). The pioneering environmental sociologist Robert D. Bullard argued that the environmental movement was originally created by—and largely benefited—wealthier, more educated people. In this part's selection from his book *Dumping in Dixie*, Bullard details how people in the poorest neighborhoods, most of whom are minorities, are consigned to endure higher rates of asthma and other environment-induced illnesses than their wealthier and whiter counterparts do. These findings fed the fire of a burgeoning movement for "environmental justice," which demands equal protection from environmental hazards for all people—regardless of race, class, gender, or geography—and insists that community members are given a say in matters that impact their health.

The inclusion of social justice, health, and urban issues broadens the parameters of environmentalism. Robert Gottlieb aims to construct a more complex history of the environmental movement by highlighting how activists from prior eras who focused on industrial and occupational hazards should be reinterpreted as being just as engaged in environmental struggles as John Muir was. In such a retelling, the poor and people of color emerge as central—even founding—figures of environmentalism; and the incorporation of social movements for better housing, closed sewers, and safer working conditions under the banner of environmentalism challenges the myth that the movement was born after the publication of *Silent Spring*.

While "mainstream" environmentalism and the environmental justice movement have achieved some notable gains over the past half century, little progress has been made in decelerating—let alone reversing—the trend that many people argue is our world's greatest threat: climate change. This has led to a lot of hand wringing and soul searching among environmental advocates, many of whom wonder why it is that the public seems so apathetic in the face of dramatic ecological devastation. In the controversial essay "The Death of Environmentalism," Ted Nordhaus and Michael Shellenberger Nordhaus claim that the conventional environmentalist narrative, personified in Rachel Carson's "doom and gloom" prediction of a ravaged future, fails to motivate the public to act on behalf of the environment because it fosters a collective sense of cynicism and fatalism. The authors argue that the story of the human race is one of overcoming and achieving, and so they urge environmentalists to replace their apocalyptic prognostications with brighter visions of prosperity and possibility premised on embracing the economy and technology rather than demonizing them.

While Nordhaus and Shellenberger hope to help inspire a new universalistic paradigm of environmentalism, the reading by the historian Ramachandra Guha complains of "environmental imperialism," whereby wealthy Western countries impose their culturally specific notion of environmentalism onto poor developing countries. It is well known that a handful of rich nations like the U.S. have contributed the most to global warming and other environmental problems by consuming natural resources and producing waste at far greater levels than the rest of the world does, and yet poor countries disproportionately suffer from the effects of ecological degradation. Guha argues that Western environmentalists' attempts to atone for their societies' sins by advocating to preserve rainforests and charismatic species (e.g., elephants) in Asia, Africa, and Latin America end up inflicting additional harm on underprivileged populations. The wilderness ideal that they espouse, he writes, redefines local farmers, herders, and hunters—who have lived in the "wild" well before it became a "park"—as the enemy who must now be evicted in the name of environmental stewardship. A more effective and egalitarian solution to environmental problems, Guha implies, is to target the unsustainable lifestyles of the most privileged.

The category of "developing" or "poor" nations, however, masks the fact that environmental movements within these countries may also be elite driven and detrimental to the poor. In this part's final selection, the developmental sociologist Amita Baviskar details how the closure of polluting factories in Delhi, India, displaced over two million poor workers at the exact time that many of them lost their homes as their squatter camps were removed to make way for landscaped urban parks. Delhi's master plan, Baviskar contends, envisaged a hygienic and orderly model city but failed to recognize that its construction could occur only through the labor of the working poor, "for whom no provision had been made in the plans." Thus, it is not only environmental problems but also environmental solutions that can produce uneven social impacts. As citizens around the world strive to protect the environment and promote sustainability, the implication is that we ought always ask, "environment for whom?"

8

Hetch Hetchy Valley

JOHN MUIR

Yosemite is so wonderful that we are apt to regard it as an exceptional creation, the only valley of its kind in the world; but Nature is not so poor as to have only one of anything. Several other yosemites have been discovered in the Sierra that occupy the same relative positions on the Range and were formed by the same forces in the same kind of granite. One of these, the Hetch Hetchy Valley, is in the Yosemite National Park about twenty miles from Yosemite and is easily accessible to all sorts of travelers by a road and trail that leaves the Big Oak Flat road at Bronson Meadows a few miles below Crane Flat, and to mountaineers by way of Yosemite Creek basin and the head of the middle fork of the Tuolumne.

It is said to have been discovered by Joseph Screech, a hunter, in 1850, a year before the discovery of the great Yosemite. After my first visit to it in the autumn of 1871, I have always called it the "Tuolumne Yosemite," for it is a wonderfully exact counterpart of the Merced Yosemite, not only in its sublime rocks and waterfalls but in the gardens, groves and meadows of its flowery park-like floor. The floor of Yosemite is about 4000 feet above the sea; the Hetch Hetchy floor about 3700 feet. And as the Merced River flows through Yosemite, so does the Tuolumne through Hetch Hetchy. The walls of both are of gray granite, rise abruptly from the floor, are sculptured in the same style and in both every rock is a glacier monument.

Standing boldly out from the south wall is a strikingly picturesque rock called by the Indians, Kolana, the outermost of a group 2300 feet high, corresponding with the Cathedral Rocks of Yosemite both in relative position and form. On the opposite side of the Valley, facing Kolana, there is a counterpart of the El Capitan that rises sheer and plain to a height of 1800 feet, and over its massive brow flows a stream which makes the most graceful fall I have ever seen. From the edge of the cliff to the top of an earthquake talus it is perfectly free in the air for a thousand feet before it is broken into cascades among talus boulders. It is in all its glory in June, when the snow is melting fast, but fades and vanishes toward the end of summer. The only fall I know with which it may fairly be compared is the Yosemite Bridal Veil; but it excels even that favorite fall both in height and airy-fairy beauty and behavior. Lowlanders are apt to suppose that mountain streams in their wild career over cliffs lose control of themselves and tumble in a noisy chaos of mist and spray. On the contrary, on no part of their travels are they more harmonious and self-controlled. Imagine yourself in Hetch Hetchy on a sunny day in June, standing waist-deep in grass and flowers (as I have often stood), while the great pines sway dreamily with scarcely perceptible motion. Looking northward across the Valley you see a plain, gray granite cliff rising abruptly out of the gardens and groves to a height of 1800 feet, and in front of it Tueeulala's silvery scarf burning with irised sun-fire. In the first white outburst at the head there is abundance of visible energy, but it is speedily hushed and concealed in divine repose, and its tranquil progress to the base of the cliff is like that of a downy feather in a still room. Now observe the fineness and marvelous distinctness of the various sun-illumined fabrics into which the water is woven; they sift and float from form to form down the face of that grand gray rock in so leisurely and unconfused a manner that you can examine their texture, and patterns and tones of color as you would a piece of embroidery held in the hand. Toward the top of the fall you see groups of booming, comet-like masses, their solid, white heads separate, their tails like combed silk interlacing among delicate gray and purple shadows, ever forming and dissolving, worn out by friction in their rush through the air. Most of these vanish a few hundred feet below the summit, changing to varied forms of cloud-like drapery. Near the bottom the width of the fall has increased from about twenty-five feet to a hundred feet. Here it is composed of yet finer tissues, and is still without a trace of disorder—air, water and sunlight woven into stuff that spirits might wear.

So fine a fall might well seem sufficient to glorify any valley; but here, as in Yosemite, Nature seems in nowise moderate, for a short distance to the eastward of Tueeulala booms and thunders the great Hetch Hetchy Fall, Wapama, so near that you have both of them in full view from the same standpoint. It is the counterpart of the Yosemite Fall, but has a much greater volume of water, is about 1700 feet in height, and appears to be nearly vertical, though considerably inclined, and is dashed into huge outbounding bosses of foam on projecting shelves and knobs. No two falls could be more unlike—Tueeulala out in the open sunshine descending like thistledown; Wapama in a jagged, shadowy gorge roaring and thundering, pounding its way like an earthquake avalanche.

Besides this glorious pair there is a broad, massive fall on the main river a short distance above the head of the Valley. Its position is something like that of the Vernal in Yosemite, and its roar as it plunges into a surging trout-pool may be heard a long way, though it is only about twenty feet high. On Rancheria Creek, a large stream, corresponding in position with the Yosemite Tenaya Creek, there is a chain of cascades joined here and there with swift flashing plumes like the one between the Vernal and Nevada Falls, making magnificent shows as they go their glacier-sculptured way, sliding, leaping, hurrahing, covered with crisp clashing spray made glorious with sifting sunshine. And besides all these a few small streams come over the walls at wide intervals, leaping from ledge to ledge with birdlike song and watering many a hidden cliff-garden and fernery, but they are too unshowy to be noticed in so grand a place.

The correspondence between the Hetch Hetchy walls in their trends, sculpture, physical structure, and general arrangement of the main rock-masses and those of the Yosemite Valley has excited the wondering admiration of every observer. We have seen that the El Capitan and Cathedral rocks occupy the same relative positions in both valleys; so also do their Yosemite points and North Domes. Again, that part of the Yosemite north wall immediately to the east of the Yosemite Fall has two horizontal benches, about 500 and 1500 feet above the floor, timbered with golden-cup oak. Two benches similarly situated and timbered occur on the same relative portion of the Hetch Hetchy north wall, to the east of Wapama Fall, and on no other. The Yosemite is bounded at the head by the great Half Dome. Hetch Hetchy is bounded in the same way though its head rock is incomparably less wonderful and sublime in form.

The floor of the Valley is about three and a half miles long, and from a fourth to half a mile wide. The lower portion is mostly a level meadow about a mile long, with the trees restricted to the sides and the river banks, and partially separated from the main, upper, forested portion by a low bar of glacier-polished granite across which the river breaks in rapids.

The principal trees are the yellow and sugar pines, digger pine, incense cedar, Douglas spruce, silver fir, the California and golden-cup oaks, balsam cottonwood, Nuttall's flowering dogwood, alder, maple, laurel, tumion, etc. The most abundant and influential are the great yellow or silver pines like those of Yosemite, the tallest over two hundred feet in height, and the oaks assembled in magnificent groves with massive rugged trunks four to six feet in diameter, and broad, shady, wide-spreading heads. The shrubs forming conspicuous flowery clumps and tangles are manzanita, azalea, spiræa, brier-rose, several species of ceanothus, calycanthus, philadelphus, wild cherry, etc.; with abundance of showy and fragrant herbaceous plants growing about them or out in the open in beds by themselves—lilies, Mariposa tulips, brodiaeas, orchids, iris, spraguea, draperia, collomia, collinsia, castilleja, nemophila, larkspur, columbine, goldenrods, sunflowers, mints of many species, honeysuckle, etc. Many fine ferns dwell here also, especially the beautiful and interesting rock-ferns—pellaea, and cheilanthes of several species—fringing and rosetting dry rock-piles and ledges; woodwardia and asplenium on damp spots with fronds six or seven feet high; the delicate maiden-hair in mossy nooks by the falls, and the sturdy, broad-shouldered pteris covering nearly all the dry ground beneath the oaks and pines.

It appears, therefore, that Hetch Hetchy Valley, far from being a plain, common, rock-bound meadow, as many who have not seen it seem to suppose, is a grand landscape garden, one of Nature's rarest and most precious mountain temples. As in Yosemite, the sublime rocks of its walls seem to glow with life, whether leaning back in repose or standing erect in thoughtful attitudes, giving welcome to storms and calms alike, their brows in the sky, their feet set in the groves and gay flowery meadows, while birds, bees, and butterflies help the river and waterfalls to stir all the air into music—things frail and fleeting and types of permanence meeting here and blending, just as they do in Yosemite, to draw her lovers into close and confiding communion with her.

Sad to say, this most precious and sublime feature of the Yosemite National Park, one of the greatest of all our natural resources for the uplifting joy and peace and health of the people, is in danger of being dammed and made into a reservoir to help supply San Francisco with water and light, thus flooding it from wall to wall and burying its gardens and groves one or two hundred feet deep. This grossly destructive commercial scheme has long been planned and urged (though water as pure and abundant can be got from outside of the people's park, in a dozen different places), because of the comparative cheapness of the dam and of the territory which it is sought to divert from the great uses to which it was dedicated in the Act of 1890 establishing the Yosemite National Park.

The making of gardens and parks goes on with civilization all over the world, and they increase both in size and number as their value is recognized. Everybody needs beauty as well as bread, places to play in and pray in, where Nature may heal and cheer and give strength to body and soul alike. This natural beauty-hunger is made manifest in the little window-sill gardens of the poor, though perhaps only a geranium slip in a broken cup, as well as in the carefully tended rose and lily gardens of the rich, the thousands of spacious city parks and botanical gardens, and in our magnificent National parks—the Yellowstone, Yosemite, Sequoia, etc.—Nature's sublime wonderlands, the admiration and joy of the world. Nevertheless, like anything else worth while, from the very beginning, however well guarded, they have always been subject to attack by despoiling gainseekers and mischief-makers of every degree from Satan to Senators, eagerly trying to make everything immediately and selfishly commercial, with schemes disguised in smug-smiling philanthropy, industriously, shampiously crying, "Conservation, conservation, panutilization," that man and beast may be fed and the dear Nation made great. Thus long ago a few enterprising merchants utilized the Jerusalem temple as a place of business instead of a place of prayer, changing money, buying and selling cattle and sheep and doves; and earlier still, the first forest reservation, including only one tree, was likewise despoiled. Ever since the establishment of the Yosemite National Park, strife has been going on around its borders and I suppose this will go on as part of the universal battle between right and wrong, however much its boundaries may be shorn, or its wild beauty destroyed.

The first application to the Government by the San Francisco Supervisors for the commercial use of Lake Eleanor and the Hetch Hetchy Valley was made in 1903, and on December 22nd of that year it was denied by the Secretary of the Interior, Mr. Hitchcock, who truthfully said:

Presumably the Yosemite National Park was created such by law because within its boundaries, inclusive alike of its beautiful small lakes, like Eleanor, and its majestic wonders, like Hetch Hetchy and Yosemite Valley. It is the aggregation of such natural scenic features that makes the Yosemite Park a wonderland which the Congress of the United States sought by law to reserve for all coming time as nearly as practicable in the condition fashioned by the hand of the Creator—a worthy object of national pride and a source of healthful pleasure and rest for the thousands of people who may annually sojourn there during the heated months.

In 1907 when Mr. Garfield became Secretary of the Interior the application was renewed and granted; but under his successor, Mr. Fisher, the matter has been referred to a Commission, which as this volume goes to press still has it under consideration.

The most delightful and wonderful camp grounds in the Park are its three great valleys—Yosemite, Hetch Hetchy, and Upper Tuolumne; and they are also the most important places with reference to their positions relative to the other great features—the Merced and Tuolumne Cañons, and the High Sierra peaks and glaciers, etc., at the head of the rivers. The main part of the Tuolumne Valley is a spacious flowery lawn four or five miles long, surrounded by magnificent snowy mountains, slightly separated from other beautiful meadows, which together make a series about twelve miles in length, the highest reaching to the feet of Mount Dana, Mount Gibbs, Mount Lyell and Mount McClure. It is about 8500 feet above the sea, and forms the grand central High Sierra camp ground from which excursions are made to the noble mountains, domes, glaciers, etc.; across the Range to the Mono Lake and volcanoes and down the Tuolumne Cañon to Hetch Hetchy. Should Hetch Hetchy be submerged for a reservoir, as proposed, not only would it be utterly destroyed, but the sublime cañon way to the heart of the High Sierra would be hopelessly blocked and the great camping ground, as the watershed of a city drinking system, virtually would be closed to the public. So far as I have learned, few of all the thousands who have seen the park and seek rest and peace in it are in favor of this outrageous scheme.

One of my later visits to the Valley was made in the autumn of 1907 with the late William Keith, the artist. The leaf-colors were then ripe, and the great godlike rocks in repose seemed to glow with life. The artist, under their spell, wandered day after day along the river and through the groves and gardens, studying the wonderful scenery; and, after making about forty sketches, declared with enthusiasm that although its walls were less sublime in height, in picturesque beauty and charm Hetch Hetchy surpassed even Yosemite.

That any one would try to destroy such a place seems incredible; but sad experience shows that there are people good enough and bad enough for anything. The proponents of the dam scheme bring forward a lot of bad arguments to prove that the only righteous thing to do with the people's parks is to destroy them bit by bit as they are able. Their arguments are curiously like those of the devil, devised for the destruction of the first garden—so much of the very best Eden fruit going to waste; so much of the best Tuolumne water and Tuolumne scenery going to waste. Few of their statements are even partly true, and all are misleading.

Thus, Hetch Hetchy, they say, is a "low-lying meadow." On the contrary, it is a high-lying natural landscape garden, as the photographic illustrations show.

"It is a common minor feature, like thousands of others." On the contrary it is a very uncommon feature; after Yosemite, the rarest and in many ways the most important in the National Park.

"Damming and submerging it 175 feet deep would enhance its beauty by forming a crystal-clear lake." Landscape gardens, places of recreation and worship, are never made beautiful by destroying and burying them. The beautiful sham lake, forsooth, should be only an eyesore, a dismal blot on the landscape, like many others to be seen in the Sierra. For, instead of keeping it at the same level all the year, allowing Nature centuries of time to make new shores, it would, of course, be full only a month or two in the spring, when the snow is melting fast; then it would be gradually drained, exposing the slimy sides of the basin and shallower parts of the bottom, with the gathered drift and waste, death and decay of the upper basins, caught here instead of being swept on to decent natural burial along the banks of the river or in the sea. Thus the Hetch Hetchy dam-lake would be only a rough imitation of a natural lake for a few of the spring months, an open sepulcher for the others.

"Hetch Hetchy water is the purest of all to be found in the Sierra, unpolluted, and forever unpollutable." On the contrary, excepting that of the Merced below Yosemite, it is less pure than that of most of the other Sierra streams, because of the sewerage of camp grounds draining into it, especially of the Big Tuolumne Meadows camp ground, occupied by hundreds of tourists and mountaineers, with their animals, for months every summer, soon to be followed by thousands from all the world.

These temple destroyers, devotees of ravaging commercialism, seem to have a perfect contempt for Nature, and, instead of lifting their eyes to the God of the mountains, lift them to the Almighty Dollar.

Dam Hetch Hetchy! As well dam for water-tanks the people's cathedrals and churches, for no holier temple has ever been consecrated by the heart of man.

9

Excerpts from *Silent Spring*

RACHEL CARSON

A Fable for Tomorrow

There was once a town in the heart of America where all life seemed to live in harmony with its surroundings. The town lay in the midst of a checkerboard of prosperous farms, with fields of grain and hillsides of orchards where, in spring, white clouds of bloom drifted above the green fields. In autumn, oak and maple and birch set up a blaze of color that flamed and flickered across a backdrop of pines. Then foxes barked in the hills and deer silently crossed the fields, half hidden in the mists of the fall mornings.

Along the roads, laurel, viburnum and alder, great ferns and wildflowers delighted the traveler's eye through much of the year. Even in winter the roadsides were places of beauty, where countless birds came to feed on the berries and on the seed heads of the dried weeds rising above the snow. The countryside was, in fact, famous for the abundance and variety of its bird life, and when the flood of migrants was pouring through in spring and fall people traveled from great distances to observe them. Others came to fish the streams, which flowed clear and cold out of the hills and contained shady pools where trout lay. So it had been from the days many years ago when the first settlers raised their houses, sank their wells, and built their barns.

Then a strange blight crept over the area and everything began to change. Some evil spell had settled on the community: mysterious maladies swept the flocks of chickens; the cattle and sheep sickened and died. Everywhere was a shadow of death. The farmers spoke of much illness among their families. In the town the doctors had become more and more puzzled by new kinds of sickness appearing among their patients. There had been several sudden and unexplained deaths, not only among adults but even among children, who would be stricken suddenly while at play and die within a few hours.

There was a strange stillness. The birds, for example—where had they gone? Many people spoke of them, puzzled and disturbed. The feeding stations in the backyards were deserted. The few birds seen anywhere were moribund; they trembled violently and could not fly. It was a spring without voices. On the mornings that had once throbbed with the dawn chorus of robins, catbirds, doves, jays, wrens, and scores of other bird voices there was now no sound; only silence lay over the fields and woods and marsh.

On the farms the hens brooded, but no chicks hatched. The farmers complained that they were unable to raise any pigs—the litters were small and the young survived only a few days. The apple trees were coming into bloom but no bees droned among the blossoms, so there was no pollination and there would be no fruit.

The roadsides, once so attractive, were now lined with browned and withered vegetation as though swept by fire. These, too, were silent, deserted by all living things. Even the streams were now lifeless. Anglers no longer visited them, for all the fish had died.

In the gutters under the eaves and between the shingles of the roofs, a white granular powder still showed a few patches; some weeks before it had fallen like snow upon the roofs and the lawns, the fields and streams.

No witchcraft, no enemy action had silenced the rebirth of new life in this stricken world. The people had done it themselves.

*   *   *

This town does not actually exist, but it might easily have a thousand counterparts in America or elsewhere in the world. I know of no community that has experienced all the misfortunes I describe. Yet every one of these disasters has actually happened somewhere, and many real communities have already suffered a substantial number of them. A grim specter has crept upon us almost unnoticed, and this imagined tragedy may easily become a stark reality we all shall know.

What has already silenced the voices of spring in countless towns in America? This book is an attempt to explain.

The Obligation to Endure

The history of life on earth has been a history of interaction between living things and their surroundings. To a large extent, the physical form and the habits of the earth's vegetation and its animal life have been molded by the environment. Considering the whole span of earthly time, the opposite effect, in which life actually modifies its surroundings, has been relatively slight. Only within the moment of time represented by the present century has one species—man—acquired significant power to alter the nature of his world.

During the past quarter century this power has not only increased to one of disturbing magnitude but it has changed in character. The most alarming of all man's assaults upon the environment is the contamination of air, earth, rivers, and sea with dangerous and even lethal materials. This pollution is for the most part irrecoverable; the chain of evil it initiates not only in the world that must support life but in living tissues is for the most part irreversible. In this now universal contamination of the environment, chemicals are the sinister and little-recognized partners of radiation in changing the very nature of the world—the very nature of its life. Strontium 90, released through nuclear explosions into the air, comes to earth in rain or drifts down as fallout, lodges in soil, enters into the grass or corn or wheat grown there, and in time takes up its abode in the bones of a human being, there to remain until his death. Similarly, chemicals sprayed on croplands or forests or gardens lie long in soil, entering into living organisms, passing from one to another in a chain of poisoning and death. Or they pass mysteriously by underground streams until they emerge and, through the alchemy of air and sunlight, combine into new forms that kill vegetation, sicken cattle, and work unknown harm on those who drink from once pure wells. As Albert Schweitzer has said, "Man can hardly even recognize the devils of his own creation."

It took hundreds of millions of years to produce the life that now inhabits the earth—eons of time in which that developing and evolving and diversifying life reached a state of adjustment and balance with its surroundings. The environment, rigorously shaping and directing the life it supported, contained elements that were hostile as well as supporting. Certain rocks gave out dangerous radiation; even within the light of the sun, from which all life draws its energy, there were short-wave radiations with power to injure. Given time—time not in years but in millennia—life adjusts, and a balance has been reached. For time is the essential ingredient; but in the modern world there is no time.

The rapidity of change and the speed with which new situations are created follow the impetuous and heedless pace of man rather than the deliberate pace of nature. Radiation is no longer merely the background radiation of rocks, the bombardment of cosmic rays, the ultraviolet of the sun that have existed before there was any life on earth; radiation is now the unnatural creation of man's tampering with the atom. The chemicals to which life is asked to make its adjustment are no longer merely the calcium and silica and copper and all the rest of the minerals washed out of the rocks and carried in rivers to the sea; they are the synthetic creations of man's inventive mind, brewed in his laboratories, and having no counterparts in nature.

To adjust to these chemicals would require time on the scale that is nature's; it would require not merely the years of a man's life but the life of generations. And even this, were it by some miracle possible, would be futile, for the new chemicals come from our laboratories in an endless stream; almost five hundred annually find their way into actual use in the United States alone. The figure is staggering and its implications are not easily grasped—500 new chemicals to which the bodies of men and animals are required somehow to adapt each year, chemicals totally outside the limits of biologic experience.

Among them are many that are used in man's war against nature. Since the mid-1940's over 200 basic chemicals have been created for use in killing insects, weeds, rodents, and other organisms described in the modern vernacular as "pests"; and they are sold under several thousand different brand names.

These sprays, dusts, and aerosols are now applied almost universally to farms, gardens, forests, and homes—nonselective chemicals that have the power to kill every insect, the "good" and the "bad," to still the song of birds and the leaping of fish in the streams, to coat the leaves with a deadly film, and to linger on in soil—all this though the intended target may be only a few weeds or insects. Can anyone believe it is possible to lay down such a barrage of poisons on the surface of the earth without making it unfit for all life? They should not be called "insecticides," but "biocides."

The whole process of spraying seems caught up in an endless spiral. Since DDT was released for civilian use, a process of escalation has been going on in which ever more toxic materials must be found. This has happened because insects, in a triumphant vindication of Darwin's principle of the survival of the fittest, have evolved super races immune to the particular insecticide used, hence a deadlier one has always to be developed—and then a deadlier one than that. It has happened also because, for reasons to be described later, destructive insects often undergo a "flareback," or resurgence, after spraying, in numbers greater than before. Thus the chemical war is never won, and all life is caught in its violent crossfire.

Along with the possibility of the extinction of mankind by nuclear war, the central problem of our age has therefore become the contamination of man's total environment with such substances of incredible potential for harm—substances that accumulate in the tissues of plants and animals and even penetrate the germ cells to shatter or alter the very material of heredity upon which the shape of the future depends.

Some would-be architects of our future look toward a time when it will be possible to alter the human germ plasm by design. But we may easily be doing so now by inadvertence, for many chemicals, like radiation, bring about gene mutations. It is ironic to think that man might determine his own future by something so seemingly trivial as the choice of an insect spray.

All this has been risked—for what? Future historians may well be amazed by our distorted sense of proportion. How could intelligent beings seek to control a few unwanted species by a method that contaminated the entire environment and brought the threat of disease and death even to their own kind? Yet this is precisely what we have done. . . .

*　*　*

The problem whose attempted solution has brought such a train of disaster in its wake is an accompaniment of our modern way of life. Long before the age of man, insects inhabited the earth—a group of extraordinarily varied and adaptable beings. Over the course of time since man's advent, a small percentage of the more than half a million species of insects have come into conflict with human welfare in two principal ways: as competitors for the food supply and as carriers of human disease.

Disease-carrying insects become important where human beings are crowded together, especially under conditions where sanitation is poor, as in time of natural disaster or war or in situations of extreme poverty and deprivation. Then control of some sort becomes necessary. It is a sobering fact, however, as we shall presently see, that the method of massive chemical control has had only limited success, and also threatens to worsen the very conditions it is intended to curb.

Under primitive agricultural conditions the farmer had few insect problems. These arose with the intensification of agriculture—the devotion of immense acreages to a single crop. Such a system set the stage for explosive increases in specific insect populations. Single-crop farming does not take advantage of the principles by which nature works; it is agriculture as an engineer might conceive it to be. Nature has introduced great variety into the landscape, but man has displayed a passion for simplifying it. Thus he undoes the built-in checks and balances by which nature holds the species within bounds. One important natural check is a limit on the amount of suitable habitat for each species. Obviously then, an insect that lives on wheat can build up its population to much higher levels on a farm devoted to wheat than on one in which wheat is intermingled with other crops to which the insect is not adapted.

The same thing happens in other situations. A generation or more ago, the towns of large areas of the United States lined their streets with the noble elm tree. Now the beauty they hopefully created is threatened with complete destruction as disease sweeps through the elms, carried by a beetle that would have only limited chance to build up large populations and to spread from tree to tree if the elms were only occasional trees in a richly diversified planting.

Another factor in the modern insect problem is one that must be viewed against a background of geologic and human history: the spreading of thousands of different kinds of organisms from their native homes to invade new territories. This worldwide migration has been studied and graphically described by the British ecologist Charles Elton in his recent book *The Ecology of Invasions*. During the Cretaceous Period, some hundred million years ago, flooding seas cut many land bridges between continents and living things found themselves confined in what Elton calls "colossal separate nature reserves." There, isolated from others of their kind, they developed many new species. When some of the land masses were joined again, about 15 million years ago, these species began to move out into new territories—a movement that is not only still in progress but is now receiving considerable assistance from man.

The importation of plants is the primary agent in the modern spread of species, for animals have almost invariably gone along with the plants, quarantine being a comparatively recent and not completely effective innovation. The United States Office of Plant Introduction alone has introduced almost 200,000 species and varieties of plants from all over the world. Nearly half of the 180 or so major insect enemies of plants in the United States are accidental imports from abroad, and most of them have come as hitchhikers on plants.

In new territory, out of reach of the restraining hand of the natural enemies that kept down its numbers in its native land, an invading plant or animal is able to become enormously abundant. Thus it is no accident that our most troublesome insects are introduced species.

These invasions, both the naturally occurring and those dependent on human assistance, are likely to continue indefinitely. Quarantine and massive chemical campaigns are only extremely expensive ways of buying time. We are faced, according to Dr. Elton, "with a life-and-death need not just to find new technological means of suppressing this plant or that animal"; instead we need the basic knowledge of animal populations and their relations to their surroundings that will "promote an even balance and damp down the explosive power of outbreaks and new invasions."

Much of the necessary knowledge is now available but we do not use it. We train ecologists in our universities and even employ them in our governmental agencies but we seldom take their advice. We allow the chemical death rain to fall as though there were no alternative, whereas in fact there are many, and our ingenuity could soon discover many more if given opportunity.

Have we fallen into a mesmerized state that makes us accept as inevitable that which is inferior or detrimental, as though having lost the will or the vision to demand that which is good? Such thinking, in the words of the ecologist Paul Shepard, "idealizes life with only its head out of water, inches above the limits of toleration of the corruption of its own environment . . . Why should we tolerate a diet of weak poisons, a home in insipid surroundings, a circle of acquaintances who are not quite our enemies, the noise of motors with just enough relief to prevent insanity? Who would want to live in a world which is just not quite fatal?"

Yet such a world is pressed upon us. The crusade to create a chemically sterile, insect-free world seems to have engendered a fanatic zeal on the part of many specialists and most of the so-called control agencies. On every hand there is evidence that those engaged in spraying operations exercise a ruthless power. "The regulatory entomologists . . . function as prosecutor, judge and jury, tax assessor and collector and sheriff to enforce their own orders," said Connecticut entomologist Neely Turner. The most flagrant abuses go unchecked in both state and federal agencies.

It is not my contention that chemical insecticides must never be used. I do contend that we have put poisonous and biologically potent chemicals indiscriminately into the hands of persons largely or wholly ignorant of their potentials for harm. We have subjected enormous numbers of people to contact with these poisons, without their consent and often without their knowledge. If the Bill of Rights contains no guarantee that a citizen shall be secure against lethal poisons distributed either by private individuals or by public officials, it is surely only because our forefathers, despite their considerable wisdom and foresight, could conceive of no such problem.

I contend, furthermore, that we have allowed these chemicals to be used with little or no advance investigation of their effect on soil, water, wildlife, and man himself. Future generations are unlikely to condone our lack of prudent concern for the integrity of the natural world that supports all life.

There is still very limited awareness of the nature of the threat. This is an era of specialists, each of whom sees his own problem and is unaware of or intolerant of the larger frame into which it fits. It is also an era dominated by industry, in which the right to make a dollar at whatever cost is seldom challenged. When the public protests, confronted with some obvious evidence of damaging results of pesticide applications, it is fed little tranquilizing pills of half truth. We urgently need an end to these false assurances, to the sugar coating of unpalatable facts. It is the public that is being asked to assume the risks that the insect controllers calculate. The public must decide whether it wishes to continue on the present road, and it can do so only when in full possession of the facts. In the words of Jean Rostand, "The obligation to endure gives us the right to know."

10

Excerpts from "Environmentalism and Social Justice"

ROBERT D. BULLARD

The environmental movement in the United States emerged with agendas that focused on such areas as wilderness and wildlife preservation, resource conservation, pollution abatement, and population control. It was supported primarily by middle- and upper-middle-class whites. Although concern about the environment cuts across racial and class lines, environmental activism has been most pronounced among individuals who have above-average education, greater access to economic resources, and a greater sense of personal efficacy.[1] . . .

An abundance of documentation shows blacks, lower-income groups, and working-class persons are subjected to a disproportionately large amount of pollution and other environmental stressors in their neighborhoods as well as in their workplaces.[2] However, these groups have only been marginally involved in the nation's environmental movement. . . .

Research on environmental quality in black communities has been minimal. Attention has been focused on such problems as crime, drugs, poverty, unemployment, and family crisis. Nevertheless, pollution is exacting a heavy toll (in health and environmental costs) on black communities across the nation. There are few studies that document, for example, the way blacks cope with environmental stressors such as municipal solid-waste facilities, hazardous-waste landfills, toxic-waste dumps, chemical emissions from industrial plants, and on-the-job hazards that pose extreme risks to their health. . . .

Many of the interactions that emerged among core environmentalists, the poor, and blacks can be traced to distributional equity questions. How are the benefits and burdens of environmental reform distributed? Who gets what, where, and why? Are environmental inequities a result of racism or class barriers or a combination of both? After more than two decades of modern environmentalism, the equity issues have not been resolved. There has been, however, some change in the way environmental problems are presented by mainstream environmental organizations. More important, environmental equity has now become a major item on the local (grassroots) as well as national civil rights agenda.[3] . . .

The problem of polluted black communities is not a new phenomenon. Historically, toxic dumping and the location of locally unwanted land uses (LULUs) have followed the "path of least resistance," meaning black and poor communities have been disproportionately burdened with these types of exter-nalities. However, organized black resistance to toxic dumping, municipal waste facility siting, and discriminatory environmental and land-use decisions is a relatively recent phenomenon.[4] Black environmental concern has been present but too often has not been followed up with action. . . .

The environmental movement of the 1960s and 1970s, dominated by the middle class, built an impressive political base for environmental reform and regulatory relief. Many environmental problems of the 1980s and 1990s, how-ever, have social impacts that differ somewhat from earlier ones. Specifically, environmental problems have had serious regressive impacts. These impacts have been widely publicized in the media, as in the case of the hazardous-waste problems at Love Canal and Times Beach. The plight of polluted minority com-munities is not as well known as the New York and Missouri tragedies. Nev-ertheless, a disproportionate burden of pollution is carried by the urban poor and minorities.[5]

Few environmentalists realized the sociological implications of the not-in-my-backyard (NIMBY) phenomenon.[6] Given the political climate of the times, the hazardous wastes, garbage dumps, and polluting industries were likely to end up in somebody's backyard. But whose backyard? More often than not, these LULUs ended up in poor, powerless, black communities rather than in afflu-ent suburbs. This pattern has proven to be the rule, even though the benefits derived from industrial waste production are directly related to affluence.[7] Pub-lic officials and private industry have in many cases responded to the NIMBY phenomenon using the place-in-blacks'-backyard (PIBBY) principle.[8] . . .

The push for environmental equity in the black community has much in common with the development of the modern civil rights movement that began in the South. That is, protest against discrimination has evolved from "organiz-ing efforts of activists functioning through a well-developed indigenous base."[9] Indigenous black institutions, organizations, leaders, and networks are coming together against polluting industries and discriminatory environmental policies. This book addresses this new uniting of blacks against institutional barriers of racism and classism.

Race versus Class in Spatial Location . . .

Houston, Texas, the nation's fourth largest city, is a classic example of an area where race has played an integral part in land-use outcomes and municipal ser-vice delivery.[10] As late as 1982, there were neighborhoods in Houston that still did not have paved streets, gas and sewer connections, running water, regular garbage service, and street markers. Black and Hispanic neighborhoods were far more likely to have service deficiencies than their white counterparts. One of the neighborhoods (Bordersville) was part of the land annexed for the bustling

Houston Intercontinental Airport. Another area, Riceville, was a stable black community located in the city's sprawling southwest corridor, a mostly white sector that accounted for nearly one-half of Houston's housing construction in the 1970s.

The city's breakneck annexation policy stretched municipal services thin. Newly annexed unincorporated areas, composed of mostly whites, often gained at the expense of older minority areas. How does one explain the service disparities in this modern Sunbelt city? After studying the Houston phenomenon for nearly a decade, I have failed to turn up a single case of a white neighborhood (low- or middle-income) in the city that was systematically denied basic municipal services. The significance of race may have declined, but racism has not disappeared when it comes to allocating scarce resources.

Do middle-income blacks have the same mobility options that are available to their white counterparts? The answer to this question is no. Blacks have made tremendous economic and political gains in the past three decades with the passage of equal opportunity initiatives at the federal level. Despite legislation, court orders, and federal mandates, institutional racism and discrimination continue to influence the quality of life in many of the nation's black communities.[11]

The differential residential amenities and land uses assigned to black and white residential areas cannot be explained by class alone. For example, poor whites and poor blacks do not have the same opportunities to "vote with their feet." Racial barriers to education, employment, and housing reduce mobility options available to the black underclass and the black middle class.[12] . . .

Environmental degradation takes an especially heavy toll on inner-city neighborhoods because the "poor or near poor are the ones most vulnerable to the assaults of air and water pollution, and the stress and tension of noise and squalor."[13] A high correlation has been discovered between characteristics associated with disadvantage (i.e., poverty, occupations below management and professional levels, low rent, and a high concentration of black residents [due to residential segregation and discriminatory housing practices]) and poor air quality.[14] Individuals that are in close proximity to health-threatening problems (i.e., industrial pollution, congestion, and busy freeways) are living in endangered environs. The price that these individuals pay is in the form of higher risks of emphysema, chronic bronchitis, and other chronic pulmonary diseases.[15]

Blacks and other economically disadvantaged groups are often concentrated in areas that expose them to high levels of toxic pollution: namely, urban industrial communities with elevated air and water pollution problems or rural areas with high levels of exposure to farm pesticides. . . .

All Americans, white or black, rich or poor, are entitled to equal protection under the law. Just as this is true for such areas as education, employment, and housing, it also applies to one's physical environment. Environmental discrimination is a fact of life. Here, environmental discrimination is defined as disparate treatment of a group or community based on race, class, or some other distinguishing characteristic. The struggle for social justice by black Americans has been and continues to be rooted in white racism. White racism is a factor in the impoverishment of black communities and has made it easier for black residential areas to become the dumping grounds for all types of health-threatening toxins and industrial pollution.

Government and private industry in general have followed the "path of least resistance" in addressing externalities as pollution discharges, waste disposal, and nonresidential activities that may pose a health threat to nearby communities.[16] Middle- and upper-class households can often shut out the fumes, noise, and odors with their air conditioning, dispose of their garbage to keep out the rats and roaches, and buy bottled water for drinking.[17] Many lower-income households (black or white) cannot afford such "luxury" items; they are subsequently forced to adapt to a lower-quality physical environment. . . .

Why has this happened and what have blacks done to resist these practices? In order to understand the causes of the environmental dilemma that many black and low-income communities find themselves in, the theoretical foundation of environmentalism needs to be explored.

The Theoretical Basis of Environmental Conflict

Environmentalism in the United States grew out of the progressive conservation movement that began in the 1890s. The modern environmental movement, however, has its roots in the civil rights and antiwar movements of the late 1960s.[18] The more radical student activists splintered off from the civil rights and antiwar movements to form the core of the environmental movement in the early 1970s. The student environmental activists affected by the 1970 Earth Day enthusiasm in colleges and universities across the nation had hopes of bringing environmental reforms to the urban poor. They saw their role as environmental advocates for the poor since the poor had not taken action on their own.[19] They were, however, met with resistance and suspicion. Poor and minority residents saw environmentalism as a disguise for oppression and as another "elitist" movement.[20]

Environmental elitism has been grouped into three categories: (1) *compositional elitism* implies that environmentalists come from privileged class strata, (2) *ideological elitism* implies that environmental reforms are a subterfuge for distributing the benefits to environmentalists and costs to nonenvironmentalists, and (3) *impact elitism* implies that environmental reforms have regressive distributional impacts.[21]

Impact elitism has been the major sore point between environmentalists and advocates for social justice who see some reform proposals creating, exacerbating, and sustaining social inequities. Conflict centered largely on the "jobs versus environment" argument. Imbedded in this argument are three competing advocacy groups (1) *environmentalists* are concerned about leisure and recreation, wildlife and wilderness preservation, resource conservation, pollution abatement, and industry regulation, (2) *social justice advocates'* major concerns include basic civil rights, social equity, expanded opportunity, economic mobility, and institutional discrimination, and (3) *economic boosters* have as their chief concerns maximizing profits, industrial expansion, economic stability, laissez-faire operation, and deregulation. . . .

The offer of a job (any job) to an unemployed worker appears to have served a more immediate need than the promise of a clean environment. There is evidence that new jobs have been created as a direct result of environmental reforms.[22] Who got these new jobs? The newly created jobs are often taken by people who already have jobs or by migrants who possess skills greater than the indigenous workforce. More often than not, "newcomers intervene between the jobs and the local residents, especially the disadvantaged."[23]

Minority residents can point to a steady stream of industrial jobs leaving their communities. Moreover, social justice advocates take note of the miserable track record that environmentalists and preservationists have on improving environmental quality in the nation's racially segregated inner cities and hazardous industrial workplaces, and on providing housing for low-income groups. Decent and affordable housing, for example, is a top environmental problem for inner-city blacks. On the other hand, environmentalists' continued emphasis on wilderness and wildlife preservation appeal to a population that can afford leisure time and travel to these distant locations. This does not mean that poor people and people of color are not interested in leisure or outdoor activities. Many wilderness areas and national parks remain inaccessible to the typical inner-city resident because of inadequate transportation. Physical isolation, thus, serves as a major impediment to black activism in the mainstream conservation and resource management activities.

Translating Concern into Action . . .

There is no single agenda or integrated political philosophy in the hundreds of environmental organizations found in the nation. The type of issues that environmental organizations choose can greatly influence the type of constituents they attract.[24] The issues that are most likely to attract the interests of black community residents are those that have been couched in a civil rights or equity framework (see Table 10.1). They include those that (1) focus on inequality and distributional impacts, (2) endorse the "politics of equity" and direct action, (3) appeal to urban mobilized groups, (4) advocate safeguards against

TABLE 10.1. Type of Environmental Groups and Issue Characteristics
That Appeal to Black Community Residents

Issue Characteristic	Type of Environmental Group			
	Mainstream	Grassroots	Social Action	Emergent Coalition
Appeal to urban mobilized groups	–	+	+	+
Concern about inequality and distributional impacts	–/+	–/+	+	+
Endorse the "politics of equity" and direct action	–/+	+	+	–/+
Focus on economic-environment trade-offs	–	–/+	+	+
Champion of the political and economic "underdog"	–	–/+	+	+

–: Group is unlikely to have characteristic.
+: Group is likely to have characteristic.
- /+: Group in some cases may have characteristic.
Source: Adapted from Richard P. Gale, "The Environmental Movement and the Left: Antagonists or Allies?," *Sociological Inquiry* 53 (Spring 1983): 194, Table 1.

environmental blackmail with a strong pro-development stance, and (5) are ideologically aligned with policies that favor social and political "underdogs."

Mainstream environmental organizations, including the "classic" and "mature" groups, have had a great deal of influence in shaping the nation's environmental policy. Classic environmentalism continues to have a heavy emphasis on preservation and outdoor recreation, while mature environmentalism is busy in the area of "tightening regulations, seeking adequate funding for agencies, occasionally focusing on compliance with existing statutes through court action, and opposing corporate efforts to repeal environmental legislation or weaken standards."[25] These organizations, however, have not had a great deal of success in attracting working-class persons, the large black population in the nation's inner cities, and the rural poor. Many of these individuals do not see the mainstream environmental movement as a vehicle that is championing the causes of the "little man," the "underdog," or the "oppressed."[26]

Recently emerged grassroots environmental groups, some of which are affiliated with mainstream environmental organizations, have begun to bridge the class and ideological gap between core environmentalists (e.g., the Sierra Club) and grassroots organizations (e.g., local activist groups in southeast Louisiana). In some cases, these groups mirror their larger counterparts at the national level in terms of problems and issues selected, membership, ideological alignment, and tactics. Grassroots groups often are organized around area-specific and single-issue problems. They are, however, more inclusive than mainstream environmental organizations in that they focus primarily on local problems.

Grassroots environmental organizations, however, may or may not choose to focus on equity, distributional impacts, and economic-environmental trade-off issues. These groups do appeal to some black community residents, especially those who have been active in other confrontational protest activities. . . .

Social action groups that take on environmental issues as part of their agenda are often on the political Left. They broaden their base of support and sphere of influence by incorporating environmental equity issues as agenda items that favor the disenfranchised. The push for environmental equity is an extension of the civil rights movement, a movement in which direct confrontation and the politics of protest have been real weapons. In short, social action environmental organizations retain much of their civil rights flavor.

Other environmental groups that have appealed to black community residents grew out of coalitions between environmentalists (mainstream and grassroots), social action advocates, and organized labor.[27] These somewhat fragile coalitions operate from the position that social justice and environmental quality are compatible goals. Although these groups are beginning to formulate agendas for action, mistrust still persists as a limiting factor. These groups are often biracial with membership cutting across class and geographic boundaries. There is a down side to these types of coalition groups. For example, compositional factors may engender less group solidarity and sense of "control" among black members, compared to the indigenous social action or grassroots environmental groups where blacks are in the majority and make the decisions. The question of "who is calling the shots" is ever present.

Environmentalists, thus, have had a difficult task convincing blacks and the poor that they are on their side. Mistrust is engendered among economically and politically oppressed groups in this country when they see environmental reforms being used to direct social and economic resources away from problems of the poor toward priorities of the affluent. For example, tighter government regulations and public opposition to disposal facility siting have opened up the Third World as the new dumping ground for this nation's toxic wastes. Few of these poor countries have laws or the infrastructure to handle the wastes from the United States and other Western industrialized nations.[28] Blacks and other ethnic minorities in this country also see their communities being inundated with all types of toxics. This has been especially the case for the southern United States (one of the most underdeveloped regions of the nation) where more than one-half of all blacks live.

Environmentalism and Civil Rights

The civil rights movement has its roots in the southern United States. Southern racism deprived blacks of "political rights, economic opportunity, social justice, and human dignity."[29] The new environmental equity movement also is centered in the South, a region where marked ecological disparities exist between black and white communities.[30] The 1980s have seen the emergence of a small cadre of blacks who see environmental discrimination as a civil rights issue. A fragile alliance has been forged between organized labor, blacks, and environmental groups as exhibited by the 1983 Urban Environment Conference workshops held in New Orleans.[31] Environmental and civil rights issues were presented as compatible agenda items by the conference organizers. Environmental protection and social justice are not necessarily incompatible goals.[32] . . .

Institutional racism continues to affect policy decisions related to the enforcement of environmental regulations. Slowly, blacks, lower-income groups, and working-class persons are awakening to the dangers of living in a polluted environment. They are beginning to file and win lawsuits challenging governments and private industry that would turn their communities into the dumping grounds for all type of unwanted substances and activities. Whether it is a matter of deciding where a municipal landfill or hazardous-waste facility will be located, or getting a local chemical plant to develop better emergency notification, or trying to secure federal assistance to clean up an area that has already been contaminated by health-threatening chemicals, it is apparent that blacks and other minority groups must become more involved in environmental issues if they want to live healthier lives.

Black communities, mostly in the South, are beginning to initiate action (protests, demonstrations, picketing, political pressure, litigation, and other forms of direct action) against industries and governmental agencies that have targeted their neighborhoods for nonresidential uses including municipal garbage, hazardous wastes, and polluting industries. The environmental "time bombs" that are ticking away in these communities are not high on the agendas of mainstream environmentalists nor have they received much attention from mainstream civil rights advocates. Moreover, polluted black communities have received little national media coverage or remedial action from governmental agencies charged with cleanup of health-threatening pollution problems. The time is long overdue for placing the toxics and minority health concerns (including stress induced from living in contaminated communities) on the agenda of federal and state environmental protection and regulatory agencies. The Commission for Racial Justice's *Toxic Wastes and Race* has at least started government officials, academicians, and grassroots activists talking about environmental problems that disproportionately affect minority communities.

NOTES

1. See Frederick R. Buttel and William L. Flinn, "Social Class and Mass Environmental Beliefs: A Reconsideration," *Environment and Behavior* 10 (1978): 433–450; Kenneth M. Bachrach and Alex J. Zautra, "Coping with Community Stress: The Threat of a Hazardous Waste

Landfill," *Journal of Health and Social Behavior* 26 (1985): 127–141; Paul Mohai, "Public Concern and Elite Involvement in Environmental-Conservation Issues," *Social Science Quarterly* 66 (1985): 820–838.

2. See Morris E. Davis, "The Impact of Workplace Health and Safety on Black Workers: Assessment and Prognosis," *Labor Studies Journal* 4 (Spring 1981): 29–40; Richard Kazis and Richard Grossman, *Fear at Work: Job Blackmail, Labor, and the Environment* (New York: Pilgrim, 1983), Ch. 1; W. J. Kruvant, "People, Energy, and Pollution," in Dorothy K. Newman and Dawn Day, eds., *The American Energy Consumer* (Cambridge, MA: Ballinger, 1975), 125–167; Robert D. Bullard, "Solid Waste Sites and the Black Houston Community," *Sociological Inquiry* 53 (Spring 1983): 273–288; Robert D. Bullard, "Endangered Environs: The Price of Unplanned Growth in Boomtown Houston," *California Sociologist* 7 (Summer 1984): 85–101; Robert D. Bullard and Beverly H. Wright, "Dumping Grounds in a Sunbelt City," *Urban Resources* 2 (Winter 1985): 37–39.

3. Robert D. Bullard and Beverly H. Wright, "Environmentalism and the Politics of Equity: Emergent Trends in the Black Community," *Mid-American Review of Sociology* 12 (Winter 1987): 21–37.

4. See Robert D. Bullard and Beverly H. Wright, "Blacks and the Environment," *Humboldt Journal of Social Relations* 14 (Summer 1987): 165–184; Robert D. Bullard, "Solid Waste Sites and the Black Houston Community," 273–288; Bullard, "Endangered Environs," 84–102.

5. Brian J. L. Berry, ed., *The Social Burden of Environmental Pollution: A Comparative Metropolitan Data Source* (Cambridge, MA: Ballinger, 1977); Sam Love, "Ecology and Social Justice: Is There a Conflict?," *Environmental Action* 4 (1972): 3–6; Julian McCaull, "Discriminatory Air Pollution: If Poor Don't Breathe," *Environment* 19 (1976): 26–32; Vernon Jordon, "Sins of Omission," *Environmental Action* 11 (1980): 26–30.

6. Denton E. Morrison, "How and Why Environmental Consciousness Has Trickled Down," in Allan Schnaiberg, Nicholas Watts, and Klaus Zimmermann, eds., *Distributional Conflict in Environmental-Resource Policy* (New York: St. Martin's, 1986), 187–220.

7. Robert D. Bullard and Beverly H. Wright, "The Politics of Pollution: Implications for the Black Community," *Phylon* 47 (1986): 71–78.

8. Bullard and Wright, "Environmentalism and the Politics of Equity," 28.

9. A. D. Morris, *The Origins of the Civil Rights Movement* (New York: Free Press, 1986), xii.

10. See Robert D. Bullard, *Invisible Houston: The Black Experience in Boom and Bust* (College Station: Texas A&M University Press, 1987), 14–31.

11. Robert D. Bullard, "Blacks and the American Dream of Housing," in Jamshid A. Momeni, ed., *Race, Ethnicity, and Minority Housing in the United States* (Westport, CT: Greenwood, 1986), 53–63; Bullard and Wright, "Environmentalism and the Politics of Equity," 21–37.

12. Robert L. Lineberry, *Equity and Urban Policy: The Distribution of Municipal Public Services* (Beverly Hills, CA: Sage, 1977), 174–175.

13. Daniel Zwerdling, "Poverty and Pollution," *Progressive* 37 (1973): 25–29.

14. Kruvant, "People, Energy, and Pollution," 125–167.

15. Douglas Lee and H. K. Lee, "Conclusions and Reservations," in Douglas Lee, ed., *Environmental Factors in Respiratory Disease* (New York: Academic, 1972), 250–251; Ronald Brownstein, "The Toxic Tragedy," in Ralph Nader, Ronald Brownstein, and John Richard, eds., *Who's Poisoning America: Corporate Polluters and Their Victims in the Chemical Age* (San Francisco: Sierra Club Books, 1982), 1–52.

16. Bullard and Wright, "Blacks and the Environment," 170–171.

17. Zwerdling, "Poverty and Pollution," 27; Bullard and Wright, "The Politics of Pollution," 71–78.

18. C. R. Humphrey and F. H. Buttel, *Environment, Energy and Society* (Belmont, CA: Wadsworth, 1982), 11–136; R. P. Gale, "The Environmental Movement and the Left," *Sociological Inquiry* 53, nos. 2–3 (1983): 179–199.

19. Samuel P. Hays, *Beauty, Health, and Permanence: Environmental Politics in the United States, 1955–1985* (Cambridge: Cambridge University Press, 1987), 269.

20. David L. Sills, "The Environmental Movement and Its Critics," *Human Ecology* 13 (1975): 1–41; Denton E. Morrison, "The Soft Cutting Edge of Environmentalism: Why and How the Appropriate Technology Notion Is Changing the Movement," *Natural Resources Journal* 20 (April 1980): 275–298; Allan Schnaiberg, "Redistributive Goals versus Distributive Politics: Social Equity Limits in Environmentalism and Appropriate Technology Movements," *Sociological Inquiry* 53 (Spring 1983): 200–219.

21. Denton E. Morrison and Riley E. Dunlap, "Environmentalism and Elitism: A Conceptual and Empirical Analysis," *Environmental Management* 10 (1986): 581–589.

22. Alan S. Miller, "Toward an Environment/Labor Coalition," *Environment* 22 (June 1980): 32–39.

23. See Barry Bluestone and Bennett Harrison, *The Deindustrialization of America* (New York: Basic Books, 1982), 90.

24. The discussion of issues that are likely to attract blacks to the environmental movement was adapted from Gale, "The Environmental Movement and the Left," 182–186.

25. Ibid, 184.

26. See Ronald A. Taylor, "Do Environmentalists Care about Poor People?," *U.S. News and World Report* 96 (April 2, 1982): 51–55; Bullard, "Endangered Environs," 98; Bullard and Wright, "The Politics of Pollution," 71–78.

27. Miller, "Toward an Environment/Labor Coalition," 32–39; Sue Pollack and JoAnn Grozuczak, *Reagan, Toxics and Minorities* (Washington, DC: Urban Environment Conference, Inc., 1984), Ch. 1; Kazis and Grossman, *Fear at Work*, 3–35.

28. Andrew Porterfield and David Weir, "The Export of Hazardous Waste," *Nation* 245 (October 3, 1987): 340–344; Jim Vallette, *The International Trade in Wastes: A Greenpeace Inventory* (Washington, DC: Greenpeace, 1989), 7–16.

29. Jack Bloom, *Class, Race and the Civil Rights Movement* (Bloomington: Indiana University Press, 1987), 18.

30. Bullard and Wright, "Environmentalism and the Politics of Equity," 32.

31. Urban Environment Conference, Inc., *Taking Back Our Health: An Institute on Surviving the Toxic Threat to Minority Communities* (Washington, DC: Urban Environment Conference, Inc., 1985), 29.

32. Bullard and Wright, "Environmentalism and the Politics of Equity," 32–33.

11

Excerpts from "Where We Live, Work, and Play"

ROBERT GOTTLIEB

There was tension and excitement in the conference room at the Washington Court Hotel on Capitol Hill when Dana Alston stepped to the podium. Alston was to address an audience of more than 650 people, including 300 delegates to the first national People of Color Environmental Leadership Summit. Delegates included grassroots environmental activists from across the country: African Americans from "cancer alley" in Louisiana; Latinos from the cities and rural areas of the Southwest; Native American activists such as the Western Shoshone, who were protesting underground nuclear testing on their lands; organizers of multiracial coalitions in places such as San Francisco and Albany, New York. The purpose of the summit, held on October 24–27, 1991, was to begin to define a new environmental politics from a multiracial and social justice perspective. The delegates sought to address questions of agenda, organizational structure, movement composition, and social vision: issues central to the definition of environmentalism in the 1990s.

The delegates had just heard speeches by leaders of two important national environmental groups. The first, Michael Fischer, executive director of the Sierra Club, admitted to the summit delegates that his organization had too often been "conspicuously missing from the battles for environmental justice," but argued that the time had come for groups to "work and look into the future, rather than to beat our breasts about the past. "We national organizations are not the enemy," Fischer claimed, warning summit participants that conflict between grassroots activists and national groups would only reinforce the divide-and-conquer approach of the Reagan and Bush administrations. "We're here to reach across the table and to build the bridge of partnership with all of you," Fischer insisted.[1]

Fischer's remarks paralleled the comments of John Adams, executive director of the Natural Resources Defense Council, a prominent, staff-based group of lawyers and other environmental professionals. Adams recited how the NRDC, during its twenty-year history, had "relentlessly confronted the massive problems associated with air, water, food and toxics" and had challenged the "disproportionate impacts on communities of color" of a wide range of environmental problems. "I believe the efforts we've engaged in are significant," Adams declared, and he offered, like Fischer, to facilitate a "partnership" between the national and grassroots groups. "You can't win this battle alone," Adams concluded, underlining Fischer's warning about the consequences of disunity.[2]

Many of the delegates felt that the speeches by these environmental chief executive officers, or CEOs, were not responsive to the criticisms they and their communities had directed at these groups. Activists had complained about the absence of people of color in leadership and staff positions of the national groups, the failure of these groups to incorporate equity or social justice considerations in selecting the issues they fought, and the disregard for local cultures and grassroots concerns in the positions these national groups took with regard to environmental conflicts. But beyond these specific complaints, delegates to the People of Color Environmental Leadership Summit were seeking a redefinition of environmentalism to place the concerns, methods of organizing, and constituencies of the grassroots groups at the center of the environmental discourse. They wanted to redefine the central issues of environmental politics, not just to join a coalition of special interest groups.

When Alston, a key organizer of the summit, took the stage to respond to the speeches of the environmental CEOs, there was a great deal of anticipation about what she might say. Alston symbolized the new kind of environmentalist the summit had sought to attract. Born in Harlem, she first became active in the mid-1960s in the black student movement, addressing issues of apartheid and the Vietnam War. Pursuing an interest in the relationship between social and economic justice issues and public health concerns, Alston completed a master's degree in occupational and environmental health at Columbia University. She subsequently took a series of jobs that extended those interests, from the Red Cross, where she dealt with the new emergency issues associated with toxics and nuclear power problems, to Rural America, where she organized conferences on pesticide issues. As part of her work, she frequently encountered staff members from the national environmental groups. It was at these meetings and strategy sessions that Alston, an African-American woman, was struck by how she was consistently the only person of color in attendance and the only participant to press such issues as farmworker health or the discriminatory effect on communities of color of the location of hazardous waste sites.

In February 1990, Alston joined the staff of the Panos Institute, an organization that deals with the intersection of environment and development issues from the perspective of Third World needs and concerns. Alston was hired to develop a program related to the rise of domestic people-of-color organizations concerned with environmental justice. In that capacity, she was invited to be part of the planning committee organizing the People of Color Environmental Leadership Summit. A thoughtful speaker, Alston had been asked to respond to the presentations of the environmental CEOs, given her background and familiarity with both the national and grassroots groups.

As she began to talk, Alston told the delegates and participants that she had decided not to respond to the speeches by Fischer and Adams. Instead, she would try to "define for ourselves the issues of the ecology and the environment, to speak these truths that we know from our lives to those participants and observers who we have invited here to join us." Alston engaged her audience, responding to their appeal for self-definition. "For us," she declared, "the issues of the environment do not stand alone by themselves. They are not narrowly defined. Our vision of the environment is woven into an overall framework of social, racial, and economic justice." As Alston spoke, many in the audience talked back to her, shouting their agreement. "The environment, for us, is where we live, where we work, and where we play. The environment affords us the platform to address the critical issues of our time: questions of militarism and defense policy; religious freedom; cultural survival; energy-sustainable development; the future of our cities; transportation; housing; land and sovereignty rights; self-determination; employment—and we can go on and on." Turning to the environmental CEOs, Alston declared that what she and the delegates wanted was not a paternalistic relationship but a "relationship based on equity, mutual respect, mutual interest, and justice." This required a vision of the future. In pursuing these goals, Alston concluded (restating a dominant theme of the summit), "we refuse narrow definitions."[3]

The question of definition lies at the heart of understanding the past, present, and future of the environmental movement. Today, the environmental movement, broadly defined, contains a diverse set of organizations, ideas, and approaches: professional groups, whose claims to power rest on scientific and legal expertise; environmental justice advocates concerned about equity and discrimination; traditional conservationists or protectionists, whose long-established organizations have become a powerful institutional presence; local grassroots protest groups organized around a single issue; direct-action groups bearing moral witness in their defense of Nature.

Environmental organizations range from multimillion-dollar operations led by chief executive officers and staffed by experts to ad hoc neighborhood associations formed to do battle concerning a local environmental issue. Some environmental groups speak the language of science; others criticize the way science is used to direct policy. There are groups concerned with improving efficiency in existing economic arrangements and those that seek to remake society; groups that promote market solutions and those that want to regulate market failures; conservative environmentalists hoping to strengthen the system and radical environmentalists interested in an agenda for social change.

Given the diverse nature of contemporary environmentalism, it is striking how narrowly the movement has been retrospectively described by historians. In all the standard environmental histories, the roots of environmentalism are presented as differing perspectives on how best to manage or preserve "Nature," meaning Nature outside the cities and the experiences of people's everyday lives. The primary figures in numerous historical texts—the romantic, unyielding, Scottish mountaineer John Muir and the German-trained, management-oriented forester Gifford Pinchot are the best-known examples—represent those perspectives to the exclusion of other figures not seen as engaged in environmental struggles because their concerns were urban and industrial. There has been no place in this history for Alice Hamilton, who helped identify the new industrial poisons and spoke of reforming the "dangerous trades"; for empowerment advocates such as Florence Kelley, who sought to reform the conditions of the urban and industrial environment in order to improve the quality of life of workers, children, women, and the poor; or for urban critics such as Lewis Mumford, who spoke of the excesses of the industrial city and envisioned environmental harmony linking city and countryside at the regional scale.

In part because of these historical omissions, scholars offer sharply divergent views about the origins, evolution, and nature of contemporary environmentalism. Most common explanations place the beginning of the current environmental movement on or around Earth Day 1970. The new movement, they emphasize, came to anchor new forms of environmental policy and management based on the cleanup and control of pollution. These histories review how this movement influenced and was shaped by legislative and regulatory initiatives focused on environmental contamination rather than on the management or protection of Nature apart from daily life. This explanation thus provides a convenient way to distinguish between an earlier conservationist epoch, when battles took place concerning national parks, forest lands, resource development, and recreational resources, and today's environmental era, when pollution and environmental hazards dominate contemporary policy agendas.

The problem with the story historians have told us is whom it leaves out and what it fails to explain. Pollution issues are not just a recent concern; people have recognized, thought about, and struggled with these problems for more than a century in significant and varied ways. A history that separates resource development and its regulation from the urban and industrial environment disguises a crucial link that connects both pollution and the loss of wilderness. If environmentalism is seen as rooted primarily or exclusively in the struggle to reserve or manage extra-urban Nature, it becomes difficult to link the changes in material life after World War II—the rise of petrochemicals, the dawning of the nuclear age, the tendencies toward overproduction and mass consumption—with the rise of new social movements focused on quality-of-life issues. And by defining contemporary environmentalism primarily in reference to its mainstream, institutional forms, such a history cannot account for the spontaneity and diversity of an environmentalism rooted in communities and constituencies seeking to address issues of where and how people live, work, and play.

Forcing the Spring offers a broader, more inclusive way to interpret the environmentalism of the past as well as the nature of the contemporary movement. This interpretation situates environmentalism as a core concept of a complex of social movements that first appeared in response to the urban and industrial changes accelerating with the rapid urbanization, industrialization, and closing of the frontier that launched the Progressive Era in the 1890s. The pressures on human and natural environments can then be seen as connected and as integral to the urban and industrial order. The social and technological changes brought about by the Depression and World War II further stimulated environmental points of view. And if Earth Day 1970 is seen not simply as the beginning of a new movement, but as the culmination of an era of protest and as prefiguring the different approaches within contemporary environmentalism, it is possible to more fully explain the commonalities and differences of today's complex environmental claims. . . .

*   *   *

When I began research for this book more than five years ago, I immediately confronted what was to emerge as my central research question: What was and is environmentalism? Which individuals and groups make up the environmental movement? Is environmentalism a new kind of social movement? In seeking to answer these questions, I encountered, among other important figures, the extraordinary Alice Hamilton, the mother of American occupational and community health, who so clearly anticipated many contemporary environmental themes. This pivotal figure is not found in any of the environmental history texts. Yet Hamilton is clearly as much an environmentalist as John Muir, the much celebrated defender of Yosemite and passionate advocate of wilderness. I mentioned my interest in Hamilton to a staff member of one of the leading mainstream environmental groups who was curious about my project. "But who's Alice Hamilton?" he asked in a puzzled manner. When I recounted the story to another friend involved in public health issues, I explained that the response by my environmentalist friend was equivalent to ignorance about John Muir. "Who's John Muir?" my public health friend replied.

Through its effort to broaden the definition of environmentalism, this book shifts environmental analysis from an argument about protection or management of the natural environment to a discussion of social movements in response to the urban and industrial forces of the past hundred years. Defining environmentalism in this broad way draws attention to the commonalities and connections among segments of complex and varied movements for change. It includes groups focused not just on wilderness or resource management but on issues affecting daily life. And while the agendas, organizational forms, and political biases of environmental groups can differ significantly, they still share a common search for a response to the dominant urban and industrial order.

Whether this search leads to a new direction and a new vision for environmentalism relates back to the question Dana Alston posed about definitions at the People of Color Environmental Leadership Summit. *Forcing the Spring* seeks to answer that question by providing a more comprehensive view of where environmentalism comes from within American experience and whether environmentalism is capable of transcending its narrow definitions to change the very fabric of social life.

NOTES

1. Michael Fisher's remarks are from his speech, "We Need Your Help," given at the session "Our Vision of the Future: A Redefinition of Environmentalism," People of Color Environmental Leadership Summit, Washington, D.C., October 26, 1991.

2. John Adam's talk was given at the session "Our Vision of the Future: A Redefinition of Environmentalism," People of Color Environmental Leadership Summit, Washington, D.C., October 26, 1991.

3. D. Alston, "Moving beyond the Barriers," speech delivered at the First National People of Color Environmental Leadership Summit, Washington, D.C., October 24–27, 1991.

12

Excerpts from "The Death of Environmentalism"

TED NORDHAUS AND MICHAEL SHELLENBERGER

Rachel Carson opened *Silent Spring*, her 1962 polemic against chemical pesticides in general and DDT in particular, with a terrible prophecy: "Man has lost the capacity to foresee and forestall. He will end by destroying the earth."[1]

Silent Spring set the template for nearly a half century of environmental writing: wrap the latest scientific research about an ecological calamity in a tragic narrative that conjures nostalgia for Nature while prophesying ever worse disasters to come, unless human societies repent for their sins against Nature and work for a return to a harmonious relationship with the natural world.

Eco-tragedies are premised on the notion that humankind's survival depends on understanding that ecological crises are a consequence of human intrusions on Nature, and that humans must let go of their consumer, religious, and ideological fantasies and recognize where their true self-interest lies.

Part of the allure of the tragic narrative for environmental writers was that it *appeared* to be responsible for motivating action on the pollution problems of the 1960s and 1970s. As we have seen, the less obvious but far more powerful drivers for antipollution laws were growing postmaterial desires, rising prosperity, and postwar optimism. In primarily crediting books like *Silent Spring* for the antipollution victories of the 1960s, environmentalists continue to preach terrifying stories of eco-apocalypse, expecting them to result in the change we need. . . .

Environmental tales of tragedy begin with Nature in harmony and almost always end in a quasi-authoritarian politics. Eco-tragic narratives diagnose human desire, aspiration, and striving to overcome the constraints of our world as illnesses to be cured or sins to be punished. They aim to short-circuit democratic values by establishing Nature as it is understood and interpreted by scientists as the ultimate authority that human societies must obey. And they insist that humanity's future is a zero-sum proposition—that there is only so much prosperity, material comfort, and modernity to go around. If too many people desire such things, we will all be ruined. *We*, of course, meaning those of us who have already achieved prosperity, material comfort, and modernity. In the end, the story told by these eco-tragedies is not that humankind cannot stand too much reality but rather that Nature cannot stand too much humanity. . . .

In the Book of Genesis, the Fall from Eden occurs because Adam and Eve eat fruit from the Tree of Knowledge. In the environmentalist's telling of our fall, human are being punished by Nature with ecological crises like global warming for our original sin of eating from the tree of knowledge—thus acting equal or superior to Nature. Our fall from Nature was triggered by our control of fire, the rise of agriculture, the birth of modern civilization, or sometimes, as in the case of *Silent Spring*, by modern science itself—which is ironic, given the privileged role the so-called natural sciences played in inventing the idea of Nature as separate from humans in the first place.

The eco-tragedy narrative imagines humans as living in a fallen world where wilderness no longer exists and a profound sadness pervades a dying earth. The unstated aspiration is to return to a time when humans lived in harmony with their surroundings. That tragic narrative is tied to an apocalyptic vision of the future—an uncanny parallel to humankind's Fall from Eden in the Book of Genesis and the end of the world in the final Book of Revelation.

Carson closes *Silent Spring* with these three grim sentences:

The "control of nature" is a phase conceived in arrogance, born of the Neanderthal age of biology and philosophy, when it was supposed that nature exists for the convenience of man. The concepts and practices of applied entomology for the most part date from the Stone Age of science. It is our alarming misfortune that so primitive a science has armed itself with the most modern and terrible weapons, and that in turning them against the insects it has also turned them against the earth.[2]

It is this reality—human agency—that most bothers environmentalists like Carson. For her, human attempts to control Nature inevitably end in tragedy.

In 1969, the microbiologist René Dubos won the Pulitzer Prize for a book calling for a new eco-religion based on the principle of harmony with nature. "Whatever form this religion takes, it will have to be based on harmony with nature as well as man, instead of the drive to mastery."[3]

The contrast between living in harmony with Nature and mastering it is what unites Carson and Dubos with virtually every strain of contemporary environmentalism. Environmentalists imagine that their values are in opposition to the Western philosophical tradition, which sees humans as separate from and superior to Nature. This is what Carson speaks of when she refers to the "Neanderthal age of biology and philosophy." But what environmentalists challenge is the ordering of the categories, not the categories themselves. Rather than dissolving the distinction between humans and Nature, environmentalists reverse the hierarchy, arguing that humans are still separate from but *subordinate* to Nature.[4]

This reversal is motivated by the view that our perfectly healthy and natural desire to control our environment is a sinful desecration of Nature. But it must be asked: can human societies exist without, in one way or another, controlling Nature? Isn't that what agriculture is all about? Virtually any attempt to alter one's surroundings—whether by gathering wood to build a fire, constructing shelter, raising livestock, growing crops, or hunting and gathering—are efforts to control Nature. Nor is doing so uniquely human: beavers build dams, ants farm aphids, and more than a few other animals use tools.

There is nothing wrong with human and nonhuman desires for control over the environment. Indeed, we wouldn't exist were it not for our ancestors' will to control. Saving the redwoods and banning DDT were no less acts of controlling Nature than were logging ancient forests and spraying toxic pesticides. The issue is not whether humans *should* control Nature, for that is inevitable, but rather *how* humans should control natures—nonhuman and human.

From beginning to end, we humans are as terrestrial as the grounds on which we walk. We are neither a cancer on, nor the stewards for, planet Earth. We are neither destined to go extinct nor destined to live in harmony. Rather, we are the first species to have any control whatsoever over how we evolve. . . .

∗ ∗ ∗

Given that most of the intellectual founders of the environmental paradigm were scientists studying nonhuman nature, it's not surprising that environmentalism has constructed its politics so literally around objectively representing Nature through Science. Each of these scientists criticizes the way the sciences have been misused to destroy the environment. But few have doubted that Science should have a privileged role in shaping politics and human society.[5]

This faith in science is often accompanied by the antiquated view that there are facts separate from values and interpretations. But the fact that there is a strong international consensus among scientists that global warming is caused almost entirely by humans does not make it any less of an interpretation. And simply deciding *what* to study, and what kind of hypotheses to form, is a value of judgment. The facts one chooses to give grater weight to in the case of global warming are deeply informed by one's values. The facts tell us that global temperatures have been rising over the last century. They tell us that human sources of pollution have probably been in some significant part responsible for those temperature increases. They tell us that global climate change and habitat destruction may be leading to the mass die-off of many plant and animal species.

But the facts also tell us that global temperatures have fluctuated wildly over the five billion years that the planet has existed; that there have been at least five previous mass extinctions during the history of the planet; that asteroids, comets, volcanoes, and ice ages have dramatically changed the climate and habitat at a planetary level; that the earth will very likely be here for billions of years after all traces of humanity have vanished from its surface; and that some form of humanity and human society will likely survive the ecological crises we face.

So which facts do we choose to focus on? Which conclusions do we draw? And what actions do we take based on those facts? These are questions as much about values as facts.

The questions before us are centrally about *how* we will survive, *who* will survive, and *how* we will live. These are questions that climatologists and other scientists can inform but not decide. For their important work, scientists deserve our gratitude, not special political authority.

What's needed today is a politics that seeks authority not from Nature or Science but from a compelling vision of the future that is appropriate for the world we live in and the crises we face. The idea that we should respect Nature implies that Nature has a particular single being (or dream) to be respected. If we define *Nature* as all things, then it is not at all clear *which natures* we should respect and which we should overcome. We are Nature and Nature is us. Nature can neither instruct our actions nor punish them. Whatever actions we choose to take or not to take in the name of the survival of the human species or human societies will be natural.

Many environmentalists imagine overcoming global warming to be about saving the planet. But the fate of the planet is not in question. The earth has survived meteorites and ice ages. It will certainly survive us. . . .

*   *   *

There is a very different story that can be told about human history, one that embraces our agency, and that is the story of constant human *overcoming*. Whereas the tragic story imagines that humans have *fallen*, the narrative of overcoming imagines that we have *risen*.

Consider how much our ancestors—human and nonhuman—overcame for us to become what we are today. For beginners, they were prey. Given how quickly and effectively humans are driving the extinction of nonhuman animal species, the notion that our ancestors were food seems preposterous. And yet, understanding that we evolved from being prey goes a long way toward understanding some of the feelings and motivations that drive us into suicidal wars and equally suicidal ecological collapses.

Against the happy accounts of harmonious premodern human societies at one with Nature, there is the reality that life was exceedingly short and difficult. Of course, life could also be wonderful and joyous. But it was hunger not obesity, oppression not depression, and violence not loneliness that were primary concerns.

Just as the past offers plenty of stories of humankind's failure, it also offers plenty of stories of human overcoming. Indeed, we can only speak of past collapses because we have survived them. There are billions more people on earth than there were when the tiny societies of the Anasazi in the American Southwest and the Norse in Greenland collapsed in the twelfth and fifteenth centuries, respectively.[6] That there are nearly several billion of us alive today is a sign of our success, not failure.

Perhaps the most powerful indictment of environmentalism is that environmentalists so often consider our long life spans and large numbers terrible tragedies rather than extraordinary achievements. The narrative of overpopulation voiced almost entirely by some of the richest humans ever to roam the earth is utterly lacking in gratitude for the astonishing labors of our ancestors.

Of course, none of this is to say that human civilizations won't collapse again in the future. They almost certainly will. Indeed, some already are collapsing. But to focus on these collapses is to miss the larger picture of rising prosperity and longer life spans. Not only have we survived, *we've thrived.* Today more and more of us are "free at last"—free to say what we want to say, love whom we want to love, and live within a far larger universe of possibilities than any other generation of humans on earth.

At the very moment that we humans are close to overcoming hunger and ancient diseases like polio and malaria, we face ecological crises of our own making, ones that could trigger drought, hunger, and the resurgence of ancient diseases.

The narrative of overcoming helps us to imagine and thus create a brighter future. Human societies will continue to stumble. Many will fall. But we have overcome starvation, disease, deprivation, oppression, and war. We can overcome ecological crises. . . .

* * *

The philosopher Richard Rorty has suggested that we borrow from the Romantics the view that in politics, as in life, imagination is more important than reason.[7] If he is right, then environmentalists have gotten their politics backwards. For too long they have demanded that Americans *Wake up!* rather than encouraging them to dream. But Rorty also recognizes that, in the end, all imagination entails a kind of rationality, and all rationality rests on a kind of imagination. The question is not "Reason or imagination?" but "*Which* reason? *Which* imagination?"

How might history have turned out if environmentalists had adopted a generous, adaptive, contingent, and *anthrophilic* politics that flowed from an affirmative and expansive vision of our natures? How might it have differed had environmentalists adopted an egalitarian vision that embraced politics, in all its untidiness and contestations, rather than simply substituting the new authorities of Science, Nature, place, and race for the old authorities of God, Man, Progress, and Tradition?

A new politics requires a new mood, one appropriate for the world we hope to create. It should be a mood of gratitude, joy, and pride, not sadness, fear, and regret. A politics of overcoming will trigger feelings of joy rather than sadness, control rather than fatalism, and gratitude rather than resentment. If we are grateful to be alive, then we must also be grateful that our ancestors overcame. It is thanks to them, and the world they made, that we live.

NOTES

1. The words belong not to her but to the man to whom she dedicated the book, Albert Schweitzer.

2. Rachel Carson, *Silent Spring* (New York: Houghton Mifflin, 1962), 2–3.

3. René Dubos, *So Human an Animal* (New York: Scribner, 1968), 7.

4. Even the language of connection between nature and humans, the geographer Jim Proctor has noted, presumes a separation. "Greater connection is not, then, needed between people and the environment," he writes, but rather the deconstruction of both categories. James D. Proctor, "Environment after Nature: Time for a New Vision," in *Envisioning Nature, Science, and Religion*, ed. James D. Proctor, 293–311 (West Conshohocken, PA: Templeton Foundation Press, 2009), 8, accessible at http://college.lclark.edu/live/ files/11226-envisioningnsr2009bpdf.

5. Dubos is a partial exception. Though he elevated an antiquated notion of nature as harmonious and spoke of our collective guilt, he also recognized the significance of changing values. "The most hopeful sign for the future is the attempt by the rebellious young to reject our social values . . . As long as there are rebels in our midst there is reason to hope that our societies can be saved." Dubos, *So Human*, 5.

6. Jared Diamond, *Collapse: How Societies Choose to Fail or Succeed* (New York: Viking, 2006).

7. Richard Rorty, *Contingency, Irony, and Solidarity* (Cambridge: Cambridge University Press, 1989), 7.

13

The Paradox of Global Environmentalism

RAMACHANDRA GUHA

The central paradox of global environmentalism is that the people who are the most vocal in defense of nature are the people who most actively destroy it. As biologists have repeatedly reminded us, the present epoch is witness to an unprecedented attack on species and habitats. The most vital as well as the most glamorous of these species and habitats are found in the poorer countries of the South, such as Brazil, Ecuador, Kenya, Tanzania, Indonesia, and India. However, the movement for their conservation is fueled principally by processes originating in the richer countries of the North, such as Norway, Australia, Germany and, preeminently, the United States.

The American wilderness movement has a history that extends back more than a century. Its two most influential and venerated figures have been John Muir (1838–1914), who founded the Sierra Club, and Aldo Leopold (1887–1948), who cofounded the Wilderness Society. Muir and Leopold advanced both scientific and ethical reasons for protecting endangered species and ecosystems. They, and their colleagues, helped inspire the creation of the National Park Service, which in turn put in place perhaps the world's best managed system of protected areas.

Until the middle decades of this century, wilderness protection in the United States was the preoccupation of precocious pioneers, whose shouts of alarm sometimes led to changes in public policy. However, when environmentalism emerged as a popular movement in the 1960s and 1970s, it principally focused on two concerns: the threats to human health posed by pollution, and the threats to wild species and wild habitats posed by economic expansion. The latter concern became, in fact, the defining motif of the movement. The dominance of wilderness protection in American environmentalism has promoted an essentially negative agenda: the protection of parks and their animals by freeing them of human habitation and productive activities. As the historian Samuel Hays points out, "natural environments which formerly had been looked upon as 'useless,' waiting only to be developed, now came to be thought of as 'useful' for filling human wants and needs. They played no less a significant role in the advanced consumer society than did such material goods as hi-fi sets or indoor gardens."[1] While saving these islands of biodiversity, American environmentalists have paid scant attention to what was happening outside them. This was especially apparent in their indifference to America's growing consumption of energy and materials.

The growing popular interest in the wild and the beautiful has thus not merely accepted the parameters of the affluent society but tends to see nature itself as merely one more good to be consumed. The uncertain commitment of most nature lovers to a more comprehensive environmental ideology is illustrated by the puzzle that they are willing to drive thousands of miles, using scarce oil and polluting the atmosphere, to visit national parks and sanctuaries—thus using anti-ecological means to marvel at the beauty of forests, swamps, or mountains protected as specimens of a "pristine" and "untouched" nature.

Consuming Nature Abroad

Crucially, the most gorgeous examples of pristine nature are located outside the United States (and outside Europe as well). The most charismatic mammals—the tiger and the elephant, the rhinoceros and the lion—are found in Asia and Africa; the most charismatic habitats, such as the rainforest, in Latin America. In the decades after World War II, and more so since the 1970s, the gaze of the North Atlantic wilderness lover has increasingly turned outward. What his or her homeland offered was not quite as exotic or attractive as what might be found overseas. And the appeal of foreign species was enhanced by new technologies, such as satellite television, which brought the beauties of the tiger or the rainforest into the living room. Meanwhile, air travel had become cheaper, more extensive, and more reliable; within days of reading about a tiger or watching it on your screen, you could be with it in its own wild habitat.

In response to a growing global market for nature tourism and driven also by strong domestic pressures, many nations in the developing South have undertaken ambitious programs to conserve and demarcate habitats and species for strict protection. For instance, when India became independent in 1947, it had less than a half-dozen wildlife reserves; it now has more than 400 parks and sanctuaries, covering 4.3 percent of the country (there are proposals to double this area). A similar expansion of territory under wilderness conservation can be observed in other Asian and African countries too. These parks are governed by two axioms: that wilderness has to be big, continuous wilderness and that *all* human intervention is bad for the retention of diversity. These axioms have led to the constitution of numerous very large sanctuaries, with a total ban on human ingress in their "core" areas. In the process, hundreds of thousands of Indian villagers have been uprooted from their homes, and millions more have had their access to fuel, fodder, and small timber restricted or cut off.

Five major groups fuel the movement for wildlife conservation in the South. The first are the city-dwellers and foreign tourists who merely season their lives, a week at a time, with the wild. Their motive is straightforward: pleasure and

fun. The second group consists of ruling elites who view the protection of particular species (for example, the tiger in India) as central to the retention or
enhancement of national prestige. The third group is composed of international
conservation organizations such as the World Conservation Union (IUCN) and
the World Wildlife Fund, whose missions are "educating" people and politicians
about the virtues of biological conservation. A fourth group consists of func-
tionaries of the state forest or wildlife service mandated by law to physically control the parks. While some officials are genuinely inspired by a love of nature,
the majority—at least in Asia and Africa—are motivated merely by the power
and spin-off benefits (overseas trips, for example) that come with the job. The
final group are biologists, who believe in wilderness and species preservation for
the sake of "science."

These five groups are united in their hostility to the farmers, herders, swid-
den cultivators, and hunters who have lived in the "wild" from well before it
became a "park" or "sanctuary." They see these human communities as having
a destructive effect on the environment, their forms of livelihood aiding the
disappearance of species and contributing to soil erosion, habitat simplification, and worse. Often their feelings are expressed in strongly pejorative language. Touring Africa in 1957, one prominent member of the Sierra Club sharply
attacked the Masai for grazing cattle in African sanctuaries. He held the Masai
to be illustrative of a larger trend, wherein "increasing population and increasing land use," rather than industrial exploitation, constituted the main threat to
the world's wilderness areas. The Masai and "their herds of economically worth-
less cattle," he remarked, "have already overgrazed and laid waste to much of
the 23,000 square miles of Tanganyika they control, and as they move into the
Serengeti, they bring the desert with them, and the wilderness and wildlife must
bow before their herds."[2]

Thirty years later, the World Wildlife Fund initiated a campaign to save
the Madagascar rainforest, the home of the ring-tailed lemur, the Madagascar
serpent eagle, and other endangered species. The group's fund-raising posters boasted spectacular sketches of the lemur and the eagle and of the half-ton
elephant bird that once lived on the island but is now extinct. Man "is a relative newcomer to Madagascar," noted the accompanying text, "but even with the
most basic of tools—axes and fire—he has brought devastation to the habitats
and resources he depends on." The posters also had a picture of a muddy river
with the caption: "Slash-and-burn agriculture has brought devastation to the
forest, and in its wake, erosion of the topsoil."

Environmental Imperialism

This poster succinctly summed up the conservationist position with regard
to the tropical rainforest. This holds that the enemy of the environment is the

hunter and farmer living in the forest, who is too short-sighted for his, and our, good. This belief (or prejudice) has informed the many projects, spread across the globe, to constitute nature parks by evicting the original human inhabitants of these areas, with scant regard for their past or future. All this is done in the name of the global heritage of biological diversity. Cynics might conclude, however, that tribal people in the Madagascar or Amazon forest are expected to move out only so that residents of London or New York can have the comfort of knowing that the lemur or toucan has been saved for posterity—evidence of which is then provided for them by way of the wildlife documentary they can watch on their television screens.

Raymond Bonner's remarkable 1993 book on African conservation, *At the Hand of Man: Peril and Hope for Africa's Wildlife*, laid bare the imperialism, unconscious and explicit, of Northern wilderness lovers and biologists working on that luckless continent. Bonner remarks that:

> Africans [have been] ignored, overwhelmed, manipulated and outmaneuvered— by a conservation crusade led, orchestrated, and dominated by white Westerners. . . . As many Africans see it, white people are making rules to protect animals that white people want to see in parks that white people visit. Why should Africans support these programs? . . . Africans do not use the parks and they do not receive any significant benefits from them. Yet they are paying the costs. There are indirect economic costs—government revenues that go to parks instead of schools. And there are direct personal costs [that is, from the ban on hunting and fuel collecting, or through physical displacement].

A Zambian biologist, E. N. Chidumayo, echoes Bonner's argument: "The only thing that is African about most conventional conservation policies is that they are practiced on African land."[3]

Bonner's book focuses on the elephant, one of approximately six animals that have come to acquire "totemic" status among Western wilderness lovers. Animal totems existed in most premodern societies, but as the Norwegian scholar Arne Kalland points out, in the past the injunction not to kill the totemic species applied only to members of the group. Hindus do not ask others to worship the cow, but those who love and cherish the elephant, seal, whale, or tiger try to impose a worldwide prohibition on its killing. No one, they say, anywhere, anytime, shall be allowed to harm the animal they hold sacred even if (as with the elephant and several species of whale) scientific evidence has established that small-scale hunting will not endanger its viable populations and will, in fact, save human lives put at risk by the expansion, after total protection, of the *lebensraum* of the totemic animal. The new totemists also insist that their species is the only true inhabitant of the ocean or forest, and ask that human beings who have lived in the same terrain (and with these animals) for many generations be sent elsewhere.[4]

Throughout Asia and Africa, the management of parks has sharply posited the interests of poor villagers who have traditionally lived in them against those of wilderness lovers and urban pleasure seekers who wish to keep parks "free of human interference"—free, that is, of humans other than themselves. This conflict has led to violent clashes between local people and government officials. At present, the majority of wildlife conservationists, domestic or foreign, seem to believe that species and habitat protection can succeed only through a punitive guns-and-guards approach. However, some Southern scientists have called for a more inclusively democratic approach to conservation, whereby tribal people and peasants can be involved in management and decision making and can be fairly compensated for the loss of their homes and livelihood.[5]

Environmentalists?

The Northern wilderness lover has largely been insensitive to the needs and aspirations of human communities that live in or around habitats they wish to "preserve for posterity." At the same time, he or she has also been insensitive to the deep asymmetries in global consumption, to the fact that it is precisely the self-confessed environmentalist who practices a lifestyle that lays an unbearable burden on the finite natural resources of the earth. The United States and the countries of Western Europe consume a share of the world's resources radically out of proportion to their percentage of the world's population. A recent study by the Wuppertal Institute for Climate, Environment, and Energy, in Wuppertal, Germany, notes that the North lays excessive claim to the South's "environmental space." The way the global economy is currently structured, it argues, "the North gains cheap access to cheap raw materials and hinders access to markets for processed products from those countries; it imposes a system [the World Trade Organization] that favors the strong; it makes use of large areas of land in the South, tolerating soil degradation, damage to regional ecosystems, and disruption of local self-reliance; it exports toxic waste; [and] it claims patent rights to utilization of biodiversity in tropical regions. . . ."

Seen "against the backdrop of a divided world," says the report, "the excessive use of nature and its resources in the North is a principal block to greater justice in the world. . . . A retreat of the rich from overconsumption is thus a necessary first step towards allowing space for improvement of the lives of an increasing number of people."

The problem thus identified, the report itemizes, in meticulous detail, how Germany can take the lead in reorienting its economy and society toward a more sustainable path. It begins with an extended treatment of overconsumption, the excessive use of the global commons by the West over the past 200 years, and the terrestrial consequences of profligate lifestyles—namely soil erosion, forest depletion, biodiversity loss, and air and water pollution. It then outlines a long-range plan for reducing the "throughput" of nature in the economy and cutting down on emissions.[6]

Consider, conversely, the approach to global environmental problems advocated by a man regarded as the "dean" of tropical biology, the American scientist Daniel Janzen. In an editorial written for the October 1988 issue of the journal *Conservation Biology*, Janzen asked his fellow biologists—professors as well as graduate students—to devote 20 percent of their funds and time to tropical conservation. He calculated that the $500 million and the 20,000 man-years thus generated would be enough to "solve virtually all neotropical conservation problems." "What can academics and researcher committees do?" asks Janzen. He offers this answer: "Significant input can be anything from voluntary secretarial work for a fund-raising drive to a megalomaniacal effort to bootstrap an entire tropical country into a permanent conservation ecosystem." Janzen assumes that money plus biologists will suffice to solve "virtually all neotropical conservation problems," although some of us think that a more effective solution would be for biologists to throw themselves into a megalomaniacal effort to bootstrap but one temperate country—Janzen's own—into living off its own resources.

Wilderness lovers like to speak of the equal rights of all species to exist. This ethical cloaking cannot hide the truth that green missionaries are possibly more dangerous, and certainly more hypocritical, than their economic or religious counterparts. The globalizing advertiser and banker works for a world in which everyone, regardless of class or color, is in an economic sense an American or Japanese—driving a car, drinking a Pepsi, owning a refrigerator and a washing machine. The missionary, having discovered Christ or Allah, wants all pagans or kaffirs also to share in the discovery. The conservationist wants to "protect the tiger or whale for posterity," yet expects other people to make the sacrifice, expects indigenous tribal people or fisherfolk to vacate the forest or the ocean so that he may enjoy his own brief holiday in communion with nature. But few among these lovers of nature scrutinize their own lifestyle, their own heavy reliance on nonrenewable resources, and the ecological footprint their consumption patterns leave on the soil, forest, waters, and air of lands other than their own.

NOTES
1. Samuel Hayes, "From Conservation to Environment: Environmental Politics in the United States since World War Two," *Environmental Review* 6, no. 1 (1982): 21f. See also Hayes's *Beauty, Health and Permanence: The American Environmental Movement, 1955–85* (Cambridge: Cambridge University Press, 1987).
2. Lee Merriam Talbot, "Wilderness Overseas," *Sierra Club Bulletin* 42, no. 6 (1957).

3. E. N. Chidumayo, "Realities for Aspiring Young African Conservationists," in Dale Lewis and Nick Carter, eds., *Voices from Africa: Local Perspectives on Conservation* (Washington, DC: World Wildlife Fund, 1993), 49.

4. Arne Kalland, "Seals, Whales and Elephants: Totem Animals and the Anti-Use Campaigns," in *Proceedings of the Conference on Responsible Wildlife Management* (Brussels: European Bureau for Conservation and Development, 1994). See also Kalland's "Management by Totemization: Whale Symbolism and the Anti-Whaling Campaign," *Arctic* 46, no. 2 (1993).

5. For thoughtful suggestions as to how the interests of wild species and those of poor humans might be made more compatible, see M. Gadgil and P. R. S. Rao, "A System of Positive Incentives to Conserve Biodiversity," *Economic and Political Weekly* (Mumbai, India), August 6, 1994; see also Ashish Kothari, Saloni Suri, and Neena Singh, "Conservation in India: A New Direction," *Economic and Political Weekly*, October 28, 1995.

6. Wolfgang Sachs, Reinhard Loske, Manfred Linz, et al., *Greening the North: A Post-Industrial Blueprint for Ecology and Equity* (London: Zed Books, 1998).

14

Excerpts from "Between Violence and Desire: Space, Power, and Identity in the Making of Metropolitan Delhi"

AMITA BAVISKAR

Introduction

Delhi, on the morning of January 30, 1995, was waking up to another winter day. In the well-to-do colony of Ashok Vihar, early risers were setting off on morning walks, some accompanied by their pet dogs. As one of these residents walked into the neighbourhood "park," the only open area in the locality, he saw a young man, poorly clad, walking away with an empty bottle in hand. Incensed, he caught the man, called his neighbours and the police. A group of enraged house-owners and two police constables descended on the youth and, within minutes, beat him to death.

The young man was 18-year-old Dilip, a visitor to Delhi, who had come to watch the Republic Day parade in the capital. He was staying with his uncle in a *jhuggi* (shanty house) along the railway tracks bordering Ashok Vihar. His uncle worked as a labourer in an industrial estate nearby which, like all other planned industrial zones in Delhi, had no provision for workers' housing. The *jhuggi* cluster with more than 10,000 households shared three public toilets, each one with eight latrines, effectively one toilet per 2083 persons. For most residents, then, any large open space, under cover of dark, became a place to defecate. Their use of the "park" brought them up against the more affluent residents of the area who paid to have a wall constructed between the dirty, unsightly *jhuggis* and their own homes. The wall was soon breached, as much to allow the traffic of domestic workers who lived in the *jhuggis* but worked to clean the homes and cars of the rich, wash their clothes, and mind their children, as to offer access to the delinquent defecators.

Dilip's death was thus the culmination of a long-standing battle over a contested space that, to one set of residents, embodied their sense of gracious urban living, a place of trees and grass devoted to leisure and recreation, and that to another set of residents, was the only available space that could be used as a toilet. If he had known this history of simmering conflict, Dilip would probably have been more wary and would have run away when challenged, and perhaps he would still be alive.[1]

This incident made a profound impression on me. During my research in central India, the site of struggles over displacement due to dams and forestry projects as well as the more gradual but no less compelling processes of impoverishment due to insecure land tenure, I had witnessed only too often state violence that tried to crush the aspirations of poor people striving to craft basic subsistence and dignity.[2] Now I was watching a similar contestation over space unfold in my own back yard. I had previously analysed struggles over the environment in rural India; now my attention was directed towards how, in an urban context, the varied meanings at stake in struggles over the environment were negotiated through different projects and practices. This concern has been strengthened over the last 2 years by two sets of processes, each an extraordinarily powerful attempt to remake the urban landscape of Delhi.

Through a series of judicial orders, the Supreme Court of India has initiated the closure of all polluting and non-conforming industries in the city, throwing out of work an estimated 2 million people employed in and around 98,000 industrial units. At the same time, the Delhi High Court has ordered the removal and relocation of all *jhuggi* squatter settlements on public lands, an order that will demolish the homes of more than 3 million people. In a city of 12 million people, the enormity of these changes is mind-boggling. Both these processes, which were set in motion by the filing of public interest litigation by environmentalists and consumer rights groups, indicate that bourgeois[3] environmentalism has emerged as an organised force in Delhi, and upper-class concerns around aesthetics, leisure, safety, and health have come significantly to shape the disposition of urban spaces.

This bourgeois environmentalism converges with the disciplining zeal of the state and its interest in creating legible spaces and docile subjects.[4] According to Alonso, "modern forms of state surveillance and control of populations as well as of capitalist organisation and work discipline have depended on the homogenising, rationalising and partitioning of space."[5] Delhi's special status and visibility as national capital has made state anxieties around the management of urban spaces all the more acute: Delhi *matters* because very important people live and visit there; its image reflects the image of the nation-state. . . .

The Logic of the Planned City . . .

Delhi's Master Plan envisaged a model city, prosperous, hygienic, and orderly, but failed to recognise that this construction could only be realised by the labours of large numbers of the working poor, for whom no provision had been made in the plans. Thus the building of planned Delhi was mirrored in the simultaneous mushrooming of unplanned Delhi. In the interstices of the Master Plan's zones, the liminal spaces along railway tracks and barren lands acquired by the DDA [Delhi Development Authority], grew the shanty towns built by construction workers, petty vendors, and artisans, and a whole host of workers whose ugly existence had been ignored in the plans. The development of slums was, then, not a violation of the Plan; it was an essential accompaniment to it, its Siamese twin. The "legal geography"[6] created by the Plan criminalised vast sections of the city's working class, adding another layer of vulnerability to their existence. At the same time, the existence of the slums over time was enabled by a series of on-going transactions: the periodic payment of bribes to municipal officials, and the intervention of local politicians. Planners' attempts to map inflexible legal geographies became a resource by which state officials and political entrepreneurs could profit, as they brokered deals that allowed slums to stay. Planners lamented the absence of "political will," the apparent impotence of the municipal authorities to enforce the law, but failed to recognise their own complicity in creating a situation where illegal practices could flourish. Erasing (through criminalising) the necessary presence of the working class was thus not an oversight but rather intrinsic to the project of producing and reproducing powerful inequalities. . . .

Environment for Whom?

Bourgeois desires for a clean and green Delhi have combined with commercial capital and the state to deny the poor their rights to the environment. Although the environment is seen as a luxury for those who can barely carve out a livelihood, attending to the struggles for work and home allows us to appreciate what the environment means across time to different groups as they are reconfigured by the contestations around place-making. The proliferation of deplorable squatter settlements, and the criminalisation of the working poor who live in them, is a direct consequence of processes of displacement written into the Master Plan. State monopoly over urban land, combined with the state's failure to build or facilitate the construction of legal low-cost housing, makes slums the only possible option. While the bourgeois gaze regards these encroachments as disfiguring the landscape, for their residents the *jhuggis* represent a tremendous investment in terms of the capital and labour that has gone into making a habitable place: coordinating with other builders, laying out plots and lanes, putting in drains, improving building materials, negotiating with the municipal authorities, petitioning for toilets, schools, and healthcare. The visible difference between relatively new and old *jhuggi* settlements makes clear the incremental efforts that go into the making of homes and habitable neighbourhoods. With the passage of time, plastic sheets and bamboo thatch shacks are replaced with more sturdy plaster and brick, roads and drains are laid out, the tentative hope of permanence signified also by the carefully cultivated rose and sacred basil plants in recycled plastic containers that are lined beside front doors.

The hope of permanence is not a foolhardy fantasy. Slum-dwellers know that if they endure the hardship and hazard of being illegal residents, the *fait accompli* of encroachment can be a powerful argument for recognition and legal status. Over time, the claims of *jhuggi*-dwellers to be regularised become stronger, with the state either legalising their settlement or granting them alternative sites in resettlement colonies on the edge of the city. Having learnt to anticipate this sequence of conflict and compromise, the poor and their political patrons willingly collaborate in the enterprise of encroachment, negotiating the risk of displacement in the hope of securing future recognition and permanent tenure. The slums, like the non-conforming and polluting industries that in the eyes of the Supreme Court are a violation of law, are for their residents the manifestation of years of compromise in which law enforcement agencies have been fully complicit. Preying upon working class hopes and dreams of a better future, these relations of conflict and compromise are embedded in profound structural violence. The collective efforts of slum-dwellers who mobilise to improve and defend their modest homes, confronting demolition crews and doggedly rebuilding after the destruction, are sabotaged by the state's promise of limited housing sites in resettlement colonies. Driven by the desire to secure legal housing and a stable foothold in the uncertain economy of the city, slum-dwellers abandon their collective struggle for individual gain. When the municipal trucks arrive to take people to the bleak resettlement sites on the city's outskirts, and the municipal officials begin handing out the slips of paper that promise a plot in these wastelands, there is a scramble to dismantle the homes painstakingly built brick by brick over the years, to be the first to board the trucks. Arriving at the resettlement sites, bare tracts of land without any services, the poor tackle once again the arduous challenge of imagining and crafting liveable places. The civilising and improving mission of the state is thus realised by the labours of the poor, their sweat and blood and dreams.

The making of Delhi's working class is also bound to the perpetuation of their identity as migrants. A migrant identity, with its implication of belonging elsewhere, keeps the poor from being recognised as full residents of Delhi entitled to the full complement of civic rights and social opportunities. Despite Delhi's history as a city of migrants, where the overwhelming majority of the population consists of first- or second-generation migrants, the fact of migration is selectively used to stigmatise certain social groups. While attempts by the bourgeoisie to construct a genealogy explaining its presence in Delhi are granted legitimacy, similar strategies are denied to the property-less. Perceiving the poor as migrants and as newly arrived interlopers on the urban scene is a strategy to disenfranchise them from civic citizenship. . . .

Conclusion: Reform or Transform Delhi?

In Delhi, the poor have responded to such disciplining attempts by adopting varied strategies of enterprise, compromise, and resistance. They have exercised their franchise as citizens (the "vote banks" that the bourgeoisie holds in contempt), used kinship networks, entered into unequal bargains with politicians and employers, mobilised collectively through neighbourhood associations, and most recently, attempted to create a coalition of slum-dwellers' organisations, trade unions, and NGOs. This coalition, called Saajha Manch (Joint Forum), has over the last three years created a powerful critique of Delhi's Master Plan, pointing to the absence of participatory processes in its formulation and highlighting the sharp inequalities in the consumption of urban resources. These multiple practices, simultaneously social and spatial, attempt to democratise urban development even as they challenge dominant modes of framing the environment-development question.

This paper has shown that planned urban development, like other modes of state-making, attempts to transform the relations between populations and spaces, in the process displacing and impoverishing large sections of the citizenry. In the case of Delhi, state-making is not only about reproducing the state nationally and internationally and securing resources for capitalist restructuring, but it also includes interventions aimed at improving the environmental quality of life for Delhi's bourgeoisie. For the bourgeoisie as well as for poor migrants, processes of place-making are marked by both violence and desire,[7] as displacement collides with dreams of a better life.

NOTES

1. The violence did not end there. When a group of people from the *jhuggis* gathered to protest against this killing, the police opened fire and killed four more people.
2. A. Baviskar, "Written on the body, written on the land: Violence and environmental struggles in central India," in N. Peluso and M. Watts, eds., *Violent Environments* (Ithaca, NY: Cornell University Press, 2001), 354–379.
3. I am using the terms "bourgeois" and "upper-class" to refer to the group that is instantly recognisable in Delhi by dress, deportment, and language: the *padhelikhe* (educated) and the propertied, white-collar professionals, and those engaged in business: the owners of material and symbolic capital.
4. J. C. Scott, *Seeing Like a State: How Certain Schemes to Improve the Human Condition Have Failed* (New Haven, CT: Yale University Press, 1996).
5. A. M. Alonso, "The politics of space, time and substance: State formation, nationalism and ethnicity," *Annual Review of Anthropology* 23 (1994): 379–405.
6. N. Sundar, "Beyond the bounds? Violence at the margins of new legal geographies," in N. Peluso and M. Watts, eds., *Violent Environments* (Ithaca, NY: Cornell University Press, 2001), 328–353.
7. L. Malkki, "National geographic: The rooting of peoples and the territorialisation of national identity among scholars and refugees," *Cultural Anthropology* 7, no. 7 (1992): 24–44.

Reading Questions and Further Readings

Reading Questions

1. What is Muir's argument for why Hetch Hetchy Valley in Yosemite ought to be preserved, and what is the image of nature that he constructs?
2. In what ways does the environmentalism documented by Gottlieb differ from Muir's preservationism?
3. What is "environmental justice," and how does this concept both complement and challenge conventional environmentalism?
4. According to Nordhaus and Shellenberger, what is wrong with environmentalism?
5. In what ways can environmental protection efforts in the developing world negatively impact the well-being of underprivileged people?

Further Readings

Dowie, Mark. *Losing Ground: American Environmentalism at the Close of the Twentieth Century.* Cambridge, MA: MIT Press, 1996.

Nixon, Rob. *Slow Violence and the Environmentalism of the Poor.* Cambridge, MA: Harvard University Press, 2013.

Speth, Gus. *America the Possible: Manifesto for a New Economy.* New Haven, CT: Yale University Press, 2012.

Thoreau, Henry David. *Walden.* Boston: Ticknor and Fields, 1854.

PART III

Population and Consumption

The relationship between the earth's human population and general environ-mental vitality has been a core concern for Environmental Studies scholars since the field first galvanized in the mid-twentieth century. In this part, we trace how the relationship between human population and natural resource consumption has animated debates and motivated critical engagement with issues of socio-economic equity, patterns of global development, and the possibility of limits to the availability of the earth's natural resources. Starting with the controversial eighteenth-century thinker Thomas Malthus, the readings outline the form and content of some key debates and trace how those debates continue to infuse contemporary inquiries into sustainability, biodiversity, and our technologi-cal future. Central themes that unfold across the readings in this part include (1) the relationship between human demography and its ecological impacts, (2) questions about our technological capacity to overcome the limits suggested by nonrenewable resource consumption, (3) our individual and collective poten-tial to limit natural resource consumption (and therefore adverse anthropogenic impacts on the planet), and (4) the role of markets and manufacturing norms in mediating the effects of consumption on environmental quality.

The term "consumption" signals this bundle of core concerns in Environmen-tal Studies. As modern history unfolded, Western environmental scholars found themselves repeatedly confronted with a debate over the relationship between human population and the ultimate ecological fate of the planet. It has taken many forms, but despite decades of research and inquiry, this debate remains fundamentally unresolved. At its heart is the following question: does envi-ronmental vitality require human beings to limit their consumption of natural resources, or does the unique quality of being human—that we can reason—free us from any limits to what the earth can provide for human use? Simply, must human beings be mindful of natural resource consumption, or will we always be able to meet an apparent natural resource limit with an alternative, a technologi-cal fix? The major aspects of this analytical puzzle hinge on understanding the dynamics of global demography, technological advancement, and the nature of human material needs and desires.

The readings assembled here elaborate this debate. On the question of how many human beings the earth can sustain and why, we read Malthus's classic, and highly controversial, argument for the concept of a human carrying capacity that regards people living in poverty as the most notorious source of environmental damage. We then read accounts from the Western environmental movement's mid-twentieth-century debates about the importance of human population, material consumption patterns, and potentially looming ecological limits. The selection from Peter J. Taylor and Frederick H. Buttel describes some of the key technical models of the era that sought to determine the precise human carrying capacity of the earth. They discuss the influential "Limits to Growth" model, for example, and show how analytical concepts of scale and regional socioeconomic dynamics might be used to call such models into question.

More recently, Marian R. Chertow reviews how Barry Commoner, Paul Ehrlich, and John Holdren tried to quantify human impact by also considering the human capacity to mitigate adverse environmental impacts through technological advancement. Their famous, but widely critiqued and elaborated, IPAT model proposed a formula for quantifying human impact on the earth by enumerating human population, levels of affluence, and technological sophistication. In this model, analysts could acknowledge vastly uneven capacities for material consumption across a large and complex global human population and in doing so consider the ecological "impact" of poorer human beings differently from those with relative wealth and material access. The model also allowed scholars to consider how human technological advancements might alleviate the adverse impacts of resource consumption or facilitate transitions to the use of relatively plentiful resources when others became scarce. Chertow shows how this conceptual history in Environmental Studies gave rise to wholly new ways of thinking in the field; in this case, a new subfield called Industrial Ecology gave scholars the analytical space to consider how conventional industrial processes might be made environmentally benign or even beneficial.

Gretchen C. Daily and Paul R. Ehrlich remind us that the analytics of equity are far more complex than socioeconomic class. In their analysis, they attempt to integrate sensitivity to a full range of indicators of relative equity. These include poverty and wealth but also gender, age, geographic location, and fertility rates. It is only in achieving increasing equality of opportunity in these arenas, they argue, that a fuller account of sustainability and carrying capacity might be taken. As Environmental Studies thinking and scholarship moved into the twenty-first century, the polarized limits-versus-technology debate gave way to more nuanced thinking about how to integrate a sensitivity to ecosystem vitality with the assumed inevitability of increasing human material consumption. Global economic development proceeded in a way that brought questions of global economic equity to the center of the consumption debate so that natural resource consumption and population problems were now framed through a lens not only of impact but also of entitlement. Influential new concepts of environmental justice demanded it.

Readings from William McDonough and Michael Braungart and from Michael Maniates give some indication of these changes in the core population-consumption debate. McDonough and Braungart signal the integrated thinking behind what eventually became the now-pervasive idea of sustainable design. Maniates outlines a quest for "consumptive resistance" that calls for responsible environmental engagement and new ideas about the relationship between material consumption and personal and collective satisfaction.

Finally, Leon Kolankiewicz infuses these questions with renewed attention to the species with which humans share the earth. Human beings may have the unique ability to employ technological fixes for the limits of natural resources, but we also have the capacity to recognize an affinity for, and interdependence with, nonhuman nature. Kolankiewicz calls on readers to consider the profound impacts of human population and consumption patterns on our capacity to conserve biodiversity. At stake are the building blocks of ecology, he argues, and no assessment of the impact of human material consumption is complete without an accounting of the costs of that consumption in biodiversity terms.

The earth's human population continues to grow, and demographic projections prepare us for even greater numbers in the decades to come. In Environmental Studies scholarship, the relationship between the earth's ecological vitality and the consumption patterns human beings produce will remain central to our analytical agenda and the politics and policy challenges that will animate our environmental future.

15

Excerpts from "An Essay on the Principle of Population"

THOMAS MALTHUS

In an inquiry concerning the improvement of society, the mode of conducting the subject which naturally presents itself, is,

1. To investigate the causes that have hitherto impeded the progress of mankind towards happiness; and,
2. To examine the probability of the total or partial removal of these causes in future.

To enter fully into this question, and to enumerate all the causes that have hitherto influenced human improvement, would be much beyond the power of an individual. The principal object of the present essay is to examine the effects of one great cause intimately united with the very nature of man; which, though it has been constantly and powerfully operating since the commencement of society, has been little noticed by the writers who have treated this subject. The facts which establish the existence of this cause have, indeed, been repeatedly stated and acknowledged; but its natural and necessary effects have been almost totally overlooked; though probably among these effects may be reckoned a very considerable portion of that vice and misery, and of that unequal distribution of the bounties of nature, which it has been the unceasing object of the enlightened philanthropist in all ages to correct.

The cause to which I allude, is the constant tendency in all animated life to increase beyond the nourishment prepared for it.

It is observed by Dr. Franklin, that there is no bound to the prolific nature of plants or animals, but what is made by their crowding and interfering with each other's means of subsistence. Were the face of the earth, he says, vacant of other plants, it might be gradually sowed and overspread with one kind only, as for instance with fennel: and were it empty of other inhabitants, it might in a few ages be replenished from one nation only, as for instance with Englishmen.[1]

This is incontrovertibly true. Through the animal and vegetable kingdoms Nature has scattered the seeds of life abroad with the most profuse and liberal hand; but has been comparatively sparing in the room and the nourishment necessary to rear them. The germs of existence contained in this earth, if they could freely develop themselves, would fill millions of worlds in the course of a few thousand years. Necessity, that imperious, all-pervading law of nature, restrains them within the prescribed bounds. The race of plants and the race of animals shrink under this great restrictive law; and man cannot by any efforts of reason escape from it.

In plants and irrational animals, the view of the subject is simple. They are all impelled by a powerful instinct to the increase of their species; and this instinct is interrupted by no doubts about providing for their offspring. Wherever therefore there is liberty, the power of increase is exerted; and the superabundant effects are repressed afterwards by want of room and nourishment.

The effects of this check on man are more complicated. Impelled to the increase of his species by an equally powerful instinct, reason interrupts his career, and asks him whether he may not bring beings into the world, for whom he cannot provide the means of support. If he attend to this natural suggestion, the restriction too frequently produces vice. If he hear it not, the human race will be constantly endeavouring to increase beyond the means of subsistence. But as, by that law of our nature which makes food necessary to the life of man, population can never actually increase beyond the lowest nourishment capable of supporting it, a strong check on population, from the difficulty of acquiring food, must be constantly in operation. This difficulty must fall somewhere, and must necessarily be severely felt in some or other of the various forms of misery, or the fear of misery, by a large portion of mankind.

That population has this constant tendency to increase beyond the means of subsistence, and that it is kept to its necessary level by these causes, will sufficiently appear from a review of the different states of society in which man has existed. But, before we proceed to this review, the subject will, perhaps, be seen in a clearer light, if we endeavour to ascertain what would be the natural increase of population, if left to exert itself with perfect freedom; and what might be expected to be the rate of increase in the productions of the earth, under the most favourable circumstances of human industry.

It will be allowed, that no country has hitherto been known, where the manners were so pure and simple, and the means of subsistence so abundant, that no check whatever has existed to early marriages from the difficulty of providing for a family, and that no waste of the human species has been occasioned by vicious customs, by towns, by unhealthy occupations, or too severe labour. Consequently in no state that we have yet known, has the power of population been left to exert itself with perfect freedom.

Whether the law of marriage be instituted, or not, the dictate of nature and virtue seems to be an early attachment to one woman; and where there were no impediments of any kind in the way of an union to which such an attachment would lead, and no causes of depopulation afterwards, the increase of the human species would be evidently much greater than any increase which has been hitherto known.

In the northern states of America, where the means of subsistence have been more ample, the manners of the people more pure, and the checks to early marriages fewer, than in any of the modern states of Europe, the population has been found to double itself, for above a century and a half successively, in less than twenty-five years.[2] Yet, even during these periods, in some of the towns, the deaths exceeded the births,[3] a circumstance which clearly proves that, in those parts of the country which supplied this deficiency, the increase must have been much more rapid than the general average.

In the back settlements, where the sole employment is agriculture, and vicious customs and unwholesome occupations are little known, the population has been found to double itself in fifteen years.[4] Even this extraordinary rate of increase is probably short of the utmost power of population. Very severe labour is requisite to clear a fresh country; such situations are not in general considered as particularly healthy; and the inhabitants, probably, are occasionally subject to the incursions of the Indians, which may destroy some lives, or at any rate diminish the fruits of industry.

According to a table of Euler, calculated on a mortality of 1 in 36, if the births be to the deaths in the proportion of 3 to 1, the period of doubling will be only 12 years and 4-5ths. And this proportion is not only a possible supposition, but has actually occurred for short periods in more countries than one.

Sir William Petty supposes a doubling possible in so short a time as ten years.[5]

But, to be perfectly sure that we are far within the truth, we will take the slowest of these rates of increase, a rate in which all concurring testimonies agree, and which has been repeatedly ascertained to be from procreation only.

It may safely be pronounced, therefore, that population, when unchecked, goes on doubling itself every twenty-five years, or increases in a geometrical ratio.

The rate according to which the productions of the earth may be supposed to increase, it will not be so easy to determine. Of this, however, we may be perfectly certain, that the ratio of their increase in a limited territory must be of a totally different nature from the ratio of the increase of population. A thousand millions are just as easily doubled every twenty-five years by the power of population as a thousand. But the food to support the increase from the greater number will by no means be obtained with the same facility. Man is necessarily confined in room. When acre has been added to acre till all the fertile land is occupied, the yearly increase of food must depend upon the melioration of the land already in possession. This is a fund; which, from the nature of all soils, instead of increasing, must be gradually diminishing. But population, could it be supplied with food, would go on with unexhausted vigour; and the increase of one period would furnish the power of a greater increase the next, and this without any limit.

From the accounts we have of China and Japan, it may be fairly doubted, whether the best-directed efforts of human industry could double the produce of these countries even once in any number of years. There are many parts of the globe, indeed, hitherto uncultivated, and almost unoccupied; but the right of exterminating, or driving into a corner where they must starve, even the inhabitants of these thinly-peopled regions, will be questioned in a moral view. The process of improving their minds and directing their industry would necessarily be slow; and during this time, as population would regularly keep pace with the increasing produce, it would rarely happen that a great degree of knowledge and industry would have to operate at once upon rich unappropriated soil. Even where this might take place, as it does sometimes in new colonies, a geometrical ratio increases with such extraordinary rapidity, that the advantage could not last long. If the United States of America continue increasing, which they certainly will do, though not with the same rapidity as formerly, the Indians will be driven further and further back into the country, till the whole race is ultimately exterminated, and the territory is incapable of further extension.

These observations are, in a degree, applicable to all the parts of the earth, where the soil is imperfectly cultivated. To exterminate the inhabitants of the greatest part of Asia and Africa, is a thought that could not be admitted for a moment. To civilise and direct the industry of the various tribes of Tartars and Negroes, would certainly be a work of considerable time, and of variable and uncertain success.

Europe is by no means so fully peopled as it might be. In Europe there is the fairest chance that human industry may receive its best direction. The science of agriculture has been much studied in England and Scotland; and there is still a great portion of uncultivated land in these countries. Let us consider at what rate the produce of this island might be supposed to increase under circumstances the most favourable to improvement.

If it be allowed that by the best possible policy, and great encouragements to agriculture, the average produce of the island could be doubled in the first twenty-five years, it will be allowing, probably, a greater increase than could with reason be expected.

In the next twenty-five years, it is impossible to suppose that the produce could be quadrupled. It would be contrary to all our knowledge of the properties of land. The improvement of the barren parts would be a work of time and labour; and it must be evident to those who have the slightest acquaintance with agricultural subjects, that in proportion as cultivation extended, the additions that could yearly be made to the former average produce must be gradually and regularly diminishing. That we may be the better able to compare the increase of population and food, let us make a supposition, which, without pretending to accuracy, is clearly more favourable to the power of production in the earth, than any experience we have had of its qualities will warrant.

Let us suppose that the yearly additions which might be made to the former average produce, instead of decreasing, which they certainly would do, were to remain the same; and that the produce of this island might be increased every twenty-five years, by a quantity equal to what it at present produces. The most enthusiastic speculator cannot suppose a greater increase than this. In a few centuries it would make every acre of land in the island like a garden.

If this supposition be applied to the whole earth, and if it be allowed that the subsistence for man which the earth affords might be increased every twenty-five years by a quantity equal to what it at present produces, this will be supposing a rate of increase much greater than we can imagine that any possible exertions of mankind could make it.

It may be fairly pronounced, therefore, that, considering the present average state of the earth, the means of subsistence, under circumstances the most favorable to human industry, could not possibly be made to increase faster than in an arithmetical ratio. . . .

. . . A promiscuous intercourse to such a degree as to prevent the birth of children, seems to lower, in the most marked manner, the dignity of human nature. It cannot be without its effect on men, and nothing can be more obvious than its tendency to degrade the female character, and to destroy all its most amiable and distinguishing characteristics. Add to which, that among those unfortunate females with which all great towns abound, more real distress and aggravated misery are, perhaps, to be found, than in any other department of human life.

When a general corruption of morals, with regard to the sex, pervades all classes of society, its effects must necessarily to be, to poison the springs of domestic happiness, to weaken conjugal and parental affection, and to lessen united exertions and ardour of parents in the care and education of their children; effects which cannot take place without a decided diminution of general happiness and virtue of society; particularly as the necessity of art in the accomplishment and conduct of intrigues, and in the concealment of their consequences, necessarily leads to many other vices.

The positive checks to population are extremely various, and include every cause, whether arising from vice or misery, which in any degree contribute to shorten the natural duration of human life. Under this head, therefore, may enumerated all unwholesome occupations, severe labour and exposure to seasons, extreme poverty, bad nursing of children, great towns, excesses of kinds, the whole train of common diseases and epidemics, wars, plague, and famine.

On examining these obstacles to the increase of population which are classed under the heads of preventive and positive checks, it will appear that they are all resolvable into moral restraint, vice, and misery.

Of the preventive checks, the restraint from marriage which is not followed by irregular gratifications, may properly be termed moral restraint.[6]

Promiscuous intercourse, unnatural passions, violations of the marriage bed, and improper arts to conceal the consequences of irregular connections, are preventive checks that clearly come under the head of vice.

Of the positive checks, those which appear to arise unavoidably from the laws of nature, may be called exclusively misery; and those which we obviously bring upon ourselves, such as wars, excesses, and many others which it would be in our power to avoid, are of a mixed nature. They are brought upon us by vice and their consequences are misery.[7]

The sum of all these preventive and positive checks, taken together, forms the immediate check to population; and it is evident that, in every country where the whole of the procreative power cannot be called into action, the preventive and the positive checks must vary inversely as each other; that is, in countries either naturally unhealthy, or subject to a great mortality, from whatever cause it may arise, the preventive check is found to prevail with considerable force, the positive check will prevail very little, or the mortality be very small.

In every country some of these checks are, with more or less force, in constant operation; yet, notwithstanding their general prevalence, the states in which there is not a constant effort in the population to increase beyond the means of subsistence. This constant effort as constantly tends to subject the lower classes of society to distress, and to prevent any great melioration of their condition.

These effects, in the present state of society, seem to be produced in the following manner. We will suppose the means of subsistence in any country just equal to the easy support of its inhabitants. The constant effort towards population, which is found to act even in the most vicious societies, increases the number of people before the means of subsistence are increased. The food, therefore, which before supported eleven millions, must now be divided among eleven millions and a half. The poor consequently must live much worse, and many of them be reduced to severe distress. The number of labourers also being above the proportion of work in the market, the price of labour must tend to fall, while the price of provisions would at the same time tend to rise. The labourer therefore must do more work to earn the same as he did before. During this season of distress, the discouragements to marriage and the difficulty of rearing a family are so great, that the progress of population is retarded. In the mean time, the cheapness of labour, the plenty of labourers, and the necessity of an increased industry among them, encourage cultivators to employ more labour upon their land, to turn up fresh soil, and to manure and improve more completely what is already in tillage, till ultimately the means of subsistence may become in the same proportion to the population as at the period from which we set out. The situation of the labourer being then again tolerably comfortable, the restraints to population are in some degree loosened; and, after a short period, the same retrograde and progressive movements, with respect to happiness, are repeated.

This sort of oscillation will not probably be obvious to common view; and it may be difficult even for the most attentive observer to calculate its periods. Yet that, in the generality of old states, some alternation of this kind does exist though in a much less marked, and in a much more irregular manner, than

I have described it, no reflecting man, who considers the subject deeply, can well doubt.

One principal reason why this oscillation has been less remarked, and less decidedly confirmed by experience than might naturally be expected, is, that the histories of mankind which we possess are, in general, histories only of the higher classes. We have not many accounts that can be depended upon, of the manners and customs of that part of mankind, where these retrograde and progressive movements chiefly take place. A satisfactory history of this kind, of one people and of one period, would require the constant and minute attention of many observing minds in local and general remarks on the state of the lower classes of society, and the causes that influenced it; and, to draw accurate inferences upon this subject, a succession of such historians for some centuries would be necessary. This branch of statistical knowledge, has, of late years, been attended to in some countries,[8] and we may promise ourselves a clearer insight into the internal structure of human society from the progress of these inquiries. But the science may be said yet to be in its infancy, and many of the objects, on which it would be desirable to have information, have been either omitted or not stated with sufficient accuracy. Among these, perhaps, may be reckoned the proportion of the number of adults to the number of marriages; the extent to which vicious customs have prevailed in consequence of the restraints upon matrimony; the comparative morality among the children of the most distressed part of the community, and of those who live rather more at their ease; the variations in the real price of labour; the observable differences in the state of the lower classes of society, with respect to ease and happiness, at different times during a certain period; and very accurate registers of births, deaths, and marriages, which are of the utmost importance in this subject.

A faithful history, including such particulars, would tend greatly to elucidate the manner in which the constant check upon population acts; and would probably prove the existence of the retrograde and progressive movements that have been mentioned; though the times of their vibration must necessarily be rendered irregular from the operation of many interrupting causes; such as, the introduction or failure of certain manufactures; a greater or less prevalent spirit of agricultural enterprise; years of plenty, or years of scarcity; wars, sickly seasons, poor-laws, emigrations and other causes of a similar nature.

A circumstance which has, perhaps, more than any other, contributed to conceal this oscillation from common view, is the difference between the nominal and real price of labour. It very rarely happens that the nominal price of labour universally falls; but we well know that it frequently remains the same, while the nominal price of provisions has been gradually rising. This, indeed, will generally be the case, if the increase of manufactures and commerce be sufficient to employ the new labourers that are thrown into the market, and to prevent the increased supply from lowering the money-price.[9] But an increased number of labourers receiving the same money-wages will necessarily, by their competition, increase the money-price of corn. This is, in fact, a real fall in the price; and during this period, the condition of the lower classes of the community must be gradually growing worse. But the farmers and capitalists are growing rich from the real cheapness of labour. Their increasing capitals enable them to employ a greater number of men; and, as the population had probably suffered some check from the greater difficulty of supporting a family, the demand for labour, after a certain period, would be great in proportion to supply, and its price would of course rise, if left to find its natural level; and thus the wages of labour, and consequently the condition of the lower classes of society, might have progressive and retrograde movements, through the price of labour might never nominally fall.

In savage life, where there is no regular price of labour, it is little to be doubted that similar oscillations take place. When population has increased nearly to the utmost limits of the food, all the preventive and the positive checks will naturally operate with increased force. Vicious habits with respect to the sex will be more general, the exposing of children more frequent, and both the probability and fatality of wars and epidemics will be considerably greater; and these causes will probably continue their operation till the population is sunk below the level of the food; and then the return to comparative plenty will again produce an increase, and, after a certain period, its further progress will again be checked by the same causes.[10]

But without attempting to establish these progressive and retrograde movements in different countries, which would evidently require more minute histories than we possess, and which the progress of civilisation naturally tends to counteract, the following propositions are intended to be proved:

1. Population is necessarily limited by the means of subsistence.
2. Population invariably increases where the means of subsistence increase, unless prevented by some very powerful and obvious checks.[11]
3. These checks, and the checks which repress the superior power of population, and keep its effects on a level with the means of subsistence, are all resolvable into moral restraint, vice and misery.

The first of these propositions scarcely needs illustration. The second and third will be sufficiently established by a review of the immediate checks to population in the past and present state of society.

NOTES
1. Benjamin Franklin, *Political, Miscellaneous and Philosophical Pieces* (London, 1779), 9.
2. It appears, from some recent calculations and estimates, that form the first settlement of America, to the year 1800, the periods of doubling have been but very little above twenty years.

3. Richard Price, *Observations on Reversionary Payments*, vol. 1, 4th ed. (London: T. Cadell, 1773), 274.

4. Ibid., 282.

5. Sir William Petty, *Political Arithmetick* (London: R. Clavel, 1690), 14.

6. It will be observed, that I here use the term *moral* in its most confined sense. By moral constraint I would be understood to mean a restraint from marriage from prudential motives, with a conduct strictly moral during the period of this constraint; and I have never intentionally deviated from this sense. When I have wished to consider the restraint from marriage unconnected with its consequences, I have either called it prudential restraint, or a part of the preventive check, of which indeed it forms the principal branch.

   In my review of the different stages of society, I have been accused of not allowing sufficient weight in the prevention of population to moral restraint; but when the confined sense of the term, which I have here explained, is adverted to, I am fearful that I shall not be found to have erred much in this respect. I should be very glad to believe myself mistaken.

7. As the general consequence of vice is misery, and as this consequence is the precise reason why an action is termed vicious, it may appear that the term *misery* alone would be here sufficient, and that it is superfluous to use both. But the rejection of the term *vice* would introduce a considerable confusion into our language and ideas. We want it particularly to distinguish those actions, the general tendency of which is to produce misery, and which are therefore prohibited by the commands of the Creator, and the precepts of the moralist, although, in their immediate or individual effects, they may produce perhaps exactly the contrary. The gratification of all our passions in its immediate effect is happiness, not misery; and, in individual instances, even the remote consequences (at least in this life) may possibly come under the same denomination. There may have been some irregular connections with women, which have added to the happiness of both parties, and have injured no one. These individual actions, therefore, cannot come under the head of misery. But they are still evidently vicious, because an action is so denominated, which violates an express precept, founded upon its general tendency to produce misery, whatever may be its individual effect; and no person can doubt the general tendency of an illicit intercourse between the sexes, to injure the happiness of society.

8. The judicious questions which Sir John Sinclair circulated in Scotland, and the valuable accounts which he has collected in that part of the island, do him the highest honour; and these accounts will ever remain an extraordinary monument of the learning, good sense, and general information of the clergy of Scotland. It is to be regretted that the adjoining parishes are not put together in the work, which would have assisted the memory both in attaining and recollecting the state of particular districts. The repetitions and contradictory opinions which occur are not, in my opinion, so objectionable; as, to the result of such testimony, more faith may be given than we could possibly give to the testimony of any individual. Even were this result drawn for us by some master hand, though much valuable time would undoubtedly be saved, the information would not be so satisfactory. If, with a few subordinate improvements, this work had contained accurate and complete registers for the last 150 years, it would have been inestimable, and would have exhibited a better picture of the internal state of a country than has yet been presented to the world. But this last most essential improvement no diligence could have effected.

9. If the new labourers thrown yearly into the market should find no employment but in agriculture, their competition might so lower the money-price of labour as to prevent the increase of population from occasioning an effective demand for more corn; or, in other words, if the landlords and farmers could get nothing but an additional quantity of

agricultural labour in exchange for any additional produce which they could raise, they might not be tempted to raise it.

10. Sir James Stuart very justly compares the generative faculty to a spring loaded with a variable weight (Sir James Stuart, *An Inquiry Into the Principles of Political Economy*, vol. 1 [London: A. Millar and T. Cadell, 1767], book 1, chap. 4, p. 20), which would of course produce exactly that kind of oscillation which has been mentioned. In the first book of his *Political Economy*, he has explained many parts of the subject of population very ably.

11. I have expressed myself in this cautious manner, because I believe there are some instances, where population does not keep up to the level of the means of subsistence. But these are extreme cases; and, generally speaking, it might be said that,

1. Population is necessarily limited by the means of subsistence.

2. Population always increases where the means of subsistence increase.

3. The checks which repress the superior power of population, and keep its effects on a level with the means of subsistence, are all resolvable into moral restrains, vice and misery.

It should be observed that, by an increase in the means of subsistence, is here meant such an increase as will enable the mass of the society to command more food. An increase might certainly take place, which in the actual state of a particular society would not be distributed to the lower classes, and consequently give no stimulus to population.

16

How Do We Know We Have Global Environmental Problems?

Science and the Globalization of Environmental Discourse

PETER J. TAYLOR AND FREDERICK H. BUTTEL

Introduction

Since scientists a generation ago detected radioactive strontium in reindeer meat and linked DDT to the non-viability of bird eggs, science has a had a central role in shaping what count as environmental problems. Over the last few years, environmental scientists and environmentalists have called attention, in particular, to analyses of carbon dioxide concentrations in polar ice, measurements of upper atmospheric ozone depletion, remote sensing assessments of tropical deforestation, and, most notably, projections of future temperature and precipitation changes drawn from computation-intensive atmosphere circulation models. This current coalition of environmental activism and "planetary science" has stimulated a rapid rise in awareness and discussion of global environmental problems.[1] A wave of natural and social scientific studies has followed on the effects of global environmental change on vegetation and wildlife, agriculture, world trade and national economic viability, and international security. We know we have global environmental problems because, in short, science documents the existing situation and ever tightens its predictions of future changes. Accordingly, science supplies the knowledge needed to stimulate and guide social-political action.

Science-centered environmentalism, is, however, vulnerable to "deconstruction." Environmental problems, almost by definition, involve multiple, interacting causes, allowing scientists to question the definitions and procedures of other scientists, promote alternative explanations and cast doubt on the certainty of predictions. In turn, people trying to make or influence policy often find the lack of scientific closure a potent weapon.[2] After an initial honeymoon period during the late 1980s, global climate modeling, estimates of biodiversity loss, and other studies of the implications of environmental change have become subject to scientific and consequent political dispute.

The purpose of this paper is not to add our own assessment of the reliability of global environmental science or of the severity of the problems this science is indicating. Instead, building on the sociology and social studies of science, we propose a different construction of the special relationship between environmental science and politics. The sociology and social study of science has, over the last 15 years, illuminated the social influences that shape what counts as scientific knowledge.[3] Truth or falsity of the science is rarely sufficient to account for its acceptance, either within science or, as will be an equally important concern to us here, within the political realm. In this light we make three propositions, each confounding the first answer above to the question of how we know we have global environmental problems:

(1) In science, certain courses of action are facilitated over others, not just in the use or misuse of science, but in its very foundation—the problems chosen, categories used, relationships investigated, and confirming evidence required.[4] Politics—in the sense of courses of social action pursued or promoted—are not merely stimulated by scientific findings; politics are *woven into* science at its "upstream" end. In the case of environmental problems, we know that they are global in part because scientists and political actors jointly construct them in global terms.

(2) In global environmental discourse, two allied views of politics—the moral and the technocratic—have been privileged. Both views of social action emphasize people's *common* interests in remedial environmental efforts while, at the same time, steering attention away from the difficult politics that result from differentiated social groups and nations having different interests in causing and alleviating environmental problems. We know we have global environmental problems, in part, because we act as if we are a unitary and not a differentiated "we."

(3) Global environmental change, simultaneously a scientific framework and a movement ideology, is particularly vulnerable to deconstruction. The point is not that appeals to common or universal interests are without efficacy as a political tactic (as, for example, human rights campaigns in times of severe repression demonstrate). Rather, inattention to the national and localized political and economic dynamics of socio-environmental change will ensure that scientists, both natural and social, and the environmentalists who invoke their findings will be continually surprised by the unpredicted conflicts and unlikely coalitions. To the extent that "we" attempt to focus on global environmental problems, to stand above the formation of such coalitions and the conduct of such conflicts, "we" are more likely to be spectators, rather than engaged participants in the shaping of our related, but different, futures.

To explore these proposition, we will begin with a reconstruction and overview of the interwoven science and politics of *The Limits to Growth* (LTG) study of the 1970s.[5] This case is convenient not only for reasons of demonstrating historical continuity; there is also a vast literature on the topic[6] and a long span of experience by which to assess its consequences. Although the study should be familiar to most readers, we believe that our interpretation of the LTG is novel. From this beginning we then make extensions to current studies of the human/ social impacts of climate change. Finally, we discuss the possible sources of deconstruction of the globalization of environmental discourse, affecting both environmental action and the planetary science upon which it draws.

Global Modeling, 1970s Style

The LTG study was funded by the Club of Rome, an elite group of Western businessmen, government leaders, and scientists, and was conducted by system dynamics (SD) modelers at MIT.[7] The predictions from World 3, an SD model of the world's population, industry and resources were for population and economic collapse unless universal (coordinated, global-level) no-growth or steady-state policies were immediately established.[8]

A major debate developed over the LTG study.[9] Environmentalists applauded the attention the LTG drew to the finiteness of the Earth's resource, and the environmental movement took up notions such as finiteness of resource, "economic growth vs. the environment," growth control, and the steady-state economy as their major ideology and agenda. Economists, however, strongly criticized the LTG's pessimism. Scarcity, signaled in price changes, they contended, would stimulate technological advance and thus push back the limits of available resources. From a different vantage point, many leftist and social-justice-oriented progressives saw the LTG worldview as being insensitive to the needs of the poor and innocent of the realities of the penetration of multinational capital across the world.[10] Others, particularly those skilled in the methodology of systems analysis, pointed to the weaknesses in the model's empirical basis, structure and validation.[11]

Some of the technical objections were addressed in a subsequent Club of Rome–sponsored modeling effort, *Mankind at the Turning Point*.[12] This study disaggregated the world into 10 regions and increased the detail of the model 1000-fold. Collapse was still predicted, but its timing and character would differ from region to region. By the time of this second report, however, the debate had cooled, an outcome that has been given divergent interpretations: the result of the unproductive polarization of pro-growth and anti-growth positions,[13] a decline in public environmental concern,[14] a shift toward greater specificity of discussion of environmental issues,[15] a quick rejection because the LTG's proposal for a steady-state economy threatened interests that were tied to economic growth and precipitated a "corporate veto."[16]

Despite the initial firestorm of criticism, the system dynamicists never conceded that their modeling was in error.[17] After the heated reaction to the LTG, they adopted a lower profile, but continued to use SD in a wide variety of modeling and educational projects,[18] most notably in the explanation of broad modes of economic behavior—business cycles, inflation, and long waves (Kondratiev cycles). We can understand their continued belief in the validity of SD if we look more closely at construction of the LTG model of the world, noting that, whilst the system dynamicists were "doing science," they were also constructing interventions in that world. Both the representation of how that world works and the interventions proposed for improving it made each other seem more real.

System dynamics, pioneered by Jay Forrester at MIT in the 1950s, was used first to model individual firms, then to explain urban decay and, by the end of the 1960s, to uncover the dynamics of the whole world. The origin of SD in the modeling of firms has significance for the subsequent applications. Managers with whom Forrester had talked (recall that the LTG model and its predecessor models were developed at the Sloan School of Management at MIT) had observed repeated cycles of running up inventories, then laying off workers, and then once again accumulating a backlog of orders, adding labor and increasing production, only to find themselves overcompensating and running up inventories again. Instead of attributing this cycle to the business cycle, Forrester concluded that the causes were endogenous to the firm. Each decision of management was rational but, when coupled together and incorporating the unavoidable time delays between setting a goal and fulfilling it, the overshoot-undershoot cycle resulted. Given that the undesirable behavior was caused by the interactions among different sectors of the firm, the firm's overall management could overcome the cycling only if there were a superintending manager in a position to override the decision of mangers in the separate sectors of the firm. For example, the sector managers could be instructed to keep larger inventories and respond more slowly to changes in the backlog of orders than they would otherwise prefer to do.

SD for firms set the pattern for the subsequent urban, global and other SD models. In general, the modeler does not rely solely on recorded data, but instead invokes common-sense knowledge of how individuals work when they face a task with the usual information available. Computer games are often employed to convince players that they would not behave any differently from the people or other entities in the models.[19] Building on this common-sense validation of the separate decisions, SD then demonstrates that these locally rational decisions, when worked through feedbacks in the system model, generate unanticipated and undesired, or pathological, outcomes.[20]

Using decision rules that look plausible to an individual, not only the LTG but almost all SD models exhibit undesirable cycles or positive-feedback-based exponential growth and collapse. These cycles are difficult to overcome by adjusting the parameter values, even if set as high as economic or technological optimists would like. SD modelers infer that this behavior is intrinsic to the structure of the system modeled, not in its detailed specifications. The actions of some individuals *within* the system cannot override the structure, even if those individuals understand the system as a whole. But in the case of the LTG "world system," unlike in firms, there is no superintending manager to enforce the required interrelated changes in or at this world level. Catastrophe is thus inevitable unless "everyone"—all people, all decision-makers, all nations—can be convinced to act in concert to change the basic structure of population and production growth. In this fashion SD models support either a moral response—everyone must change to avert catastrophe!—or a technocratic response—only a superintending agency able to analyze the system as a whole can direct the changes needed. There is no paradox here—moral and technocratic responses are alike in attempting to bypass the political terrain in which different groups experience problems differently and act accordingly. Forrester has argued that global questions, such as the "feasibility" of continued growth of the world's population, capital stock and resource usage, require global models.[21] When we examine, however, how events would develop if population growth proved "infeasible," a politicized alternative to the LTG's diagnosis becomes apparent. Consider two hypothetical countries. Country A has a relatively equal land distribution; country B has a typical 1970s Central American land distribution: 2% of the people own 60% of the land; 70% own 2%. In other respects, these countries are similar: they have the same amount of arable land, the same population, the same level of capital availability and scientific capacity, and the same population growth rate, say, 3%. If we follow through the calculations of rates of population growth, food production increase, levels of poverty, and the like, we find that five generations before anyone is malnourished in country A, all of the poorest 70% in country B already are.[22] Food shortages linked to inequity in land distribution would be the likely level at which they, and by implication most of the world's population, would first experience "population pressure." Aggregation of the world's population and resources into the LTG's global model obscured the fact that crises will not emerge according to a strictly global logic, much less in any global form as such.[23] The *spatial* disaggregation in *Mankind at the Turning Point*[24] does not resolve this issues. Land-starved peasants share regions and nations with their creditors, landlords and employers. The socio-political responses of the peasants and, by extension, the ramifications of such local responses through national, regional and international political and economic linkages, will be (and already have been) *qualitatively* different from those highlighted by the LTG.[25]

The simple example does not tell us how to analyze the politics of localities, nations, regions, or a world in which people contribute differentially to environmental problems. Our point here is simply to highlight the political dimension excluded by the science of SD in its analysis of global limits to growth. The LTG's moral and technocratic emphasis is, of course, by no means a unique characteristic of their study. Our critique of the LTG's science-politics can be extended to the current globalization of environmental discourse. Before doing so, let us first say a little more about this moral-technocratic alliance that such discourse generally presupposes.

In technocratic formulations, objective, scientific and (typically) qualitative analyses are employed to identify the policies that society (in the case of the LTG, humanity) needs in order to restore order or ensure its sustainability or survival—polices to which individuals, citizens, and countries would then submit. In the LTG these policies are deduced from the model structure, which is held to reveal a dynamic that the ordinary citizen, politician or businessperson would not have recognized or specified. Moral formulations, in contrast, reject coercion and rely on each individual making the change needed to maintain valued social or natural qualities of life. Yet, in many senses the moral and technocratic are allied. The solutions appeal to *common, undifferentiated* interests as a corrective to corrupt, self-serving, naïve or scientifically ignorant governance. Moreover, appearances notwithstanding, special places in the proposed social transformation are reserved for their exponents—the technocrat as analyst/policy advisor; the moralist as guide.[26]

Revealingly, the LTG report at numerous junctures combined managerial language and moral recruitment: "Until the underlying structures of our socio-economic systems are thoroughly analyzed, they cannot be managed effectively";[27] "The economic preferences of society are [to be] shifted more toward services";[28] "We cannot say with certainty how much longer mankind can postpone initiating deliberate control of his growth";[29] "The two missing ingredients are a realistic, long-term goal that can guide mankind . . . and the human will to achieve that goal."[30] In short, the global society needs management to achieve control; mankind, like an individual person, needs a goal and a will to change.

Global Modeling Today

We are moving into a period of chronic, global, and extremely complex syndromes of ecological and economic interdependence. These emerging syndromes threaten to constrain and even reverse progress in human development. They will be manageable—if at all—only with a commitment of resources and consistency of purpose that transcends normal cycles and boundaries of scientific research and political action.[31]

Global climate models—or, more precisely, general circulation models (GCMs) of the atmosphere—have, especially since the hot dry summer of 1988 in the United States, provided a new scientific basis for projections of imminent global environmental crisis. The actual modeling technique bears no similarity to the system dynamics, but, as the diagnosis of environmental scientists Clark and Holling illustrates, the language of the LTG lives on. More importantly for our argument, the science of global environmental change continues to reflect, and in turn reinforce, the moral-technocratic formulation of global environmental problems. Two observations about contemporary research will serve to illustrate this point and to remind us of alternative formulations that, as in the LTG case, tend to be obscured by globalized discourse.

First, consider the high premium that is currently being placed on reducing uncertainty about physical processes in GCMs. To date, GCMs concur in predicting an average global warming, but the projected magnitude of the increase varies among the models. Moreover, at the level of regional predictions, larger uncertainties and inconsistencies among the GCMs are evident. Indirect climatic feedbacks, creating new uncertainty, have now been added to the research agenda.[32]

Tightening long-term projections or highlighting their severity is not, however, the only means by which policy responses to climate change could by catalyzed. As Glantz has observed, extreme climate-related events, such as droughts, storms and floods, already elicit socio-political responses that can be relatively easily studied.[33] Recent and historical cases of climatic-related "natural hazards" shed light on the impact of different emergency plans, investment in infrastructure and its maintenance and reconstruction schemes. Policy makers, from the local level up, can learn "by analogy" from experience and prepare for future crises. This approach is valuable whether or not these crises increase in frequency (or are already increasing in frequency) as a result of global climate change. Instead of emphasizing the investigation of physical processes and waiting for uncertainty to be eliminated before action is taken from the top, this approach calls for systematic analysis of effective vs. vulnerable institutional arrangements.[34] Such discussion of specific, local responses to climate change is not absent. Nevertheless, the vast majority of funds for global climate research is currently being devoted to improving GCMs and allied climatic studies.

This dominance of physical climate research over institutional analysis points to the second issue, the hierarchy of the physical over the life and social sciences. This hierarchy constitutes an environmental determinism: the physics and chemistry of climate change set the parameters for environmental and biological change; societies must then adjust as best they can to the chance in their environment. The hierarchy is evident in the conceptual and temporal relationships of GCMs to other areas of environmental change research. GCM research is over two decades old. Building on the prominence given to GCMs in the late 1980s, a second tier of research arose which has generated scenarios of agricultural, vegetation and wildlife changes. This research models the interaction of projected temperature and precipitation changes with regional soils, watersheds, timing of snowmelts, wildlife susceptibility, coastal upwelling, and so on.[35] Following shortly after, a third tier of research was added which has been devoted to assessing the economic or security consequences of these biotic chances or of the more direct consequences of climate change, such as a rise in sea level.[36] Modes of geopolitical response to the global climate change threat then began to be discussed by political scientists.[37] Finally, and most recently, social scientists and humanists have begun investigating popular understanding of global climate change,[38] furnishing the bottom rung on the ladder from the hard and physical down to the soft and personal.

Of course, global climate researchers know that climate change is a social problem, since it is through industrial production, transport and electrical generation systems and tropical deforestation that societies generate greenhouse gases. Nonetheless, it is *physical change*—the mechanical and inexorable greenhouse effect—that is invoked to promote policy responses and social change.[39] Moreover, the research undertaken often belies the stated awareness of the social dimension of environmental problems. Natural scientists, Harte et al., for example, recognize that "designing conservation policies without considering the role of existing institutions or societal responses to climatic change will likely lead to failure."[40] Yet the same authors advise that "models work best for predicting change when the important underlying [physical and biological] mechanisms are well understood."[41] Natural scientists have benefited from the prestige and funding that have flowed down from the high-status climate stimulations, fueling their confidence that political affairs can be influenced by technical knowledge without (or prior to) analysis of existing social arrangements. Harte et al.'s research reflects this sense of politics, not the earlier caveat.

Again, the physical-natural-social scientific hierarchy is not necessary in the construction of environmental problems. Over the last 15 years, fields such as geography, anthropology and international development studies have become increasingly sophisticated at analyzing environmental change as *socio-*environmental change. Processes such as deforestation, drought, land degradation and migration of "environmental refugees" are shown to be, in their causes and their effects, social and environmental at one and the same time.[42] The social dynamics are most apparent on the economic level: resource distribution determines whether and *for whom* a bad year becomes a drought. Inequities in land tenure and rural political power ensure that the rural poor will exploit land vulnerable to erosion, migrate to carve new plots from the forest, or add to the margins of burgeoning cities well before the resources of their original locale are exhausted. Industrialization and other opportunities for off-farm income can result in insufficient labor remaining to keep up traditional conservation practices.[43] Such economic observations readily lead us to consider local particularity and historical contingency—in some areas traditional practices have resisted disruption by linkage into global markets and have instead contributed to environmental sustainability, while in other areas social organization has been rapidly restructured with significant environmental consequences.[44] What accounts for such differences? We are also led to consider the contextuality of local change—local particularity is influenced by institutions, processes, and activities well beyond the immediate locale and, in turn, can have distant ramifications; e.g., the neglect of old terraces can lead to accelerated erosion and thus to siltation of reservoirs downstream.[45] In a rich sense of the word social, environmental problems require social diagnosis and response. This will continue to be the case as climate change deepens and extends already existing crises.

Sites of "Deconstruction" of Global Environmental Change

In highlighting the moral technocratic construction of global environmental problems, we hope to steer the attention of scientists and environmentalists towards the differentiated politics and economics of socio-environmental change. There are, of course, other sources of opposition to global and political formulations of environmental issues which threaten to render global environment discourse, like science-centered environmentalism in general, vulnerable to deconstruction. In this section we review some major places where globalization is disputed. Most of this opposition, it should be noted, centers more on disparities among nations than on the differentiated economic and political conditions within nations—a particular construction in its own right.

As we have observed elsewhere,[46] global change knowledge was appropriated within the environmental activist community and employed to mobilize support for the movement's goals. The selective promotion of global change/warming increased support among prospective environmental supporters, and minimized opposition among the political and corporate officialdoms in the advanced industrial countries.[47] The popularization of the global warming notion was accompanied by, if not substantially based on, disproportionate stress on Third World sources of greenhouse gases, particularly tropical rainforest destruction. Tropical rainforest destruction probably accounts for less than 15% of global greenhouse gases,[48] and is a relatively minor source compared with industrial, transport and other greenhouse gas emissions from the developed countries.[49] The "rainforest connection" has, however, been central in the scientific and popular construction of global change knowledge. At the level of environmental science, it has led to greater stress on the conservation biology of rainforest biodiversity, not only as a subordinate theme within the global environmental change framework,[50] but also as a glamour topic in its own right. The rainforest theme has been of even greater importance at the political level. Thrupp,[51] for example, has stressed that the pre-climate-change preoccupations of environmental scientists and environmental groups (e.g., preservation of primary rainforests and other "sensitive" zones, protection of wildlife and endangered species) were instrumental in the formulation of the environmentalists' global change worldview and program.

As awareness of global climate change and the biodiversity implications of rainforest destruction grew in tandem, environmentalists came to focus the bulk of their efforts at two interrelated levels: on one hand, considerable activity was focused on the UN System (particularly UNEP) and other "international regimes" in order to forge international conventions on climate change, biodiversity and forest management (which were under investigation in preparation for a hoped-for ratification at the 1992 UN-sponsored "Earth Summit" in Rio de Janeiro); on the other, environmental groups have sought to influence, and to employ the influence of, the international development finance and assistance establishment, particularly the World Bank/IMF, because of the important role of these institutions in affecting economic activity in the tropics. Within both of these fora, as well as among the international development intelligentsia and NGOs, environmental groups have played an important role in shaping understandings and policies with regard to "sustainable development."[52] In particular, there is a very strong stress on rainforest environments and biodiversity in sustainable development doctrine.[53]

The rise of global-change-led international environmentalism occurred during a significant shift of the political center of gravity of the industrial world toward neo-conservative regimes. Modern environmentalism has accommodated itself surprisingly readily to the global free-market resurgence. While international environmental groups yet reserve the right to criticize the World Bank and related institutions about the environmental destruction that results *from particular projects or types of projects*,[54] environmental groups have generally worked with the Bank/IMF in a surprisingly harmonious manner in implementing conservation/preservation policies and programs in the Third World.[55] There is a key coincidence of interest in the environmental group / World Bank / IMF relationship: the Bank and IMF gain legitimacy in the eyes of the citizens and political officialdoms of the advanced (increasingly "green"-oriented) countries by helping to implement environmental and conservation policies, while the implied threat of Bank of IMF termination of bridging, adjustment and project loans is useful in securing developing-country compliance with environmental initiatives. Given this relationship, most environmental organizations have been disinclined to take on the world debt crisis, the net South-North capital drain, and the international monetary order[56] as being fundamental contributors to environmental degradation. Postel and Flavin of the Worldwatch Institute, however, have recently stated, in a bold manner quite uncharacteristic of the international environmental community's silence on debt, that the net South-North capital drain (now approaching $50 billion annually) and the environmentally destructive imperatives of Bank- and IMF-supervised structural adjustment and debt repayment programs must be addressed before implementing global environmental policies.[57]

Nonetheless, the disproportionate stress given to the rainforest component of global climate change, along with the political alliance of environmental groups with the dominant countries and institutions of global society, have proven to be major catalysts of developing-country opposition to the global climate and related treaties that were to be considered at the Earth Summit.[58] In part, Third World opposition to these institutional practices follows a well-trodden path. International development policy has long been conflictual, involving struggles between official development agencies on one hand and developing- and developed-country groups critical of the performance and consequences of development projects and policies on the other.[59] One of the most important concomitants of the conservative drift of Western politics has been the implementation of "structural adjustment" doctrine, which is widely regarded as a response to the global debt crisis, the uncertainty of Third World repayment of private loans, and the resulting international monetary instability.[60]

The political economy of debt in the overall context of a stagnant world economy has become the principal parameter affecting both Third World development prospects and its environmental performance. It has largely been through the "debt regime" that environmental agendas have been grafted onto Third World development planning. Only heavily indebted countries, for example, have debt that is sufficiently discounted on the secondary debt market to be attractive to environmental groups for purchase in debt-for-nature swaps. Likewise, heavily indebted countries are most subject to joint environmental group and development agency pressures to protect the environment. But as much as external debt has facilitated the implementation of environmental conservation policies, debt also serves to *exacerbate* environmental degradation. Third World countries that are most "debt-stressed," and thus which are most in need of hard-currency export revenues, are most likely to see little alternative but to aggressively "develop" their tropical rainforests and other sensitive habitats in order to maintain their balance of payments and service their debts. Environmental activism through the debt regime is thus likely to be a standoff: two steps forward, and one or two steps back.

Given these political and economic conditions, it is not surprising that a strong force for deconstruction of global change/discourse is that of the growing Third World reaction to "environmental colonialism." Developing-country opposition to international environmental regulation is increasingly seen as being likely to frustrate, if not prevent, the appearance or reality of meaningful international environmental conventions.[61] This Third World reaction is surprisingly broad based. Growing quarters of the Third World intelligentsia and the NGO community stress, for example, that international environmental organizations have exaggerated the Third World contribution to global warming, and that Western calculations of developing-country contributions to greenhouse gas emissions have failed to note a fundamental First World / Third World difference in the nature of these emissions: that between the "survival emissions" of the South and the "luxury emissions" of the North.[62] But Third World criticism of global environmental regulation policies as "environmental colonialism" also includes increasingly forceful opposition by Third World politicians and business leaders to proposed global change conventions on the ground of their being an unjust violation of "national sovereignty."[63] As the Earth Summit drew near, there were strong indications that it would be dominated by North-South acrimony as much as by environmental science.

Deconstruction of the science and the action program of global climate change is by no means confined to dissenting Third World voices or to those who speak for the interests of the world's poor. Spurred by contrary evidence within Western planetary science (particularly the Marshall Institute report),[64] dissent on the part of the propertied and powerful has also been expressed; e.g., the Bush Administration in the U.S.A. has largely remained a bulwark against rushing into a global climate change convention, invoking the lack of conclusive scientific evidence that there will be significant global warming to justify their position.

Parallel to and complementing the dissident planetary science claims and the Bush Administration's "inconclusive evidence" posture in the delegitimation of the dominant construction of global climate change has been the "adaptationist" view. The adaptationist perspective has been promulgated in particular by the influential Yale economist William Nordhaus, who has served as the chair of the Adaptation Panel of the National Academy of Sciences study on global climate change.[65] Interestingly, Nordhaus and colleagues have responded to the global climate change position not by embracing a contrary planetary science position that global warming will be slight or modest—a posture that would be just as or more subject to deconstruction as the received position. Instead, Nordhaus and colleagues have premised their work on acceding to the middle to low-middle range of received global warming parameters. With this point of departure, they have striven to demonstrate empirically that the technological capabilities and the institutional and economic versatility afforded by markets and human migration within modern industrial civilization will be more than sufficient for societies to "adapt" to expected levels of global warming and climatic disruption in a cost-effective way. The Nordhaus Panel has also contended, by way of discounting and net present value analysis, that major expenditures undertaken now to mitigate global warming will be more costly than making adaptationist investments at some later point. Nordhaus, like the economists who criticized the LTG, clearly preserves a privileged role for himself in steering policy makers against the new advice of planetary scientists and their environmental allies. Science and politics again turn out to be interwoven.

Conclusion

The current globalization of environmental discourse, like the LTG debate in the 1970s, steers attention away from the differentiated politics and economics of socio-environmental change. As should be evident from this commentary, we believe both the science and politics involving environmental change would benefit from a reversal of this trend. In drawing attention to the moral-technocratic construction of global environmental problems, we have also been promoting a sociological perspective on science, namely that interpretations and action, both scientific and social, are bound together, jointly reinforced by the formulation of problems, the tools available, the audiences being addressed and enlisted to act, the support (financial and otherwise) elicited, and so on. It follows that any reconstruction of science and politics must be a multi-faceted process[66] drawing upon many more strands than simply a reconceptualization such as ours of the relationship between the knowledge claims and the views about desirable social action. Nevertheless, the critical perspectives we have introduced allow us to anticipate some ways in which global environmental discourse, although powerful, remains vulnerable to dispute and open to transformation.

NOTES

This research was supported in part by funds from a grant from Environmental Protection Agency to the Ecosystems Research Center, Cornell University.

1. In its earlier years, modern environmentalism promoted other science-based, global formulations of environmental problems, for example, Ehrlich's "population bomb" (building on population biology) and Meadows et al.'s "limits to growth" (derived from the application of system dynamics to population and resources). P. R. Ehrlich, *The Population Bomb* (New York: Ballantine Books, 1968); D. H. Meadows, D. L. Meadows, J. Randers and W. W. Behrens III, *The Limits to Growth* (New York: Universe Books, 1972). The current manifestation is, however, more broad based.

2. S. Jasanoff, "Science Politics and the Renegotiation of Expertise at EPA," *Osiris* 7 (1992).

3. E.g., R. Collins and S. Restivo, "Development, Diversity and Conflict in the Sociology of Science," *Sociol. Q.* 24 (1983): 185–200; S. Star, "Introduction: The Sociology of Science and Technology," *Social Problems* 35 (1988): 197–205; S. Woolgar, *Science: The Very Idea* (London: Tavistock, 1988).

4. P. J. Taylor, "Technocratic Optimism, H. T. Odum and the Partial Transformation of Ecological Metaphor after World War II," *J. Hist. Biol.* 21 (1988): 213–244; P. J. Taylor, "Reconstructing Socio-Ecologies: System Dynamics Modeling of Nomadic Pastoralists in Sub-Saharan Africa," in A. Clarke and J. Fujimura, eds., *The Right Tool for the Job: At Work in the Twentieth-Century Life Sciences* (Princeton, NJ: Princeton University Press, 1992).

5. See also for a more detailed sociological analysis of a closely related system dynamics environmental modeling project, Taylor, "Reconstructing Socio-Ecologies."

6. See especially B. Bloomfield, *Modeling the World: The Social Constructions of Systems Analysis* (Oxford, UK: Basil Blackwell, 1986); R. McCutcheon, *Limits of a Modern World* (London: Butterworth, 1979).

7. D. H. Meadows, D. L. Meadows, J. Randers and W. W. Behrens III, *The Limits to Growth* (New York: Universe Books, 1972).

8. Note, however, that the specific results of the study varied according to assumptions about resource availability, technological advance, and so on. In the baseline model, the model predicted a global "crash" due primarily to exhaustion of nonrenewable resources. Under more optimistic assumptions, the "crash" behavior of the model persisted, but the mechanisms of the crash were those relating to pollution and food production shortfalls.

9. See the useful summary in R. McCutcheon, *Limits of a Modern World* (London: Butterworth, 1979).

10. F. Sandbach, "The Rise and Fall of the Limits to Growth Debate," *Social Stud. Sci.* 8 (1978): 495–520; D. L. Sills, "The Environmental Movement and Its Critics," *Hum. Ecol.* 3 (1975): 1–41.

11. See especially H. S. D. Cole, C. Freeman, M. Jahoda and K. L. R. Pavitt, *Models of Doom: A Critique of the Limits of Growth* (New York: Universe Books, 1973).

12. M. Mesarovic and E. Pestel, *Mankind at the Turning Point* (New York: Dutton, 1974).

13. L. Gordon, "Limits to the Growth Debate," *Resources* 52 (1976): 1–6.

14. F. Sandbach, "The Rise and Fall of the Limits to Growth Debate," *Social Stud. Sci.* 8 (1978): 495–520.

15. C. R. Humphrey and F. H. Buttel, *Environment, Energy and Society* (Belmont, CA: Wadsworth, 1982), 110.

16. F. H. Buttel, A. Hawkins, and A. G. Power, "From Limits to Growth to Global Change: Contrasts and Contradictions in the Evolution of Environmental Science and Ideology," *Global Envir. Change* 1 (1990): 57–66.

17. D. H. Meadows, D. L. Meadows, J. Randers and W. W. Behrens, "A Response to Sussex," in H. S. D. Cole, C. Freeman, M. Jahoda and K. L. R. Pavitt, eds., *Models of Doom: A Critique of the Limits to Growth* (New York: Universe Books, 1973); B. Bloomfield, *Modeling the World: The Social Constructions of Systems Analysis* (Oxford, UK: Basil Blackwell, 1986).

18. E.g., J. W. Forrester, "Educational Implications of Responses to System Dynamics Models," in C. W. Churchman and R. O. Mason, eds., *World Modeling: A Dialogue* (New York: American Elsevier, 1976).

19. J. Sterman, "Testing Behavioral Simulation Models by Direct Experiment," *Mgmt. Sci.* 33 (1987): 1572–1592.

20. It should be noted that the science here is not exceptional; all model-making ultimately depends on certain assumptions being accepted on the basis of their plausibility, rather than on tight correspondence with empirical data. P. J. Taylor, "Revising Models and Generating Theory," *Oikos* 54 (1989): 121–126. Economists, in particular, are unapologetic about this. M. Friedman, *Essays in Positive Economics* (Chicago: University of Chicago Press, 1953).

21. J. W. Forrester, "Educational Implications of Responses to System Dynamics Models," in C. W. Churchman and R. O. Mason, eds., *World Modeling: A Dialogue* (New York: American Elsevier, 1976); see also D. H. Meadows, D. L. Meadows, J. Randers and W. W. Behrens, "A Response to Sussex," in H. S. D. Cole, C. Freeman, M. Jahoda and K. L. R. Pavitt, eds., *Models of Doom: A Critique of the Limits to Growth* (New York: Universe Books, 1973), 238.

22. See W. D. Durham, *Scarcity and Survival in Central America* (Stanford, CA: Stanford University Press, 1979), for a detailed analysis of this issue in a non-hypothetical case, that of El Salvador. The hypothetical scenario is derived originally from J. Vandermeer, "Ecological Determinism," in *Biology as a Social Weapon* (Minneapolis, MN: Burgess, 1977), 108–122.

23. Comparable observations about the differential incidence among social groups can be made with respect to the LTG's treatment of pollution, which was modeled as an aggregate world pollution level. Toxic pollutants, however, are generally relatively localized in their origins and consequences, and differentially distributed by class and race. R. Bullard and B. H.

Wright, "Environmentalism and the Politics of Equity: Emergent Trends in the Black Community," *Mid-Am Rev. Sociol.* 12 (1987): 21–38.

24. M. Mesarovic and E. Pestel, *Mankind at the Turning Point* (New York: Dutton, 1974).

25. E.g., W. W. Murdoch, *The Poverty of Nations* (Baltimore: Johns Hopkins University Press, 1980); A. K. Sen, *Poverty and Famines* (Oxford: Oxford University Press, 1981).

26. P. J. Taylor, "Technocratic Optimism, H. T. Odum and the Partial Transformation of Ecological Metaphor after World War II," *J. Hist. Biol.* 21 (1988): 213–244.

27. D. H. Meadows, D. L. Meadows, J. Randers and W. W. Behrens III, *The Limits to Growth* (New York: Universe Books, 1972), 181.

28. Ibid., 181.

29. Ibid., 183.

30. Ibid., 184.

31. W. C. Clark and C. S. Holling, "Sustainable Development of the Biosphere: Human Activities and Global Change," in T. F. Malone and J. G. Roderer, ed., *Global Change* (Cambridge: Cambridge University Press, 1985), 477.

32. D. Lashof, "The Dynamic Greenhouse: Feedback Processes That May Influence Future Concentrations of Atmospheric Trace Gases," *Clim. Change* 14 (1989): 213–242.

33. Michael Glantz, ed., *Societal Responses to Regional Climate Change: Forecasting by Analogy* (Boulder, CO: Westview, 1989).

34. As we will note later, when couched in terms of general optimism about society's adaptability to climate change, this kind of response counteracts the political momentum behind international efforts to deal with climate change.

35. P. Gleick, "The Implications of Global Climatic Change for International Security," *Clim. Change* 15 (1989): 309–325; J. Harte, M. Torn and D. Jensen, "The Nature and Consequences of Indirect Linkages between Climate Change and Biological Diversity," in R. L. Peters and T. E. Lovejoy, eds., *Consequences of the Greenhouse Effect for Biological Diversity* (New Haven, CT: Yale University Press, 1992).

36. United States Department of Energy, *The Economics of Long-Term Global Climate Change: A Preliminary Assessment*, Report of an Interagency Task Force, September 1990; T. F. Homer-Dixon, "On the Threshold: Environmental Changes as the Causes of Acute Conflict," *International Security* 16 (1991): 76–116.

37. *Policies Studies Journal* 19 (Spring 1991); *Evaluation Review* 15 (February 1991).

38. W. Kempton, "Lay Perspectives on Global Environmental Change," *Global Envir. Change* 1 (1991): 183–208; A. Ross, "Is Global Culture Warming Up?," *Social Text* 28 (1991): 3–30.

39. E.g., B. L. Turner, R. E. Kasperson, W. B. Meyer, K. M. Dow, D. Golding, J. X. Kasperson, R. C. Mitchell and S. J. Ratick, "Two Types of Global Environmental Change: Definitional and Spatial-Scale Issues in Their Human Dimensions," *Global Envir. Change* 1 (1990): 15.

40. J. Harte, M. Torn and D. Jensen, "The Nature and Consequences of Indirect Linkages between Climate Change and Biological Diversity," in R. L. Peters and T. E. Lovejoy, eds., *Consequences of the Greenhouse Effect for Biological Diversity* (New Haven, CT: Yale University Press, 1992).

41. Ibid.

42. M. Watts, "On the Poverty of Theory: Natural Hazards Research in Context," in K. Hewitt, ed., *Interpretations of Calamity from the Viewpoint of Human Ecology* (Boston: Allen and Unwin, 1983), 231–262; P. Blaikie and H. Brookfield, *Land Degradation* (London: Methuen, 1987).

43. A. De Janvry and R. Garcia-Barrios, *Rural Poverty and Environmental Degradation in Latin America: Causes, Effects and Alternative Solutions* (Rome: International Fund for Agricultural Development, 1988); R. Garcia Barrios and L. Garcia Barrios, "Environmental and

Technological Degradation in Peasant Agriculture: A Consequence of Development in Mexico," *Wld. Dev.* 18 (1990): 1569–1585.

44. P. Little, "Land Use Conflicts in the Agricultural/Pastoral Borderlands: The Case of Kenya," in P. Little, M. Horowitz and A. Nyerges, eds., *Lands at Risk in the Third World: Local-Level Perspectives* (Boulder, CO: Westview, 1987), 195–212; P. Richards, *Indigenous Agricultural Revolution* (London: Hutchinson, 1985).

45. A. De Janvry and R. Garcia-Barrios, *Rural Poverty and Environmental Degradation in Latin America: Causes, Effects and Alternative Solutions* (Rome: International Fund for Agricultural Development, 1988).

46. F. H. Buttel, A. Hawkins, and A. G. Power, "From Limits to Growth to Global Change: Contrasts and Contradictions in the Evolution of Environmental Science and Ideology," *Global Envir. Change* 1 (1990): 57–66; F. H. Buttel and P. J. Taylor, "Environmental Sociology and Global Environmental Change: A Critical Assessment," *Soc. Nat. Res.* 5 (1992).

47. The popularization of global warming also tended to stress the need to shift to alternative energy sources or to use energy more efficiently rather than to prescribe strict energy-use reduction. The global warming message thus avoided threatening the living standards of the citizens and firms of the advanced industrial societies.

48. R. B. Norgaard, "Sustainability as Intergenerational Equity: The Challenge to Economic Thought and Practice," discussion paper in the Asia Regional Series (Washington, DC: World Bank, 1991).

49. Of course, if the ambitious energy and overall development plans of developing countries are implemented, there will be a very considerable expansion of their greenhouse gas emissions over the next few decades.

50. R. L. Peters and T. E. Lovejoy, eds., *Consequences of the Greenhouse Effect for Biological Diversity* (New Haven, CT: Yale University Press, 1992).

51. L. A. Thrupp, "Politics of the Sustainable Development Fad: Do the Prevailing 'Panda-Huggers' Think about Poor People?," unpublished manuscript (Berkeley: University of California Energy and Resources Group, 1989).

52. L. Timberlake, "The Role of Scientific Knowledge in Drawing Up the Brundtland Report," in S. Andresen and W. Ostreng, eds., *International Resource Management* (London: Belhaven, 1989).

53. The political Achilles' heel of LTG doctrine from the Third World perspective was that it proscribed the kind of vibrant economic growth felt to be required to achieve progress in living standards there. The notion of "sustainable development," however, obviated the LTG's problematic destruction. This conceptual-political advance, however, was not merely a concession to Third World growth aspirations. It was also a recognition that the most environmentally sensitive zones in the tropics will be endangered as long as grinding mass poverty impels the poor to exploit the rainforests. World Commission on Environment and Development (WCED), *Our Common Future* (New York: Oxford University Press, 1987).

54. Especially dam and road construction and mining projects; D. Lewis, "G7 Gets a Roasting on Environmental Record," *New Scientist*, July 20, 1991; P. Hunt and O. Sattaur, "World Bank's Conservation Record under Fire," *New Scientist*, October 19, 1991.

55. P. Parker, "The Advance of the Green Guards," *New Scientist*, August 3, 1991.

56. Which is substantially regulated by the World Bank and IMF; R. E. Wood, *From Marshall Plan to Debt Crisis* (Berkeley: University of California Press, 1986).

57. S. Postel and C. Flavin, "Reshaping the Global Economy," in L. R. Brown, C. Flavin, and S. Postel, eds., *State of the World* (New York: Norton, 1991).

58. F. Pearce, "North-South Rift Bars Path to Summit," *New Scientist*, November 23, 1991.

59. R. E. Wood, *From Marshall Plan to Debt Crisis* (Berkeley: University of California Press, 1986).

60. On the processes by which multilateral international development agencies are influenced by the world's dominant financial and industrial powers, see W. L. Canak, ed., *Lost Promises* (Boulder, CO: Westview, 1989); L. Sklair, *Sociology of the Global System* (Baltimore: Johns Hopkins University Press, 1991); R. E. Wood, *From Marshall Plan to Debt Crisis* (Berkeley: University of California Press, 1986).

61. F. Pearce, "North-South Rift Bars Path to Summit," *New Scientist*, November 23, 1991.

62. E.g., A. Agarwal and S. Nartain, "Global Warming in an Unequal World: A Case of Environmental Colonialism," *Earth Isl. J.* (Spring 1991): 39–40. Though there are no systematic data on this point, we are nonetheless struck by the frequency of reports by colleagues who work in rainforest areas about the degree to which rainforest activists are *persona non grata* in the eyes, not only of timber concessionaires, multinational mining executives, local forest products merchants and so on, but also of peasants, forest product workers and local political officials. This growing public opposition may have influenced the Third World intelligentsia and political-economic officialdoms to speak out against the biodiversity, forest management, and other conventions being prepared for Earth Summit ratification.

63. F. Pearce, "North-South Rift Bars Path to Summit," *New Scientist*, November 23, 1991. International relations scholars who focus on "international regimes" see the chief resource that Third World countries have in negotiating favorable treatment within these regimes as being the principle of national sovereignty. E.g., S. D. Krasner, *Structural Conflict* (Berkeley: University of California Press, 1985). The national sovereignty principle increases Third World bargaining power within international organizations because it lends itself to one-country, one-vote procedures. The principle can inhibit unwanted intrusion by industrial-country governments or their multi-national firms into developing-countries' national policy decisions. Nevertheless, the imposition of "structural adjustment" policies on Third World debtor countries has led to an erosion of developing-country national sovereignty that has been unprecedented since World War II. Developing-country concern about the erosion of the national sovereignty principle, and of the advantages it affords in international regimes and organizations, was no doubt very instrumental in Third World sensitivity about Earth Summit–associated global climate and other environmental conventions.

64. Marshall Institute, *Scientific Perspectives on the Greenhouse Problem* (Washington, DC: Marshall Institute, 1989).

65. See W. D. Nordhaus, "Greenhouse Economics: Count before You Leap," *Economist*, July 7, 1990.

66. P. J. Taylor, "Reconstructing Socio-Ecologies: System Dynamics Modeling of Nomadic Pastoralists in Sub-Saharan Africa," in A. Clarke and J. Fujimura, eds., *The Right Tool for the Job: At Work in the Twentieth-Century Life Sciences* (Princeton, NJ: Princeton University Press, 1992).

17

Excerpts from "The IPAT Equation and Its Variants"

MARIAN R. CHERTOW

Introduction

In a provocative article, Rockefeller University researcher Jesse Ausubel asks: "Can technology spare the earth?"[1] It is a modern rendering of an epochal question concerning the relationship of humanity and nature, and, especially since Malthus and Darwin, of the effect of human population on resources. Surely, technology does not offer, on its own, the answer to environmental problems. Sustainability is inextricably linked with economic and social considerations that differ across cultures. This article, however, discusses the imperative of technological change and the role it can play in human and environmental improvement, particularly in the United States.

The vehicle used to begin the discussion of technological change, though phrased mathematically, is largely a conceptual expression of what factors create environmental impact in the first place. This equation represents environmental impact, (I), as the product of three variables, (1) population, (P); (2) affluence, (A); and (3) technology, (T). The IPAT equation and related formulas were born, along with the modern environmental movement, circa 1970. Although first used to quantify contributions to unsustainability, the formulation has been reinterpreted to assess the most promising path to sustainability. This revisionism can be seen as part of an underlying shift among many environmentalists in their attitudes toward technology. This article examines the conversion of the IPAT equation from a contest over which variable was worst for the planet to an expression of the profound importance of technological development in Earth's environmental future. . . .

Origins of the IPAT Equation

The relationship between technological innovation and environmental impact has been conceptualized mathematically, as noted above, by the IPAT equation. IPAT is an identity simply stating that environmental impact (I) is the product of population (P), affluence (A), and technology (T).

$$I = PAT$$

Generally credited to Ehrlich, it embodies simplicity in the face of a multitude of more complex models, and has been chosen by many scholars[2] as a starting point for investigating interactions of population, economic growth, and technological development.

The IPAT identity has led, in turn, to the master equation in industrial ecology.[3] These have been followed by two concepts in sustainability research: the Factor 10 Club (1994) and Factor Four.[4] The first two references, IPAT and the master equation, state relationships about technology and environmental impact, whereas the use of Factor Four, Factor 10, and even the Factor X debate described later, are attempts to quantify potential impacts.

In reviewing the literature, an interesting history emerges. The original formulation presented by Ehrlich and Holdren[5] was intended to refute the notion that population was a minor contributor to the environmental crisis.[6] Rather, it makes population—which the authors call "the most unyielding of all environmental pressures"—central to the equation by expressing the impact of a society on the ecosystem as:

$$I = P \times F$$

where I = total impact, P = population size, and F is impact per capita. As the authors explain, impact increases as either P or F increases, or if one increases faster than the other declines. Both variables have been growing rapidly and are much intertwined. To show that the equation is nonlinear and the variables interdependent, Ehrlich and Holdren then expanded their equation as follows:[7]

$$I = P \, (I, F) \times F$$

This variant shows that F is also dependent on P, and P depends on I and F as well. For example, rapid population growth can inhibit the growth of income and consumption, particularly in developing countries. On the other hand, cornucopians such as Julian Simon maintain that greater population is the key to prosperity.[8] Ehrlich and Holdren comment extensively on the tangled relationship of these factors and note that almost no factor has been thoroughly studied.[9]

Technology, at this stage, is not expressed as a separate variable, but is discussed in relationship to F, per capita impact. First, F is related to per capita consumption—of, for example, food, energy, fibers, and metals. Then, it is related to the technology used to make the consumption possible and whether that technology creates more or less impact. The authors note that "improvements in technology can sometimes hold the per capita impact, F, constant or even decrease it, despite increases in per capita consumption."[10] Although this statement recognizes the positive role technology can play, Ehrlich and Holdren generally conclude that technology can delay certain trends but cannot avert them.

Commoner also plays an important role in the formulation of the IPAT equation. Commoner's work in his popular 1971 book *The Closing Circle*, and much of his scientific analysis during the period of 1970–1972, were concerned with measuring the amount of pollution resulting from economic growth in the United States during the postwar period. To do so, he and his colleagues became the first to apply the IPAT concept with mathematical rigor. In order to operationalize the three factors that influence *I*, environmental impact, Commoner further defines *I* as "the amount of a given pollutant introduced annually into the environment." His equation, published in a 1972 conference proceedings,[11] is:

$$I = Population \times \frac{Economic\ good}{Population} \times \frac{Pollutant}{Economic\ good}$$

Population is used to express the size of the U.S. population in a given year or the change in population over a defined period. *Economic good* is used to express the amount of a particular good produced or consumed during a given year or the change over a defined period and is referred to as "affluence." *Pollutant* refers to the amount of a specific pollutant released and is thus a measure of "the environmental impact (i.e., amount of pollutant) generated per unit of production (or consumption), which reflects the nature of the productive technology."[12] Used in this way, the equation takes on the characteristics of a mathematical identity. On the right-hand side of the equation, the two *Populations* cancel out, the two *Economic goods* cancel out, and what remains is: *I = Pollutant*.

$$I = \cancel{Population} \times \frac{\cancel{Economic\ good}}{\cancel{Population}} \times \frac{Pollutant}{\cancel{Economic\ good}}$$

Thus, for Commoner, environmental impact is simply the amount of pollutant released rather than broader measures of impact; for example, the amount of damage such pollution created or the amount of resource depletion the pollution caused.[13] His task, then, is to estimate the contribution of each of the three terms to total environmental impact. . . .

The Transition to Technological Optimism

If the approach of the environmental movement of the 1970s was to juxtapose the gains of economic growth with the devastating reality of pollution, this approach changed in the 1980s. The Brundtland Commission report in 1987 concluded that if humanity were to have a positive future, then economy and environment had to be made more compatible. "Sustainable" was paired with "development" to describe this state, and since that time there has been increasing acceptance that economy and environment can be mutually compatible.[14]

At the same time, the IPAT equation makes us keenly aware of our limited choices. The year following the Brundtland report, one environmental chieftain, under the heading "A Luddite Recants," conceded that "economic growth has its imperatives; it will occur . . . Seen this way, reconciling the economic and environmental goals societies have set for themselves will occur only if there is a transformation in technology—a shift, unprecedented in scope and pace, to technologies, high and low, soft and hard, that facilitate economic growth while sharply reducing the pressures on the natural environment."[15] Here, James Gustave Speth, then president of the World Resources Institute [WRI], converts the 1970s suspicion, as expressed by Commoner's condemnation of "ecologically faulty technology," into an expression of hope for transformed technology.[16]

This line of argument is presented in a publication of the World Resources Institute called "Transforming Technology: An Agenda for Environmentally Sustainable Growth in the 21st Century."[17] Heaton and colleagues recite a variant of the IPAT equation resurrected by Speth in a background paper a year earlier[18] and explain its critical importance to the future of the environment in a section of their article titled "Why Technological Change Is Key." They write:

> Human impact on the natural environment depends fundamentally on an interaction among population, economic growth, and technology. A simple identity encapsulates the relationship:

$$Pollution = \frac{Pollution}{GNP} \times \frac{GNP}{Population} \times Population$$

> Here, pollution (environmental degradation generally) emerges as the product of population, income levels (the GNP per capita term) and the pollution intensity of production (the pollution/GNP term).
>
> In principle, pollution can be controlled by lowering any (or all) of these three factors. In fact, however, heroic efforts will be required to stabilize global population at double today's level, and raising income and living standards is a near-universal quest. Indeed, economic growth is a basic goal for at least 80 percent of the world's population. These powerful forces give economic expansion forward momentum. In this field of forces, the pollution intensity of production looks to be the variable easiest to manipulate, which puts the burden of change largely on technology. In fact, broadly defined to include both changes within economic sectors and shifts among them, technological change is essential just to halt backsliding: Even today's unacceptable levels of pollution will rise unless the percentage of annual growth in global and economic output is matched by an annual *decline* in pollution intensity.[19]

Thus, whereas the IPAT equation can send us in several directions, recent interpretations cast the *T* term in a very positive light. In essence, a new generation of technological optimists finds that experiments in changing human behavior to vary the course of *P* and *A* are highly uncertain. Stated another way, Walter Lynn, while Dean of the Cornell University faculty, cited the lack of progress in "social engineering," and the success, even if temporary, of technological fixes. He observed: "Currently, technology provides the only viable means by which our complex interdependent society is able to address these environmental problems."[20]

Enter Industrial Ecology

The concepts of the IPAT equation are at the core of the emerging field of industrial ecology. Industrial ecology has been described as the "marriage of technology and ecology" and examines, on the one hand, the environmental impacts of the technological society, and, on the other hand, the means by which technology can be effectively channeled toward environmental benefit. According to the first textbook in this new field,[21] industrial ecology has adopted the following IPAT variant as its "master equation":

$$Environmental\ impact = Pollution \times \frac{GDP}{person} \times \frac{Environmental\ impact}{unit\ of\ per\ capita\ GDP}$$

This master equation incorporates the same relationships as the WRI equation, with some changes in terminology. Once again we see the variation between defining pollution, *P*, as WRI has done, in contrast to defining environmental impact, as in the master equation. WRI states the equation as an identity—the populations cancel and the GNPs cancel—so that *Pollution = Pollution*. The master equation is not strictly stated as an identity. Also, Graedel and Allenby use gross domestic product (GDP) for the affluence term rather than WRI's gross national product (GNP), which reflects a shift by the United States in 1991 to the use of GDP in order to conform to the practices of most other countries. Although GNP is defined as the total final output produced by a country using inputs owned by the residents of that country, GDP counts the output produced with labor and capital *located inside* the given country, whoever owns the capital.[22] . . .

The Call of the Optimist

Just as all ecological problems are contextual, so too are the issues confronted by IPAT, which may shed light on why it has multiple interpretations. Since it was introduced in the early days of the environmental movement, much has changed, in large part because of the alarm sounded in the post–*Silent Spring* era by Ehrlich and Holdren, Commoner, and many other thoughtful researchers and policy makers. Still, much of the change was motivated by pessimism, captured in this statement of Holdren and Ehrlich at the end of their 1974 article:

> Ecological disaster will be difficult enough to avoid even if population limitation succeeds: if population growth proceeds unabated, the gains of improved technology and stabilized per capita consumption will be erased, and averting disaster will be impossible.[23]

Indeed, these are still controversial issues. Environmentalists of the 1970s who were pessimistic continued to sound alarms in the 1990s.[24] Deep ecologists will not wake to find themselves warm to technological optimism. But a great deal more consciousness about environmental issues exists internationally, especially among global institutions such as the United Nations, the World Bank, and other financial players.[25] Indeed, the notion that environmental problems can be addressed and even advanced through technical and procedural innovation has achieved its own name in the European environmental sociology literature— "ecological modernization."[26] In the United States, environmental policy, for all its warts, has made an enormous contribution at the end of the pipe, and is slowly migrating toward more integrative policy. Similarly, in corporate environmental policy, researchers can now measure the early stages of a change in emphasis from regulatory compliance toward overall process efficiency,[27] even at the expense of sales in the traditional end-of-pipe environmental industry.[28]

Upon reflection, I believe that Commoner anticipates the work defined by Ausubel, WRI, and industrial ecologists. He calls for a "new period of technological transformation of the [U.S.] economy, which reverses the counperecological trends developed since 1946"—a transformation that reconnects people and their ecosystems:

> Consider the following simple transformation of the present, ecologically faulty, relationship among soil, agricultural crops, the human population and sewage. Suppose that the sewage, instead of being introduced into surface water as it is now, whether directly or following treatment, is instead transported from urban collection systems by pipeline to agricultural areas, where—after appropriate sterilization procedures—it is incorporated into the soil. Such a pipeline would literally reincorporate the urban population into the soil's ecological cycle, restoring the integrity of that cycle . . . Hence the urban population is then no longer external to the soil cycle . . . But note that this rate of zero environmental impact is not achieved by a return to "primitive" conditions, but by an actual technological advance.[29]

Conclusions

This article underscores that technology, although associated with both disease and cure for environmental harms, is a critical factor in environmental improvement. Thus, important reasons can be found to continue to develop frameworks such as industrial ecology that focus on cures. The overall shift from pessimism to optimism, captured here through changing interpretations of the IPAT equation and its variants, is shown to be partly fatalistic, in that few alternatives exist to the imperative established by the Brundtland Commission; partly pragmatic, in that technological variables often seem easier to manage than human behavior; and partly a continued act of faith, at least in the United States, in the power of scientific advance.

NOTES

1. J. H. Ausubel, "Can Technology Spare the Earth?," *American Scientist* 84 (1996): 166–178.
2. B. Commoner, M. Corr, and P. J. Stamler, *The Closing Circle: Nature, Man, and Technology* (New York: Knopf, 1971); T. Dietz and E. Rosa, "Rethinking the Environmental Impacts of Population, Affluence and Technology," *Human Ecology Review* 1 (1994): 277–300; T. Dietz and E. Rosa, "Environmental Impacts of Population and Consumption," in P. C. Stern, T. Dietz, V. W. Ruttan, R. H. Socolow, and J. L. Sweeney, eds., *Environmentally Significant Consumption: Research Directions* (Washington, DC: Committee on the Human Dimensions of Global Change, National Research Council, 1997); T. Dietz and E. Rosa, "Climate Change and Society: Speculation, Construction and Scientific Investigation," *International Sociology* 13, no. 4 (1998): 421–455; Billie Lee Turner, class notes (Tempe: Arizona State University, 1996); I. Wernick, P. Waggoner, and J. Ausubel, "Searching for Leverage to Conserve Forests," *Journal of Industrial Ecology* 1, no. 3 (1997): 125–145.
3. G. Heaton, R. Repetto, and R. Sobin, *Transforming Technology: An Agenda for Environmentally Sustainable Growth in the 21st Century* (Washington, DC: World Resources Institute, 1991); T. Graedel and B. Allenby, *Industrial Ecology* (Englewood Cliffs, NJ: Prentice Hall, 1995).
4. E. von Weizsäcker, A. B. Lovins, and L. Lovins, *Factor Four: Doubling Wealth, Halving Resource Use* (London: Earthscan, 1997).
5. P. Ehrlich and J. Holdren, "Impact of Population Growth," *Science* 171 (1971): 1212–1217; P. Ehrlich and J. Holdren, "Impact of Population Growth," in R. G. Riker, ed., *Population, Resources, and the Environment* (Washington, DC: U.S. Government Printing Office, 1972), 365–377; P. Ehrlich and J. Holdren, "A Bulletin Dialogue on the 'Closing Circle': Critique: One Dimensional Ecology," *Bulletin of the Atomic Scientists* 28, no. 5 (1972): 16–27.
6. A precursor to the IPAT formulation by sociologist Dudley Duncan in 1964 was the POET model (population, organization, environment, technology). According to Deitz and Rosa ("Rethinking the Environmental Impacts"), the model showed that each of these components are interconnected but did not specify quantifiable relationships.
7. To trace the origins of these equations accurately is challenging. The IPAT ideas emerged in 1970 and 1971. Particularly relevant was an exchange by Commoner and Ehrlich and Holdren in the *Saturday Review* during 1970 followed by a meeting at the President's Commission on Population Growth and the American Future held on November 17, 1970, the findings of which were not published until 1972. However, Ehrlich and Holdren and Commoner produced numerous publications in the meantime, which helped steer the IPAT debate in

new directions. Ehrlich and Holdren used $I = P(I,F)\ F(P)$ in the 1972 findings, but a slightly different version, $I = P\ F(P)$ in their earlier article from *Science* in March of 1971, which is otherwise almost identical to the 1972 conference report. Both equations try to express a similar point, that P and F are interactive and can increase faster than linearly.

8. J. Simon, "Resources, Population, Environment: An Oversupply of False Bad News," *Science* 208 (1980): 1431–1437.

9. Ehrlich and Holdren, "Impact of Population Growth" (1972).

10. Ibid., 372.

11. B. Commoner, "The Environmental Cost of Economic Growth," in R. G. Ridker, ed., *Population, Resources, and the Environment* (Washington, DC: U.S. Government Printing Office, 1972), 339–363.

12. Ibid., 346.

13. In fact, one of the reviewers of this article suggested that this use of the IPAT equation might better be called "EPAT," showing the emphasis on "*E*" for "Emissions" rather than the totality of *I* for all impacts.

14. World Commission on Environment and Development, *Our Common Future* (Oxford: Oxford University Press, 1987).

15. J. Speth, "The Greening of Technology," *Washington Post*, November 20, 1988.

16. J. Speth, "Needed: An Environmental Revolution in Technology," paper for the WRI/OECD symposium Toward 2000: Environment, Technology and the New Century, Annapolis, MD, June 13–15, 1990; J. Speth, "EPA Must Help Lead an Environmental Revolution in Technology," *Environmental Law* 21 (1991): 1425–1460.

17. Heaton, Repetto, and Sobin, *Transforming Technology*.

18. Speth, "Needed," 1990.

19. Heaton, Repetto, and Sobin, *Transforming Technology*, 1.

20. W. Lynn, "Engineering Our Way Out of Endless Environmental Crises," in J. H. Ausubel and H. E. Sladovich, eds., *Technology and Environment* (Washington, DC: National Academy Press, 1989), 186.

21. Graedel and Allenby, *Industrial Ecology*.

22. P. Samuelson and W. Nordhaus, *Economics*, 16th ed. (Boston: Irwin/McGraw-Hill, 1998).

23. J. Holdren and P. Ehrlich, "Human Population and the Global Environment," *American Scientist* 62 (1974): 291.

24. P. Ehrlich and A. Ehrlich, *The Population Explosion* (New York: Simon and Schuster, 1990); B. Commoner, *Making Peace with the Planet* (New York: New Press, 1992); D. Meadows, D. Meadows, and J. Randers, *Beyond the Limits: Confronting a Global Collapse, Envisioning a Sustainable Future* (Post Mills, VT: Chelsea-Green, 1992).

25. S. Schmidheiny and F. Zorraquin, *Financing Change: The Financial Community, Eco-Efficiency, and Sustainable Development* (Cambridge, MA: MIT Press, 1996); A. Mol and D. Sonnenfeld, "Ecological Modernization around the World: An Introduction," *Environmental Politics* 9, no. 1 (2000): 3–16.

26. M. Hajer, "Ecological Modernisation as Cultural Politics," in S. Lash, B. Szerszynski, and B. Wynne, eds., *Risk, Environment and Modernity: Towards a New Ecology* (London: Sage, 1996).

27. R. Florida, "Lean and Green: The Move to Environmentally Conscious Manufacturing," *California Management Review* 39, no. 1 (1996): 80–105.

28. U.S. Department of Commerce, *The U.S. Environmental Industry* (Office of Technology Policy, October 1998).

29. Commoner, "Environmental Cost of Economic Growth."

18

Excerpts from "Socioeconomic Equity, Sustainability, and Earth's Carrying Capacity"

GRETCHEN C. DAILY AND PAUL R. EHRLICH

Introduction

A doubling of human population size portends a more than doubling of human impacts because humanity has sequentially exploited the most accessible of its essential resources. It may be difficult even for drastic changes in consumption and technology to offset the increase in environmental deterioration associated with projected population growth. This makes capitalizing on human behavioral flexibility and ingenuity absolutely critical as avenues for transforming global society into a sustainable enterprise. Government policy—economic policy in particular—can be a powerful tool for influencing fertility and consumption patterns as well as rates and directions of technological and cultural innovation.

Government policy that promotes equity is especially worth examination. Not only do many consider morally undesirable the gross inequities that presently characterize most societies and the world as a whole, but various lines of evidence suggest that these inequities are biophysically (as well as socially) *unsustainable*[1] for two reasons. First, the inequities themselves help perpetuate poverty, which generates vicious cycles[2] involving deleterious and sometimes irreversible impacts on biophysical components of Earth's life-support systems. Second, they hinder cooperation among parties of differing socioeconomic status—cooperation purportedly required for averting potentially disastrous population- and environment-related problems.[3] In this paper we restrict our focus to the first of these.

We and others have tended to shy away from investigating the social, economic, and political dimensions of carrying capacity (CC) because of their complexity and political sensitivity. But we believe that concentrating solely on the biophysical dimensions is the equivalent of the drunk searching for lost keys under a lamppost rather than where they were lost "because the light is better" there. We subscribe to the view that a more equitable world—in terms of opportunity—would be a better world. But would a more equitable world be more sustainable, have a higher CC, or more easily adjust population sizes to local CCs?

In this paper, we explore the relationships between equity and sustainability and between equity and biophysical dimensions of CC. Our analysis extends across various levels of social organization, spanning the spectrum from individuals within a household (gender- and age-related equity) to relations among regions, nations, and groups of nations. Intertemporal equity is inherently subsumed in consideration of sustainability and CC. In evaluating sustainability and CC, we necessarily focus on the two aspects of the human enterprise for which there are the best and most relevant data: patterns of food production and fertility.

The relationships we seek to characterize are dauntingly complex, and our inquiry admittedly is only a beginning. Our preliminary analysis treats various aspects of the relations equity has with sustainability and CC in isolation; in reality, the interactions are often strong and complex. For example, investment, especially in human capital or research and development, clearly impacts intergenerational equity; it may, however, involve difficult trade-offs if, as Kuznets[4] implies, unequal income is associated with higher savings rates. Finally, we draw upon a vast literature, illustrating points with a few of the best-studied and most important cases rather than by attempting a comprehensive review. In spite of these constraints, the inquiry seems worthwhile and potentially illuminating.

Definition of Terms

For the purposes of this paper, terms are defined as follows:

Equity is a measure of the relative similarity among individuals or groups in opportunity to enjoy socio-political rights, material resources, technologies, health, education, and other ingredients of human well-being.[5]

Sustainability characterizes any process or condition that can be maintained indefinitely without interruption, weakening, or loss of valued qualities. Sustainability is a necessary and sufficient condition for a population to be at or below carrying capacity.[6] Carrying capacity (CC) always embodies the concept of sustainability.

Biophysical carrying capacity is the maximum population size that an area can sustain under given technological capabilities.

Social carrying capacity is the maximum population size that an area can sustain under a given social system, with particular reference to associated patterns of resource consumption. Under any set of technologies, social CC is necessarily smaller than biophysical CC because of inefficiencies inherent in resource-distribution systems.[7] Thus biophysical CC is an upper bound on social CC. . . .

Conclusions

Our characterization of aspects of the complex relationships between equity and sustainability and between equity and carrying capacity (CC) leads to the following general conclusion: *increasing equity at all levels of organization above conditions prevailing today would indeed enhance sustainability and CC.* There is one major exception to our general conclusion, however. Equity in consumer lifestyle between and within nations *cannot be achieved globally by leveling up consumption from the bottom.* The Brundtland Report's[8] notion that the scale of the human enterprise can be expanded 5 to 10 times over the next decades reveals only a profound misunderstanding of the biophysical limits to CC.[9] Considerable de-development of the overdeveloped countries will be required—that is, controlling runaway consumption in order to reduce the physical throughput of their economies.[10]

Fortunately many important types of equity do not require increasing aggregate consumption—or at least not increasing it very much. Examples include genuine land reform, improved access to education and job opportunities, and guaranteeing access by all sexually active persons to modern contraception and safe abortion.

Finally, we end on a caveat regarding the "fog of politics."[11] It is clear that, in the foreseeable future, the world will suffer (as it always has) from incompetent and venal leaders, bureaucratic deadlock, populist whims, incomplete information, etc. Plans for such measures as inserting environmental safeguards into trade treaties so as to increase CC[12] are often not going to work out in practice. Civilization, in our view, is unlikely to persist if its major strategy is to fine-tune the present system in the hope that something approaching perpetual growth can be achieved. Instead it is imperative to find ways of reducing the scale of the human enterprise and to build in forms of insurance so that a high frequency of ecological "mistakes," even serious ones, will not bring the entire edifice crashing down. That is no small order in itself. Those who are struggling today to increase equity in various ways can take heart that they are probably both helping to increase CC and empowering *Homo sapiens* to end today's overshoot and return to sustainable numbers and lifestyles.[13]

NOTES

1. J. P. Holdren, G. C. Daily, and P. R. Ehrlich, "The Meaning of Sustainability: Bio-Geophysical Aspects," *Defining and Measuring Sustainability: The Bio-Geophysical Foundations* (Washington, DC: World Bank, 1995).

2. P. Dasgupta, *An Inquiry into Well-Being and Destitution* (Oxford, UK: Clarendon, 1993).

3. G. C. Daily, A. H. Ehrlich, and P. R. Ehrlich, "Socioeconomic Equity: A Critical Element in Sustainability," *Ambio* 24 (1995): 58–59.

4. S. Kuznets, "Economic Growth and Income Inequality," *American Economic Review* 6 (1955): 127–142.

5. Dasgupta, *Inquiry into Well-Being and Destitution.*

6. G. C. Daily and P. R. Ehrlich, "Population, Sustainability and Carrying Capacity," *Bioscience* 42 (1992): 761–771.

7. E.g., G. Hardin, "Cultural Carrying Capacity: A Biological Approach to Human Problems," *BioScience* 36 (1986): 599–606; see Daily and Ehrlich, "Population, Sustainability and Carrying Capacity," for an elaboration of biophysical and social dimensions of carrying capacity.

8. G. H. Brundtland, *Our Common Future* (Oxford: Oxford University Press, 1987).

9. E.g., W. Clark, "Managing Planet Earth," *Scientific American*, September 1989, 47–54; P. R. Ehrlich and A. H. Ehrlich, *Healing the Planet* (New York: Addison-Wesley, 1991); H. Daly and R. Goodland, "An Ecological-Economic Assessment of Deregulation of International Commerce under GATT," *Ecological Economics* 9 (1994): 73–92.

10. K. E. Boulding, "The Economics of the Coming Spaceship Earth," in *Environmental Quality in a Growing Economy*, ed. H. Jarrett, 3–14. (Baltimore: Johns Hopkins University Press, 1966); P. R. Ehrlich and A. H. Ehrlich, *Population, Resources, Environment: Issues in Human Ecology* (San Francisco: W. H. Freeman, 1970).

11. Ehrlich and Ehrlich, *Healing the Planet*.

12. E.g., M. D. Young, "Ecologically-Accelerated Trade Liberalization: A Set of Disciplines for Environment and Trade Agreements," *Ecological Economics* 9 (1994): 43–51.

13. Daily, Ehrlich, and Ehrlich, "Socioeconomic Equity."

19

The NEXT Industrial Revolution

William McDonough and Michael Braungart

"Eco-efficiency," the current industrial buzzword, will neither save the environment nor foster ingenuity and productivity, the authors say. They propose a new approach that aims to solve rather than alleviate the problems that industry makes.

In the spring of 1912 one of the largest moving objects ever created by human beings left Southampton and began gliding toward New York. It was the epitome of its industrial age—a potent representation of technology, prosperity, luxury, and progress. It weighed 66,000 tons. Its steel hull stretched the length of four city blocks. Each of its steam engines was the size of a townhouse. And it was headed for a disastrous encounter with the natural world.

This vessel, of course, was the *Titanic*—a brute of a ship, seemingly impervious to the details of nature. In the minds of the captain, the crew, and many of the passengers, nothing could sink it.

One might say that the infrastructure created by the Industrial Revolution of the nineteenth century resembles such a steamship. It is powered by fossil fuels, nuclear reactors, and chemicals. It is pouring waste into the water and smoke into the sky. It is attempting to work by its own rules, contrary to those of the natural world. And although it may seem invincible, its fundamental design flaws presage disaster. Yet many people still believe that with a few minor alterations, this infrastructure can take us safely and prosperously into the future.

During the Industrial Revolution resources seemed inexhaustible and nature was viewed as something to be tamed and civilized. Recently, however, some leading industrialists have begun to realize that traditional ways of doing things may not be sustainable over the long term. "What we thought was boundless has limits," Robert Shapiro, the chairman and chief executive officer of Monsanto, said in a 1997 interview, "and we're beginning to hit them."

The 1992 Earth Summit in Rio de Janeiro, led by the Canadian businessman Maurice Strong, recognized those limits. Approximately 30,000 people from around the world, including more than a hundred world leaders and representatives of 167 countries, gathered in Rio de Janeiro to respond to troubling symptoms of environmental decline. Although there was sharp disappointment afterward that no binding agreement had been reached at the summit, many industrial participants touted a particular strategy: eco-efficiency. The machines of industry would be refitted with cleaner, faster, quieter engines.

Prosperity would remain unobstructed, and economic and organizational structures would remain intact. The hope was that eco-efficiency would transform human industry from a system that takes, makes, and wastes into one that integrates economic, environmental, and ethical concerns. Eco-efficiency is now considered by industries across the globe to be the strategy of choice for change.

What is eco-efficiency? Primarily, the term means "doing more with less"—a precept that has its roots in early industrialization. Henry Ford was adamant about lean and clean operating policies; he saved his company money by recycling and reusing materials, reduced the use of natural resources, minimized packaging, and set new standards with his time-saving assembly line. Ford wrote in 1926, "You must get the most out of the power, out of the material, and out of the time"—a credo that could hang today on the wall of any eco-efficient factory. The linkage of efficiency with sustaining the environment was perhaps most famously articulated in *Our Common Future*, a report published in 1987 by the United Nations' World Commission on Environment and Development. *Our Common Future* warned that if pollution control were not intensified, property and ecosystems would be threatened, and existence would become unpleasant and even harmful to human health in some cities. "Industries and industrial operations should be encouraged that are more efficient in terms of resource use, that generate less pollution and waste, that are based on the use of renewable rather than non-renewable resources, and that minimize irreversible adverse impacts on human health and the environment," the commission stated in its agenda for change.

The term "eco-efficiency" was promoted five years later, by the Business Council (now the World Business Council) for Sustainable Development, a group of forty-eight industrial sponsors including Dow, DuPont, ConAgra, and Chevron, who brought a business perspective to the Earth Summit. The council presented its call for change in practical terms, focusing on what businesses had to gain from a new ecological awareness rather than on what the environment had to lose if industry continued in current patterns. In *Changing Course*, a report released just before the summit, the group's founder, Stephan Schmidheiny, stressed the importance of eco-efficiency for all companies that aimed to be competitive, sustainable, and successful over the long term. In 1996 Schmidheiny said, "I predict that within a decade it is going to be next to impossible for a business to be competitive without also being 'eco-efficient'— adding more value to a good or service while using fewer resources and releasing less pollution."

As Schmidheiny predicted, eco-efficiency has been working its way into industry with extraordinary success. The corporations committing themselves to it continue to increase in number, and include such big names as Monsanto, 3M, and Johnson & Johnson. Its famous three Rs—reduce, reuse, recycle—are steadily gaining popularity in the home as well as the workplace. The trend stems in part from eco-efficiency's economic benefits, which can be considerable: 3M, for example, has saved more than $750 million through pollution-prevention projects, and other companies, too, claim to be realizing big savings. Naturally, reducing resource consumption, energy use, emissions, and wastes has implications for the environment as well. When one hears that DuPont has cut its emissions of airborne cancer-causing chemicals by almost 75 percent since 1987, one can't help feeling more secure. This is another benefit of eco-efficiency: it diminishes guilt and fear. By subscribing to eco-efficiency, people and industries can be less "bad" and less fearful about the future. Or can they?

Eco-efficiency is an outwardly admirable and certainly well-intended concept, but, unfortunately, it is not a strategy for success over the long term, because it does not reach deep enough. It works within the same system that caused the problem in the first place, slowing it down with moral proscriptions and punitive demands. It presents little more than an illusion of change. Relying on eco-efficiency to save the environment will in fact achieve the opposite—it will let industry finish off everything quietly, persistently, and completely.

We are forwarding a reshaping of human industry—what we and the author Paul Hawken call the Next Industrial Revolution. Leaders of this movement include many people in diverse fields, among them commerce, politics, the humanities, science, engineering, and education. Especially notable are the businessman Ray Anderson; the philanthropist Teresa Heinz; the Chattanooga city councilman Dave Crockett; the physicist Amory Lovins; the environmental studies professor David W. Orr; the environmentalists Sarah Severn, Dianne Dillon Ridgley, and Susan Lyons; the environmental product developer Heidi Holt; the ecological designer John Todd; and the writer Nancy Jack Todd. We are focused here on a new way of designing industrial production. As an architect and industrial designer and a chemist who have worked with both commercial and ecological systems, we see conflict between industry and the environment as a design problem—a very big design problem.

Many of the basic intentions behind the Industrial Revolution were good ones, which most of us would probably like to see carried out today: to bring more goods and services to larger numbers of people, to raise standards of living, and to give people more choice and opportunity, among others. But there were crucial omissions. Perpetuating the diversity and vitality of forests, rivers, oceans, air, soil, and animals was not part of the agenda.

If someone were to present the Industrial Revolution as a retroactive design assignment, it might sound like this:

Design a system of production that
- puts billions of pounds of toxic material into the air, water, and soil every year
- measures prosperity by activity, not legacy

- requires thousands of complex regulations to keep people and natural systems from being poisoned too quickly
- produces materials so dangerous that they will require constant vigilance from future generations
- results in gigantic amounts of waste
- puts valuable materials in holes all over the planet, where they can never be retrieved
- erodes the diversity of biological species and cultural practices

Eco-efficiency instead
- releases *fewer* pounds of toxic material into the air, water, and soil every year
- measures prosperity by *less* activity
- *meets* or *exceeds* the stipulations of thousands of complex regulations that aim to keep people and natural systems from being poisoned too quickly
- produces *fewer* dangerous materials that will require constant vigilance from future generations
- results in *smaller* amounts of waste
- puts *fewer* valuable materials in holes all over the planet, where they can never be retrieved
- standardizes and homogenizes biological species and cultural practices

Plainly put, eco-efficiency aspires to make the old, destructive system less so. But its goals, however admirable, are fatally limited.

Reduction, reuse, and recycling slow down the rates of contamination and depletion but do not stop these processes. Much recycling, for instance, is what we call "downcycling," because it reduces the quality of a material over time. When plastic other than that found in such products as soda and water bottles is recycled, it is often mixed with different plastics to produce a hybrid of lower quality, which is then molded into something amorphous and cheap, such as park benches or speed bumps. The original high-quality material is not retrieved, and it eventually ends up in landfills or incinerators.

The well-intended, creative use of recycled materials for new products can be misguided. For example, people may feel that they are making an eco-logically sound choice by buying and wearing clothing made of fibers from recycled plastic bottles. But the fibers from plastic bottles were not specifically designed to be next to human skin. Blindly adopting superficial "environmen-tal" approaches without fully understanding their effects can be no better than doing nothing.

Recycling is more expensive for communities than it needs to be, partly because traditional recycling tries to force materials into more lifetimes than they were designed for—a complicated and messy conversion, and one that itself expends energy and resources. Very few objects of modern consumption were designed with recycling in mind. If the process is truly to save money and materials, products must be designed from the very beginning to be recycled or even "upcycled"—a term we use to describe the return to industrial systems of materials with improved, rather than degraded, quality.

The reduction of potentially harmful emissions and wastes is another goal of eco-efficiency. But current studies are beginning to raise concern that even tiny amounts of dangerous emissions can have disastrous effects on biological systems over time. This is a particular concern in the case of endocrine disrupters—industrial chemicals in a variety of modern plastics and consumer goods which appear to mimic hormones and connect with receptors in human beings and other organisms. Theo Colborn, Dianne Dumanoski, and John Peterson Myers, the authors of *Our Stolen Future* (1996), a groundbreaking study on certain synthetic chemicals and the environment, assert that "astoundingly small quantities of these hormonally active compounds can wreak all manner of biological havoc, particularly in those exposed in the womb."

On another front, new research on particulates—microscopic particles released during incineration and combustion processes, such as those in power plants and automobiles—shows that they can lodge in and damage the lungs, especially in children and the elderly. A 1995 Harvard study found that as many as 100,000 people die annually as a result of these tiny particles. Although regulations for smaller particles are in place, implementation does not have to begin until 2005. Real change would be not regulating the release of particles but attempting to eliminate dangerous emissions altogether—by design.

Applying Nature's Cycles to Industry

"Produce more with less," "Minimize waste," "Reduce," and similar dictates advance the notion of a world of limits—one whose carrying capacity is strained by burgeoning populations and exploding production and consumption. Eco-efficiency tells us to restrict industry and curtail growth—to try to limit the creativity and productiveness of humankind. But the idea that the natural world is inevitably destroyed by human industry, or that excessive demand for goods and services causes environmental ills, is a simplification. Nature—highly industrious, astonishingly productive and creative, even "wasteful"—is not efficient but *effective*.

Consider the cherry tree. It makes thousands of blossoms just so that another tree might germinate, take root, and grow. Who would notice piles of cherry blossoms littering the ground in the spring and think, "How inefficient and wasteful"? The tree's abundance is useful and safe. After falling to the ground, the blossoms return to the soil and become nutrients for the surrounding environment. Every last particle contributes in some way to the health of a thriving ecosystem. "Waste equals food"—the first principle of the Next Industrial Revolution.

The cherry tree is just one example of nature's industry, which operates according to cycles of nutrients and metabolisms. This cyclical system is powered by the sun and constantly adapts to local circumstances. Waste that stays waste does not exist.

Human industry, on the other hand, is severely limited. It follows a one-way, linear, cradle-to-grave manufacturing line in which things are created and eventually discarded, usually in an incinerator or a landfill. Unlike the waste from nature's work, the waste from human industry is not "food" at all. In fact, it is often poison. Thus the two conflicting systems: a pile of cherry blossoms and a heap of toxic junk in a landfill.

But there is an alternative—one that will allow both business and nature to be fecund and productive. This alternative is what we call "eco-effectiveness." Our concept of eco-effectiveness leads to human industry that is regenerative rather than depletive. It involves the design of things that celebrate interdependence with other living systems. From an industrial-design perspective, it means products that work within cradle-to-cradle life cycles rather than cradle-to-grave ones.

Waste Equals Food

Ancient nomadic cultures tended to leave organic wastes behind, restoring nutrients to the soil and the surrounding environment. Modern, settled societies simply want to get rid of waste as quickly as possible. The potential nutrients in organic waste are lost when they are disposed of in landfills, where they cannot be used to rebuild soil; depositing synthetic materials and chemicals in natural systems strains the environment. The ability of complex, interdependent natural ecosystems to absorb such foreign material is limited if not nonexistent. Nature cannot do anything with the stuff *by design*: many manufactured products are intended not to break down under natural conditions.

If people are to prosper within the natural world, all the products and materials manufactured by industry must after each useful life provide nourishment for something new. Since many of the things people make are not natural, they are not safe "food" for biological systems. Products composed of materials that do not biodegrade should be designed as technical nutrients that continually circulate within closed-loop industrial cycles—the technical metabolism.

In order for these two metabolisms to remain healthy, great care must be taken to avoid cross-contamination. Things that go into the biological metabolism should not contain mutagens, carcinogens, heavy metals, endocrine disrupters, persistent toxic substances, or bio-accumulative substances. Things that go into the technical metabolism should be kept well apart from the biological metabolism.

If the things people make are to be safely channeled into one or the other of these metabolisms, then products can be considered to contain two kinds of materials: *biological nutrients* and *technical nutrients.*

Biological nutrients will be designed to return to the organic cycle—to be literally consumed by microorganisms and other creatures in the soil. Most packaging (which makes up about 50 percent by volume of the solid-waste stream) should be composed of biological nutrients—materials that can be tossed onto the ground or the compost heap to biodegrade. There is no need for shampoo bottles, toothpaste tubes, yogurt cartons, juice containers, and other packaging to last decades (or even centuries) longer than what came inside them.

Technical nutrients will be designed to go back into the technical cycle. Right now anyone can dump an old television into a trash can. But the average television is made of hundreds of chemicals, some of which are toxic. Others are valuable nutrients for industry, which are wasted when the television ends up in a landfill. The reuse of technical nutrients in closed-loop industrial cycles is distinct from traditional recycling, because it allows materials to retain their quality: high-quality plastic computer cases would continually circulate as high-quality computer cases, instead of being downcycled to make soundproof barriers or flowerpots.

Customers would buy the service of such products, and when they had finished with the products, or simply wanted to upgrade to a newer version, the manufacturer would take back the old ones, break them down, and use their complex materials in new products.

First Fruits: A Biological Nutrient

A few years ago we helped to conceive and create a compostable upholstery fabric—a biological nutrient. We were initially asked by Design Tex to create an aesthetically unique fabric that was also ecologically intelligent—although the client did not quite know at that point what this would mean. The challenge helped to clarify, both for us and for the company we were working with, the difference between superficial responses such as recycling and reduction and the more significant changes required by the Next Industrial Revolution.

For example, when the company first sought to meet our desire for an environmentally safe fabric, it presented what it thought was a wholesome option: cotton, which is natural, combined with PET (polyethylene terephthalate) fibers from recycled beverage bottles. Since the proposed hybrid could be described with two important eco-buzzwords, "natural" and "recycled," it appeared to be environmentally ideal. The materials were readily available, market-tested, durable, and cheap. But when the project team looked carefully at what the manifestations of such a hybrid might be in the long run, we discovered some disturbing facts. When a person sits in an office chair and shifts around, the fabric beneath him or her abrades; tiny particles of it are inhaled or swallowed by the user and other people nearby. PET was not designed to be inhaled. Furthermore, PET would prevent the proposed hybrid from going back into the soil safely, and the cotton would prevent it from re-entering an industrial cycle. The hybrid would still add junk to landfills, and it might also be dangerous.

The team decided to design a fabric so safe that one could literally eat it. The European textile mill chosen to produce the fabric was quite "clean" environmentally, and yet it had an interesting problem: although the mill's director had been diligent about reducing levels of dangerous emissions, government regulators had recently defined the trimmings of his fabric as hazardous waste. We sought a different end for our trimmings: mulch for the local garden club. When removed from the frame after the chair's useful life and tossed onto the ground to mingle with sun, water, and hungry microorganisms, both the fabric and its trimmings would decompose naturally.

The team decided on a mixture of safe, pesticide-free plant and animal fibers for the fabric (ramie and wool) and began working on perhaps the most difficult aspect: the finishes, dyes, and other processing chemicals. If the fabric was to go back into the soil safely, it had to be free of mutagens, carcinogens, heavy metals, endocrine disrupters, persistent toxic substances, and bio-accumulative substances. Sixty chemical companies were approached about joining the project, and all declined, uncomfortable with the idea of exposing their chemistry to the kind of scrutiny necessary. Finally one European company, Ciba-Geigy, agreed to join.

With that company's help the project team considered more than 8,000 chemicals used in the textile industry and eliminated 7,962. The fabric—in fact, an entire line of fabrics—was created using only thirty-eight chemicals.

The director of the mill told a surprising story after the fabrics were in production. When regulators came by to test the effluent, they thought their instruments were broken. After testing the influent as well, they realized that the equipment was fine—the water coming out of the factory was as clean as the water going in. The manufacturing process itself was filtering the water. The new design not only bypassed the traditional three-R responses to environmental problems but also eliminated the need for regulation.

In our Next Industrial Revolution, regulations can be seen as signals of design failure. They burden industry, by involving government in commerce and by interfering with the marketplace. Manufacturers in countries that are less hindered by regulations, and whose factories emit more toxic substances, have an economic advantage: they can produce and sell things for less. If a factory is not emitting dangerous substances and needs no regulation, and can thus compete directly with unregulated factories in other countries, that is good news environmentally, ethically, and economically.

A Technical Nutrient

Someone who has finished with a traditional carpet must pay to have it removed. The energy, effort, and materials that went into it are lost to the manufacturer; the carpet becomes little more than a heap of potentially hazardous petrochemicals that must be toted to a landfill. Meanwhile, raw materials must continually be extracted to make new carpets.

The typical carpet consists of nylon embedded in fiberglass and PVC. After its useful life a manufacturer can only downcycle it—shave off some of the nylon for further use and melt the leftovers. The world's largest commercial carpet company, Interface, is adopting our technical-nutrient concept with a carpet designed for complete recycling. When a customer wants to replace it, the manufacturer simply takes back the technical nutrient—depending on the product, either part or all of the carpet—and returns a carpet in the customer's desired color, style, and texture. The carpet company continues to own the material but leases it and maintains it, providing customers with the service of the carpet. Eventually the carpet will wear out like any other, and the manufacturer will reuse its materials at their original level of quality or a higher one.

The advantages of such a system, widely applied to many industrial products, are twofold: no useless and potentially dangerous waste is generated, as it might still be in eco-efficient systems, and billions of dollars' worth of valuable materials are saved and retained by the manufacturer.

Selling Intelligence, Not Poison

Currently, chemical companies warn farmers to be careful with pesticides, and yet the companies benefit when more pesticides are sold. In other words, the companies are unintentionally invested in wastefulness and even in the mishandling of their products, which can result in contamination of the soil, water, and air. Imagine what would happen if a chemical company sold intelligence instead of pesticides—that is, if farmers or agro-businesses paid pesticide manufacturers to protect their crops against loss from pests instead of buying dangerous regulated chemicals to use at their own discretion. It would in effect be buying crop insurance. Farmers would be saying, "I'll pay you to deal with boll weevils, and you do it as intelligently as you can." At the same price per acre, everyone would still profit. The pesticide purveyor would be invested in not using pesticide, to avoid wasting materials. Furthermore, since the manufacturer would bear responsibility for the hazardous materials, it would have incentives to come up with less-dangerous ways to get rid of pests. Farmers are not interested in handling dangerous chemicals; they want to grow crops. Chemical companies do not want to contaminate soil, water, and air; they want to make money.

Consider the unintended design legacy of the average shoe. With each step of your shoe the sole releases tiny particles of potentially harmful substances that may contaminate and reduce the vitality of the soil. With the next rain these particles will wash into the plants and soil along the road, adding another burden to the environment.

Shoes could be redesigned so that the sole was a biological nutrient. When it broke down under a pounding foot and interacted with nature, it would nourish the biological metabolism instead of poisoning it. Other parts of the shoe might be designed as technical nutrients, to be returned to industrial cycles. Most shoes—in fact, most products of the current industrial system—are fairly primitive in their relationship to the natural world. With the scientific and technical tools currently available, this need not be the case.

Respect Diversity and Use the Sun

A leading goal of design in this century has been to achieve universally applicable solutions. In the field of architecture the International Style is a good example. As a result of the widespread adoption of the International Style, architecture has become uniform in many settings. That is, an office building can look and work the same anywhere. Materials such as steel, cement, and glass can be transported all over the world, eliminating dependence on a region's particular energy and material flows. With more energy forced into the heating and cooling system, the same building can operate similarly in vastly different settings.

The second principle of the Next Industrial Revolution is "Respect diversity." Designs will respect the regional, cultural, and material uniqueness of a place. Wastes and emissions will regenerate rather than deplete, and design will be flexible, to allow for changes in the needs of people and communities. For example, office buildings will be convertible into apartments, instead of ending up as rubble in a construction landfill when the market changes.

The third principle of the Next Industrial Revolution is "Use solar energy." Human systems now rely on fossil fuels and petrochemicals, and on incineration processes that often have destructive side effects. Today even the most advanced building or factory in the world is still a kind of steamship, polluting, contaminating, and depleting the surrounding environment, and relying on scarce amounts of natural light and fresh air. People are essentially working in the dark, and they are often breathing unhealthful air. Imagine, instead, a building as a kind of tree. It would purify air, accrue solar income, produce more energy than it consumes, create shade and habitat, enrich soil, and change with the seasons. Oberlin College is currently working on a building that is a good start: it is designed to make more energy than it needs to operate and to purify its own wastewater.

Equity, Economy, Ecology

The Next Industrial Revolution incorporates positive intentions across a wide spectrum of human concerns. People within the sustainability movement have found that three categories are helpful in articulating these concerns: equity, economy, and ecology.

Equity refers to social justice. Does a design depreciate or enrich people and communities? Shoe companies have been blamed for exposing workers in factories overseas to chemicals in amounts that exceed safe limits. Eco-efficiency would reduce those amounts to meet certain standards; eco-effectiveness would not use a potentially dangerous chemical in the first place. What an advance for humankind it would be if no factory worker anywhere worked in dangerous or inhumane conditions.

Economy refers to market viability. Does a product reflect the needs of producers and consumers for affordable products? Safe, intelligent designs should be affordable by and accessible to a wide range of customers, and profitable to the company that makes them, because commerce is the engine of change.

Ecology, of course, refers to environmental intelligence. Is a material a biological nutrient or a technical nutrient? Does it meet nature's design criteria: Waste equals food, Respect diversity, and Use solar energy?

The Next Industrial Revolution can be framed as the following assignment:

Design an industrial system for the next century that:
- introduces no hazardous materials into the air, water, or soil
- measures prosperity by how much natural capital we can accrue in productive ways
- measures productivity by how many people are gainfully and meaningfully employed
- measures progress by how many buildings have no smokestacks or dangerous effluents
- does not require regulations whose purpose is to stop us from killing ourselves too quickly
- produces nothing that will require future generations to maintain vigilance
- celebrates the abundance of biological and cultural diversity and solar income

Albert Einstein wrote, "The world will not evolve past its current state of crisis by using the same thinking that created the situation." Many people believe that new industrial revolutions are already taking place, with the rise of cybertechnology, biotechnology, and nanotechnology. It is true that these are powerful tools for change. But they are only tools—hyperefficient engines for the steamship of the first Industrial Revolution. Similarly, eco-efficiency is a valuable and laudable tool, and a prelude to what should come next. But it, too, fails to move us beyond the first revolution. It is time for designs that are creative, abundant, prosperous, and intelligent from the start. The model for the Next Industrial Revolution may well have been right in front of us the whole time: a tree.

20

Excerpts from "In Search of Consumptive Resistance: The Voluntary Simplicity Movement"

MICHAEL MANIATES

It's quiet, countercultural, potentially subversive, but also mainstream. It flies low, usually hidden amidst reports of increasing productivity, rising consumer confidence, expanding personal debt, and the dizzying array of new products promising to make life easier, faster, more productive, and more rewarding. Unpromisingly rooted in an apolitical and consumerist response to social ills, it also sows the seeds of collective challenge to fundamental dysfunctions of industrial society. Focused as it is on the quality of work and quest for personal control of one's time and one's life, it resonates with the American deification of individual freedom. But inevitable connections to questions of environmental quality, workplace control, and civic responsibility lend it more complicated hues.

Some call it *simple living*, evoking images of earlier, more prudent times. Others prefer *downsizing, downshifting*, or *simplifying*. The popular press and many scholars know it as *voluntary simplicity*. Call it what you will, but don't lose sight of the irony: at a time when policymakers, pundits, and corporations around the world embrace ever more deeply the assumption that consumption and happiness are joined at the hip, frugality appears to be back in fashion. Through their words and by their deeds, a seemingly large and growing number of people are claiming that "we can work less, want less, and spend less, and be happier in the process."[1]

According to Rich Hayes of UC Berkeley's Energy and Resources Group, these people generally "value moderation over excess, spiritual development over material consumption, cooperation over competition, and nature over technology."[2] Historian David Shi is more pointed—to him, practitioners of the simple life "self-consciously subordinate the material to the ideal."[3] Alan Durning, a former analyst with the World-watch Institute, an environmental think tank, quotes a voluntary simplifier who says that "simple living has come to mean spending more time attending to our lives and less time attending to our work; devoting less time earning more money and more time to the daily doings of life."[4] Noted sociologist Amitai Etzioni brings a finer-grained analysis to bear by offering three categories, each more intense, of "voluntary simplifiers": (1) "downshifters," who reduce their consumption and income without deeply altering their way of living, (2) "strong simplifiers," who significantly restructure their lives, and (3) "holistic simplifiers," whose consistent rejection of consumerism flows from a coherent philosophy.[5] And Cecile Andrews, whose book *The Circle of Simplicity* is often credited with catalyzing the "voluntary simplicity movement" (VSM) in the United States in the 1990s,[6] advances an understanding of voluntary simplicity that zeros in on the pace of life:

> A lot of people [are] rushed and frenzied and stressed. They have no time for their friends; they snap at their family; they're not laughing very much. But a growing number of people aren't content to live this way. They are looking for ways to simplify their lives—to rush less, work less, and spend less. They are beginning to slow down and enjoy life again.
>
> There's a movement associated with this—it's called the voluntary simplicity movement. Around the country, thousands of people are simplifying their lives. They are questioning the standard definitions that equate success with money and prestige and the accumulation of things. They are returning to the good life.[7]

Andrews's estimate of "thousands of people" is likely conservative, though no observer of the U.S. simplicity movement really knows for sure. Writing in 1981, Daniel Elgin, author of the simplicity bible *Voluntary Simplicity: An Ecological Lifestyle That Promotes Personal and Social Renewal*,[8] claimed 10 million dedicated converts to the cause. But this figure, an estimate for the United States alone, struck some as "optimistic."[9] And yet a later study commissioned by the Merck Family Fund[10] concluded that from 1990 to 1995, 28 percent of Americans (or over 60 million) voluntarily reduced their income and their consumption in conscious pursuit of new personal or household priorities. In 1998, Harvard economist Juliet Schor found that some 20 percent of Americans (roughly 50 million people) have, in the past several years, permanently chosen to live on significantly less and are happy with the change. Most of these "permanent downshifters" are not rich, says Schor; half reside in households with annual incomes of $35,000 or less prior to any downshifting.[11] More recently, Gerald Celente at the Trends Research Institute claimed that 15 percent of the nation's 77 million baby boomers were significantly engaged in the simplicity movement.[12]

Perhaps because of their growing numbers, simplifiers have become the subject of increasing media attention. In 1993, for example, readers of major U.S. newspapers[13] would have learned little if anything about the VSM, since relevant stories rarely made it into print. This had changed by 1996; that year, an average of just over two articles or features per paper appeared.

By 1998 the number of stories or features had jumped threefold to fivefold, depending on the newspaper, and as of this writing (mid-2001) there is no indication that this pace of coverage is slowing. Strikingly, relevant articles are finding their way into marquee venues. The *Washington Post Magazine* ran a cover story on "voluntary simplicity" just before Christmas 1998. The *New York Times* ran four pieces, three in its prestigious Op-Ed section, that connect directly to the subject in three months between November 1998 and late January 1999. Major West Coast newspapers ran feature pieces on simplicity and frugality throughout 1999 and into 2000. (Ubiquitous "resolutions for the new millennium" pieces frequently highlighted the need to simplify and "slow down.") The U.S. Public Broadcast System aired two programs—*Affluenza* and *Escape from Affluenza*—on consumerism and its cure.[14] The producer of the programs characterizes the outpouring of public interest in them as "astonishing" and "completely unexpected." The distributor of the videos describes them as the biggest production his company has distributed in years.[15] *Affluenza* has been translated into four languages for distribution to 17 countries of the former Soviet Union, and similar plans are in the works for *Escape from Affluenza*.

Curiously, the media's fascination with simplicity has not been matched by attention in more scholarly circles. One reason, I explain shortly, is that voluntary simplicity butts up against mainstream environmentalism's understanding of "sustainability" and "the consumption problem" and is marginalized as a result. By its very existence, the VSM insists that real reductions in consumption—at least for some, framed in particular ways—bring real net benefits to be enjoyed rather than sacrifices to be endured. But mainstream environmentalists—not to mention policymakers, planners, and many academics—find it difficult to entertain and give voice to the distinct possibility that doing with less could mean doing better and being happier. Locked in a calculus of sacrifice, activists and academics and policymakers alike tend first to romanticize the VSM, and then to marginalize it to the domain of "fringe" activity that is oddly interesting but fundamentally irrelevant to the practical politics of sustainability.

This is unfortunate because, as I later argue, the relevance of the VSM to larger struggles for sustainability remains ambiguous. Simplicity is teetering between a self-absorbed subculture looking to effect social change by, say, using soap more sparingly, and a nascent social movement capable of fostering lasting change in how work, play, and consumption are organized in industrial society. Writing off voluntary simplicity as the newest incarnation of yuppie "self-help" claptrap[16] runs the risk of slighting a potentially defining element of an alternative politics of environmentalism. On the other hand, framing voluntary simplicity as the next major trend to sweep industrial society[17] dangerously glosses over simple living's worrisome deficiencies of politics and analysis.

Simplifiers are apparently a diverse group, drawn to simplicity for myriad reasons and networking in novel ways. The VSM is a kaleidoscope, and thus answers to reasonable questions become both straightforward and confusing. Is voluntary simplicity a growing social movement for reducing consumption and fostering environmental sustainability, as many of its adherents maintain? One could make a convincing argument. Is it instead a passing fad, the contemporary manifestation of a long-observed pattern of yearning for simplicity that is then followed by a burst of hyperconsumption? This case can also be made. Does the VSM stand as evidence of broadly shared disquiet with the direction and pace of industrial society; is it, in other words, a subversive attempt to strike at the heart of consumer capitalism? Or is it an irrelevant subculture, one that has enjoyed more than its share of media attention, which brings together two social groups: burnt-out urban professionals earning $200,000 year whose choice to "downshift" to a "simple" lifestyle has them trading in the Mercedes for a Honda, and everyday citizens hiding from the rough-and-tumble world of environmental politics by quietly focusing on simple acts of frugality like clipping coupons at home?

One way of slicing through this confusion is to return to the organizing themes of this book: production as consumption; the chain of material provisioning and resource use; the myth of consumer sovereignty; the politics of consumption and the struggle to build new institutional mechanisms that more fully communicate the full range of the costs of consumerism, commodification, and overconsumption; the consumption juggernaut. When applied to simple living, what do these concepts tell us about the strengths and weakness of the VSM and its potential as a social movement that confronts consumption? The concluding paragraphs of this chapter take on this question. . . .

*   *   *

Robert Frank lives in Ithaca and teaches at Cornell University, while Dipak Gyawali calls Kathmandu his home. From opposite sides of the planet, living in very different worlds, both point to the same dynamic: satisfaction with one's material life is significantly influenced by how much one spends and consumes relative to others. Beyond some threshold of minimally acceptable consumption, deprivation and abundance become relative conditions, easily altered by the consumption levels and decisions of benchmark social groups.[18]

Examples abound in the industrial North. A cartoon from the *New Yorker* appears in Frank's book, for example, in which a commuter, calling home from a roadside stop, says "I was sad because I had no onboard fax until I saw a man who had no mobile phone." A silly cartoon, yes, but on the mark too regarding relative satisfaction. Products constantly are becoming more "luxurious"— faster computers, larger cars, more spacious houses, smaller cell phones, better backpacks, higher-performance athletic shoes. Their introduction into the marketplace makes previously acceptable consumption choices pedestrian, and sometimes unacceptable, by comparison. "Recent changes in the spending environment," writes Frank,

> affect the kinds of gifts you must give at weddings and birthdays, and the amounts you must spend for anniversary dinners; the price you must pay for a house in

a neighborhood with a good school; the size your vehicle must be if you want your family to be relatively safe from injury; the kinds of sneakers your kids will demand; the universities they'll need to attend if you want them to face good prospects after graduation; the kinds of wine you'll want to serve to mark special occasions; and the kind of suit you'll choose to wear to a job interview.[19]

Part of this dynamic is inescapably structural: if everyone else is wearing an expensive suit to that job interview, you have got to as well if you hope to send the right message, regardless of your feelings about consumerism or your leanings toward frugality. Likewise, the presence of more behemoth vehicles on the road ratchets up the arms race on the highway; standing a fighting chance against one in a collision requires buying a larger vehicle yourself, no matter how much joy or dismay driving it brings. But a big chunk of the problem is rooted in less obvious changes in the "benchmark" group against which much of the middle class compares itself.

"Keeping up with the Joneses" used to mean measuring up to the consumption choices of neighbors of similar means and aspirations. More and more of the middle class, however, are now gauging their material prosperity against the top 10 percent of consumers nationally—the yardstick has taken a quantum leap upward. Many interlocking forces are responsible, including easy credit, vigorous marketing, the increasing concentration of income, the marked rise in casual mixing among economic classes as corporations embrace "open offices" and "team-building" management techniques,[20] a spate of television shows and movies that portray consumption levels of the top 10 percent of Americans as those of an "average" middle-class family, and real-life consumption by the upper economic strata that is conspicuous and seemingly guilt free.[21] At work too, Juliet Schor reminds us, is the "Diderot effect,"[22] whereby a new acquisition—a new bookcase for the living room, for instance—suddenly makes the couch look rather shabby and that, when then replaced, highlights the sudden inadequacy of the curtains. Together, this trinity of forces—the structural, the rising benchmark, the Diderot-driven "upward creep of desire"—conspire to keep the consumer escalator moving upward. It is luxury fever: more people are consuming more, in terms of both quality and quantity; they are going more deeply into debt as a result; and if analysts like Robert Frank or Robert Lane (whose book *The Loss of Happiness in Market Democracies* marshals an impressive array of data)[23] can be believed, they are finding life less fulfilling and less secure as a result of this relentless ratcheting up of standards.

Take this "luxury fever" and extend it to the so-called Third World, and you get Gyawali's "tyranny of expectations."[24] Globalization brings sophisticated advertising to the Third World, spreads Western television programs that glamorize consumption (*Dallas, Melrose Place, Santa Barbara*), and facilitates the impossible-to-miss emergence of a conspicuously consuming Third World elite more at home in New York, Los Angeles, or London than in Delhi, Salvador, or Nairobi. Masses of poor in these countries feel poorer by comparison and come to expect more. Though their economies have a hand in fueling rising expectations, these economies cannot satisfy people's expectations quickly, if at all—huge numbers of people are involved, after all, whose understanding of what minimally constitutes the "good life" is rising exponentially. These expectations, heightened by media and elite behavior, go unmet; an enduring sense of "missing out" and being left behind takes root; and frustration, dissatisfaction, and social instability surface. Ironically, as prudent societies of the Third World are transformed into consumer societies, mass frustration and an overall decline in felt material satisfaction ensues, even as real incomes among the poor slowly inch higher. For scholars like Canadian historian and filmmaker Gwynne Dyer, this still-unfolding dynamic is nothing less than "a bomb under the world."[25]

This is not an argument to keep poor people poor because "we" somehow know that increased consumption will not make "them" much happier. Many around the world must be lifted from the depths of poverty, and nothing in Gyawali or Frank's analysis should be heard to suggest otherwise. But if the 4 billion or more global underconsumers are to raise their consumption levels to some minimally rewarding and secure level, the 1 billion or so global overconsumers will first have to limit and then reduce their overall level of consumption to make ecological room. At this point, mainstream environmentalism stalls out, for it has not yet focused on developing a politically practical language for analyzing and confronting the explosion of wants, much less for thinking about how to foster the "downshifting" of democratic industrial societies. Indeed, the sustainable-development lens, which removes from view any political or sociological analysis of consumption, has become the metaphorical AIDS of the environmental movement. The movement's ability to recognize and respond to the core threats to the environment has been compromised because it cannot engage an ethos of frugality.

What instead emerges is simplistic moralizing about consumption that little advances the intellectual analysis or collective action necessary for taking on the consumption question. This moralizing takes three forms: rhetorical lambasting of advertising, condemnation of the immorality of overconsumption, and a rosy-eyed, apolitical romanticization of the joys of simple living. Though each response makes some sense, in that each echoes credible analyses of the current human condition (corporations do spend egregious amounts on advertising, gluttony is fraught with ethical implications, frugality and self-denial can bring personal satisfaction), none alone hold much promise for undermining luxury fever or thwarting the tyranny of expectations. . . .

Stepping Back: The Consumption Angle and Consumptive Resistance

The "consumption angle" as described in this book is at once an analysis of our currently unsustainable ways and a call for a new kind of struggle for sustainability. Production, the consumption angle reminds us, is also consumption. Consumption, it insists, occurs not just at the end point of consumer demand, but up and down a chain of material provisioning and resource use. Nodes of production and consumption along this chain come with their own structures, dynamics, and power relations, which interact with one another to cement into place patterns of production and consumption fundamentally at odds with life processes of the planet. Political and economic power—typically concentrated, often highly so, and frequently concealed from easy view—scaffolds the chains and obscures the telling costs of rampant consumerism. The prevailing dogma of consumer sovereignty, which fixes all responsibility for "overconsumption" on individual consumers ostensibly making "free" marketplace choices, blunts efforts to hold the powerful accountable for their central role in the upscaling of desire. A new environmentalism that recognizes these myriad factors and confronts the engines of consumption has never been more necessary.

Held up against the consumption angle, the VSM remains an uncertain and ambiguous thing. To their credit, downshifters and simplifiers, and those who imagine themselves joining their ranks, understand more than most the extent to which *production is consumption*—consumption of time, creativity, control over work, and a sustaining personal sense of fulfillment and accomplishment. Get past the politeness and apolitical sensibilities of the VSM, and what you have is a clear call for production systems that consume far less in terms of human and environmental capital. Of course, since "work" is a politically charged topic, and since most simplifiers come from households for whom ideas of workplace organizing are unfamiliar or taboo, simplifiers displace their anxieties about work onto consumption, thinking that by reducing expenses one can eventually resolve workplace problems. But make no mistake: for a large swath of simplifiers, the core issue is the nature of production and the organization of the workplace, and the dehumanizing and frustrating changes in both over the past two decades. The growing ranks of simplifiers are thus a potentially potent ally in any effort to bring a consumption perspective to bear on our current economic and environmental woes.

For the VSM, *the chain of material provisioning and resource use* is more problematic. Because it dislikes talking comprehensively about work, simplicity becomes a zealous conversation about what one buys and why, to the exclusion of almost everything else. As it grows in mass appeal, downshifting will almost certainly act as a drag on any systematic "up and down the chain" thinking that remains critical to the effective analysis and resolution of the consumption problem—and the winners will be those most adept at manipulating the chain of material provisioning and resource use to their own benefit, at longer-term cost to the environment and human well-being.

The VSM's current trajectory, in other words, spells trouble for those who would facilitate a deeper public understanding of the consumption problem. Rather than continuing to ignore the VSM or imagining it to be a fringe phenomenon, it would be far better to set to work widening existing cracks in the VSM's single-minded preoccupation with the end point of the consumption chain. One chink is the VSM's deep-seated suspicion of corporate power, which is displaced onto consumption and onto a sometimes-obsessive preoccupation with frugality. Another is its recently discovered and currently expanding disdain for processes of commodification and co-optation driven by the explosion of glossy magazines and slick ads extolling the virtues of simplicity. A third is its nascent environmental sensibilities; while not environmentalist per se, many simplifiers recognize that living more frugally also means living more "ecologically." These three elements, and others as well, continuously prod the movement to glance "upstream" at production, packaging, employment, and marketing practices that drive and frame final consumer behavior. Now is the time to accentuate and strengthen these prodding forces through dialogue and coalition building with the primary voices and major networking tools of the movement. The standing ovation showered on Kalle Lasn and his call at the 1998 simplicity meetings for a "pincer movement" that would mesh personal frugality with collective action suggest a large, untapped reservoir of support within the downshifting community for such efforts.

On the question of *power*, finally, the VSM remains on life support. It has not yet developed consistently coherent ways of talking about the distribution and exercise of consumption-amplifying power in any but the most unsophisticated ways. Locked into a rhetoric of the individualization of responsibility, it propagates the all-too-familiar "plant a tree, save the world" environmental mentality. This mentality imagines consumers to be immune to the marketer's ability to tap into environmental concern to sell a host of environmentally unfriendly products, and draws attention away from inequalities in power and responsibility that occupy the center of the environmental crisis. Ironically, the VSM risks aiding and abetting the very cultural and political forces it philosophically opposes.

But all this could change in a hurry. The U.S. economy is driven by consumer demand, which the simplicity movement hopes to throttle. Fulfillment of this hope would lead to production cutbacks, plant closings, and job loss, at especially great cost to the working poor, who have no savings or deep well of job skills from which to draw in a slowing economy. In theory, simplicity sounds attractive; in practice, it would, absent significant policy change, balance a greater measure of middle-class tranquility on the backs of the bottom 20 percent of American households.

Until recently, the full-employment economy made the class-warfare quality of the VSM easy for all to ignore. Times were good and jobs were plentiful. But all indicators point to an end of these economically anomalous times and a return to the more familiar economic terrain of significant unemployment and, perhaps, onerous levels of inflation. At that point, simplifiers, who are easily plunged into bouts of self-recrimination by critiques of VSM's class bias, will no longer be able to discount the macroeconomic dynamics that convert their good intentions into pain and suffering for others. That will make them uncomfortable indeed.

Signs of that discomfort, and a strategy for ameliorating it, are already appearing. One is Juliet Schor's epilogue ("Will Consuming Less Wreck the Economy?") to her book *The Overspent American*. Schor answers her wrecking-the-economy question with a firm no. "A gradual reduction in consumer spending will not cause much unemployment," she argues, because "the trend towards buying less is likely to be associated with a trend towards working less." "Fewer people would want jobs," she continues, "and the hours worked *per* job could fall."[26] The result: lower economic growth, yes, but more free time, less stress, and less useless stuff. It is a future, Schor insists, worth striving for.

But getting to this future requires an evolution of the rules, norms, and practices that govern work in the United States—dropping the hours worked per job in ways that preserve job security and benefits packages will not happen without determined and spirited national conversation. We are back to the centrality of a "time movement," something that Schor acknowledges in her epilogue but curiously sidesteps. Absent such a movement, a few simplifiers will get to work in better, more accommodating, and more flexible jobs, and they will consume less in the process. Meanwhile, many others without such options will pay the price of economic downturn; their lot will be an involuntary simplicity. The high-minded ideals of the simplicity movement will be soiled in the process, sufficiently so, perhaps, to derail its momentum.

As the economy softens, the VSM will face these sad possibilities, and the press, undoubtedly, will fully illuminate them. There will be the inevitable op-ed pieces and magazine columns charging the VSM with elitist insensitivity to the plight of the poor. The VSM will not take this lying down—if its reaction to the appropriation of "simplicity" by glossy magazines tells us anything, it is that adherents of simplicity are a prickly lot; they take their goals and ideals seriously, and they will react strongly to charges that their behavior only amplifies recessionary economic pressures. But what will they say in their defense? Anything beyond a tacit admission of complicity with economic forces that leave millions of households scrambling for cover will require of them an expanded understanding of power in American society and a newfound willingness to engage the nuts-and-bolts of workplace reform. Pressures already are building in the movement to move in just this direction—and if the economy does head south, these pressures will grow. Advocates of a consumption angle who appreciate how a "time movement" might foster the capacity for restraint would be wise to reflect on the role they might play in hurrying along these changes in the VSM.

As a social movement the VSM remains inchoate. Its numbers are large and growing, but its core ideology with respect to the many challenges posed by the consumption angle remains embryonic. As a social phenomenon it suggests that consumptive resistance flows from the dual pressures of workplace stress and the anxiety that accompanies a methodical upscaling of desire. In many ways simplifiers are no different from the average American; they work hard, raise families, live in both rural and urban areas, hold down a variety of jobs, and express a deep desire to escape the rat race. What makes them a deserving object of continued scrutiny is their choice and ability to act on these desires of escape, desires fueled and facilitated by generally high levels of education, which both raise expectations for "fulfilling work" and confer power to reform or withdraw from the workplace.

Simplifiers are like the proverbial canary in the mine, but with a twist. Like the dying canary that points to the presence of poisonous mine gas, downshifters offer an early warning that something serious is amiss. Moreover, in pursuit of their often narrow and self-interested agenda, they are also whittling away at workplace norms while modeling for others ways of divorcing personal success from willing participation in the work-and-spend cycle that structures much of our lives. Despite its reluctance to think about broader institutional dynamics, the VSM points toward a more coherent time movement—a movement that, by restructuring workplace expectations and structures, could allow millions more to discover the joys of cultivating a capacity for restraint.

NOTES

1. Linda Breen Pierce, *Choosing Simplicity: Real People Finding Peace and Fulfillment in a Complex World* (Carmel, CA: Gallagher, 2000).

2. Rich Hayes, "A Survey of People Attending a Conference on Voluntary Simplicity," unpublished manuscript (Berkeley: Energy and Resources Group, University of California, 1999).

3. David Shi, *The Simple Life: Plain Living and High Thinking in American Culture* (New York: Oxford University Press, 1985), 3.

4. Alan Durning, *How Much Is Enough?* (New York: Norton, 1992), 141.

5. Amitai Etzioni, "Voluntary Simplicity: Characterization, Select Psychological Implications, and Societal Consequences," *Journal of Economic Psychology* 19, no. 5 (1998): 619–643.

6. Cecile Andrews, *The Circle of Simplicity: Return to the Good Life* (New York: Harper Perennial, 1997). Andrews's book spawned the "The Simplicity Circles Project," which fosters small-group discussions of simplicity and links more than 500 "simplicity circles" across the United States.

7. Andrews, *The Circle of Simplicity*, xiv.

8. Daniel Elgin, *Voluntary Simplicity: An Ecological Lifestyle That Promotes Personal and Social Renewal* (New York: Bantam Books, 1981).

9. For example, Durning, *How Much Is Enough?*, 139.

10. Harwood Group, *Yearning for Balance: Views of Americans on Consumption, Materialism, and Environment* (Tacoma, MD: Merck Family Fund, 1995).

11. Juliet Schor, *The Overspent American: Upscaling, Downshifting, and the New Consumer* (New York: Basic Books, 1998).

12. See Gerald Celente, *Trends 2000* (New York: Warner Books, 1997), and later reports from the Trends Research Institute at http://www.trendsresearch.com/.

13. For the purposes of this chapter, indexes covering the following major newspapers were reviewed: *New York Times*, *Washington Post*, *Los Angeles Times*, *San Jose Mercury News*, *Chicago Tribune*, and *Atlanta Journal-Constitution*.

14. The first was *Affluenza*, which characterized consumerism as a sickness and explored ways of opposing it, both individually and collectively. The follow-up production was *Escape from Affluenza*, which focused in detail on the voluntary simplicity movement. John de Graaf and Vivia Boe, *Affluenza*, a coproduction of KCTS/Seattle and Oregon Public Broadcasting (Oley, PA: Bullfrog Films, 1997); John de Graaf and Vivia Boe, *Escape from Affluenza*, a coproduction of KCTS/ Seattle and John de Graaf (Oley, PA: Bullfrog Films, 1998).

15. In two separate communications with John Hoskyns-Abrahall of Bullfrog Films, he describes *Affluenza* as "by far our biggest hit in years." Hoskyns-Abrahall notes that "the interesting thing is that the video (*Affluenza*) is that rarity that cuts across all boundaries—geographical, political, age, you name it. We've had a lot of calls from the South (which is rare for us); a lot from older folks; a lot from congregations of all kinds including fundamentalists; and of course the simple living types. *Affluenza* is being used by Consumer Credit Counseling service nationwide. Even some enterprising Deans have used it as part of freshman orientation."

16. As, for example, James Twitchell does in his *Lead Us into Temptation: The Triumph of American Materialism* (New York: Columbia University Press, 1999), 358, notes to pages 199–202.

17. As Duane Elgin sometimes does in his *Promise Ahead: A Vision of Hope and Action for Humanity's Future* (New York: Morrow, 2000).

18. Robert Frank, *Luxury Fever: Money and Happiness in an Era of Excess* (Princeton, NJ: Princeton University Press, 1999). Though Gyawali's views on issues of environment, development, and democracy in South Asia are well documented, both in his own writings and from varied conference proceedings and speeches, his published work has not explored the tyranny of expectations. I owe my use of the term [tyranny of expectations] to a series of 1987 conversations with Gyawali in Kathmandu.

19. Frank, *Luxury Fever*, 4.

20. See for example John Freie, *Counterfeit Community: The Exploitation of Our Longings for Connectedness* (Lanham, MD: Rowman & Littlefield, 1998), especially chap. 5.

21. See Frank, *Luxury Fever*; and Juliet Schor, *The Overworked American: The Unexpected Decline of Leisure* (New York: Basic Books, 1991).

22. See Schor, *The Overspent American*, especially chap. 6. The "upward creep of desire" in the following sentence is Schor's phrase.

23. Robert Edwards Lane, *The Loss of Happiness in Market Democracies* (New Haven, CT: Yale University Press, 2000).

24. This is not Gyawali's argument alone; scholars of South Asia have drawn special attention to this phenomenon. See for example Ponna Wignaraja and Akmal Hussain, eds., *The Challenge in South Asia: Development, Democracy, and Regional Cooperation* (New Delhi: Sage, 1989), especially their "Editorial Overview: The Crisis and Promise of South Asia," pages 18–24.

25. National Film Board of Canada, *The Bomb under the World* (video recording), produced by Green Lion Productions in association with the Canadian Broadcasting Corporation; director, Werner Volkmer; writer, Gwynne Dyer; producers, Catherine Mullins and Marrin Cannelo (Oley, PA: Bullfrog Films, 1994).
26. Schor, *The Overspent American*, 169–173.

21

Excerpts from "Overpopulation versus Biodiversity"

LEON KOLANKIEWICZ

A Short History of Humans and Wildlife in North America

When it comes to wildlife, North America was already depauperate—biologically bereft—millennia before aggressive Europeans and Euro-Americans began to tame the continent and expel or exterminate its indigenous inhabitants. Boasting iconic Pleistocene (Ice Age) megafauna such as mammoths, mastodons, giant sloths, giant beavers, giant condors, giant polar bears, dire wolves, saber-toothed cats, and the like, much of North America must have been as rich as Africa's Serengeti before the arrival of the first Paleo-Indian *Homo sapiens* some thirteen thousand to twenty thousand years ago.

In recent decades, circumstantial evidence has strongly implicated Paleolithic migrants to North America in the well-documented extinctions of mammalian megafauna. Not long after their initial arrival, overall mammalian diversity plummeted at least 15–42 percent below the diversity baseline that had endured millions of years.[1] More than half of large mammals vanished in an unrivaled "cataclysmic extinction wave" at the close of the Pleistocene due to the direct effects of human predation.[2] For eons, the great mammals had survived epochal climatic shifts, vast glacial and interglacial cycles sweeping away biomes like so many autumn leaves. These shifts rivaled those of the late Pleistocene, but this time the great shaggy beasts could not survive the spears, strategies, and supreme tenacity of this cunning new predator.

What causes consternation, in the context of the IPAT formula, is that this tragic, unprecedented, and irrevocable loss of North American biodiversity occurred at a time when human population, affluence, and technology were all relatively minuscule. Yet given enough time, our primordial ancestors were apparently capable of wreaking havoc on biodiversity even before the advent of agriculture. They were pursuing immediate survival imperatives rather than stewardship and sustainability, doing what comes naturally to any organism.

In seeming contrast to this dreadful story of extinction and loss is the recent history of qualified success in wildlife conservation in America. This story too cautions against assigning blame for the demise of North American biodiversity solely to overpopulation in too hasty or facile a manner. Since President Theodore Roosevelt set aside Pelican Island National Wildlife Refuge in Florida in

1903, more than 550 national wildlife refuges have been established throughout the country, conserving more than ninety-five million acres under the slogan of "Wildlife Comes First." While a dedicated conservationist himself, Roosevelt was also responding to popular outrage at the carnage taking place in America at the time.

A century ago, whitetail deer had been all but extirpated from many states, wild turkeys were scarce, and market hunters were threatening edible waterfowl for their flesh and elegant wading birds for their feathers. Populations of charismatic birds like the California condor, ivory-billed woodpecker, trumpeter swan, and whooping crane were in freefall. The passenger pigeon, the single most abundant bird in North America and perhaps the world—Audubon described their immense migrating flocks darkening the sky for hours—was on the verge of extinction. The American buffalo (bison) had barely escaped this fate as railroads and ruthless gunners pushed westward into Indian Territory. Mountain lions, wolves, elk, and bison had been eliminated in the East, and the grizzly bear all but wiped out of the Golden Bear State (California). The fabled and ferocious plains grizzly that had chased fearless Meriwether Lewis into the Missouri River was no more. In the second half of the twentieth century, majestic birds of prey—the American bald eagle, peregrine falcon, brown pelican, and osprey—were all threatened with annihilation from the widespread use of DDT and its chemical cousins.

Today, several decades after the worst pesticides were banned and ambient concentrations have diminished, these raptors have all rebounded. Whitetail deer have become so numerous that they are considered a scourge to gardeners and a hazard to motorists in many places. Protected wading birds (herons and egrets) and managed waterfowl are far more abundant and enjoy stable populations. Bison, grizzly, and wolf populations, while not regaining their former glory, have at least stabilized and reclaimed some old haunts; their continued survival, for the time being, seems secure.

With major investments of money, technology, expertise, and tender loving care, the California condor and whooping crane have taken impressive if tenuous steps away from the brink of extinction; they still number only in the hundreds, but this beats numbering in the dozens, single digits, or zero.

What a change a genuine commitment to conservation makes! America took to heart the moving prose and pleas of venerated activists and authors like John Muir, John Burroughs, Aldo Leopold, Olaus Murie, Rachel Carson, and many other naturalists. Of course, one also needs the economic means and technical wherewithal to convert commitment into action and results. And in the twentieth century, as America grew wealthy and better educated, specialized new fields like wildlife management and conservation biology were able to develop into full-fledged (if not always well-funded) professions.

An Incomplete Formula, Other Measures, and a Broad Consensus

The point is that, here in the United States, the current status and future prospects for many prominent and beloved species of wildlife have improved even as the three main drivers of environmental degradation—population, affluence, and technology, according to the IPAT formula—have all expanded enormously. U.S. population alone more than quadrupled from 75 million in 1900 to 332 million in 2011. According to at least a superficial understanding of IPAT, the status of wildlife should have worsened considerably, not improved. What gives? Is IPAT mistaken, or incomplete? Is a perpetually growing human population compatible with the wildlife and biodiversity conservation after all?

The short answer is, "No way." While IPAT is a useful concept, it is an oversimplification of a complex reality. It is also easily misunderstood and misinterpreted. In particular, regarding wildlife, to assume that the affluence and technology factors are necessarily negative is mistaken. Rather, they are a mixed blessing, and sometimes they can mitigate rather than exacerbate a larger population. The greater resource consumption and waste generation that accompany affluence are indeed generally inimical to wildlife. Yet under the right circumstances and inspired leadership, affluence and associated higher educational levels can nurture an enlightened ethics that values wildlife and biodiversity. This is crucial to the generous support that successful public and private wildlife conservation programs require. And while some technologies like pesticides, chainsaws, coal-fired power plants, and bulldozers are generally damaging, many others, such as radio telemetry, satellite imagery, binoculars, computers, artificial insemination, and geographic information systems (GIS) can be crucial tools in modern wildlife management and endangered species recovery.

The bottom line is that we humans can wipe out wildlife even at relatively low levels of population size, affluence, and technological power, if good stewardship is not a priority. That was the case, for example, in the United States until about a century ago. However, once we become better stewards, attempting to avoid adverse impacts on wildlife, allowing for only sustainable rather than uncontrolled harvests and so forth, then we have the potential to really improve the situation, as we have to some degree. However, at that point we then reach another plateau in the pursuit of sustainable conservation. For no matter how well we excel as managers, the amount of wildlife we can save is constrained by the amount of uncontaminated, uncompromised, intact habitat remaining. If we continue to increase the number of people, levels of consumption, and aggregate demands on the land, we will inevitably decrease the abundance and diversity of wildlife. Enlightened management and cutting-edge technologies can only do so much; they cannot work miracles, hoodwink nature, or cram a thousand species onto the head of a pin. We can't game the system.

Furthermore, the situation is not as sanguine for wildlife as the heartwarming examples above might suggest. The dire state of biodiversity in besieged, over-populated California is instructive. California has more native species than any other state, as well as more endemics, those unique life forms found nowhere else on Earth. This extraordinary biodiversity is already stressed by the state's enormous population, thirty-nine million and counting, and further threatened by continuing growth.

The 2007 report *California Wildlife: Conservation Challenges*[3] by the Califor-nia Department of Fish and Game (CDFG) tallied more than eight hundred imperiled species, including half of all mammals and a third of all birds. *Califor-nia Wildlife* identified the major "stressors" impacting California's wildlife and habitats. It emphasized: "Increased needs for housing, services, transportation, and other natural resources." Of course, all of these are linked directly to popula-tion size. California's bloated population surged by nearly 50 percent from 1970 to 1990 and swelled another 14 percent in the 1990s; official projections—more a nightmare than a preview—foresee 60 million residents by 2050. The relentless spread of one life form, our own, is riding roughshod over hundreds of fellow living creatures that have been part of California far longer than humans have.

Globally, the situation is just as grim. In a dismal 2010 report card based on thirty-one indicators, the journal *Science* reported that "the rate of biodiversity loss does not appear to be slowing," in spite of the 2002 Convention on Biologi-cal Diversity in which world leaders had committed to achieving a significant reduction in the rate of biodiversity loss by 2010.[4]

Eminent biologists and authors like Jared Diamond, E. O. Wilson, and Nor-man Myers all agree that population growth and the forces it multiplies or unleashes on the landscape are devastating biodiversity. Diamond refers to an "Evil Quartet" of habitat destruction, fragmentation, overharvesting, and intro-duced species.[5] Wilson touts the acronym HIPPO—habitat destruction, inva-sive species, pollution, population, and overharvesting. He estimates that at least twelve thousands wild species are going extinct annually.[6]

Wilson has written: "The pattern of human population growth in the 20th century was more bacterial than primate. When *Homo sapiens* passed the six-billion mark we had already exceeded by perhaps as much as 100 times the bio-mass of any large animal species that ever existed on the land. We and the rest of life cannot afford another 100 years like that."[7] There is no more stinging indict-ment of human hypergrowth, or what Wilson calls "our reproductive folly."

The Conservation Measures Partnership is a collaborative effort of a dozen prominent conservation groups dedicated to improving international wildlife conservation. The partnership's "Threats Taxonomy" lists direct threats to bio-logical diversity. Their main categories include (1) residential and commercial development; (2) agriculture and aquaculture; (3) energy production and min-ing; (4) transportation and service corridors; (5) biological resource use; and

(6) human intrusions and disturbance. Clearly, each of these is a direct function of population size and, of course, affluence.[8] . . .

HANPP [human appropriation of net primary production] is what is harvested by people for our own consumption, plus the share of natural production lost to environmental degradation. Another portion of HANPP occurs from development that replaces living vegetation with pavement and buildings. In effect, the share of NPP co-opted by humans is unavailable to wildlife, which must then survive on less. Growing numbers of omnivorous human consumers with growing appetites are leaving only the leftovers—and ever fewer leftovers, at that—for the millions of creatures that inhabit the biosphere with us. The bottom line is that if HANPP continues to increase, more species will be snuffed out of existence permanently. The collective appetite of *Homo colossus*, as William Catton dubs our resource-devouring species, has grown so titanic that not even all of Planet Earth can satisfy it anymore.

This is also borne out by ecological footprint (EF) analysis. EF measures aggregate human demands, or the human load imposed on the biosphere. According to EF analysis, since about the 1980s, humanity as a whole has been living beyond the ecological means of the biosphere; we are already in ecological overshoot, an unstable and unsustainable condition. Yet even as our aggregate EF continues to increase, the Earth's biocapacity—the ability of available terrestrial and aquatic areas to provide ecological services—is being degraded by excessive human activity. We humans now use the equivalent of about 1.4 Earths.[9] But how is it even possible to use more than one Earth at any one time? Only by using up vast stores of fossil fuels, which were created over tens of millions of years. In effect, temporarily, we have at our disposal more than one planet, or what Catton refers to as the "ghost acreage" of "phantom carrying capacity."[10] Our biodiversity-besieging population and economic explosion have been ignited by a one-time jolt of "ancient sunshine."

EF analysis also helps reveal the extent to which massive numbers of massively consuming Americans are impacting the entire biosphere. In 2006, our EF was 22.3 global acres per capita, while our biocapacity was only 10.9 global acres per capita, for an ecological deficit of 11.3 global acres per capita.[11] This means that we have exceeded the nation's carrying capacity. In other words, we gargantuan Americans are living beyond our ecological means, boosting our vaunted living standards and nonnegotiable lifestyles only by drawing down biocapacity and degrading irreplaceable biodiversity across the planet. We may be able to preserve remnants of nature here, but by importing carrying capacity we are exporting nature destruction around the world. The only way out of this conundrum without risking economic unraveling or collapse—which all too many unemployed Americans have gotten a bitter taste of in the "Great Recession" of 2008–2010 and afterward—is by transitioning to a steady-state economy and sharply increasing the efficiency with which we consume energy and resources.

Even then, it will take ingenuity and innovation to avoid the Jevons paradox, the counterintuitive result by which aggregate resource consumption increases even as efficiency improves.[12]

Limiting Our Numbers to Help Other Species Survive

In analyzing environmental effects, environmental scientists refer to direct, indirect, and cumulative impacts. Human population growth impacts biodiversity in all three ways. The most significant direct impact occurs when we eliminate or fragment wildlife habitat by converting it into farmland to feed ever more humans and livestock (fed to humans "moving up the food chain") and, increasingly, vehicles (i.e., ethanol and biodiesel derived from crops). Overpopulation-related habitat damage and loss also occur with the construction of reservoirs, power lines, roads, mines, logging operations, overgrazing, bottom trawling, and urban sprawl.

Invasive species that accompany growing, spreading human populations exemplify an indirect impact. Exotic plants, animals, and microbes unleashed by humans are wreaking havoc on native terrestrial and aquatic ecosystems, flora, and fauna around the world. More people are linked to more travel, more trade, and more spread of exotics. The notorious zebra mussel, native to Eastern Europe, hopped across the Atlantic in 1988 when an ocean-going vessel dumped ballast water into Lake St. Clair, between Lake Huron and Lake Erie. It now infests and is radically altering the ecology of lakes and rivers across vast swaths of North America. This aggressive, adaptable bivalve also threatens our already besieged native freshwater mussel diversity, the richest in the world, at about three hundred species.

The ultimate cumulative impact is anthropogenic climate change, from the steady accumulation of human-emitted greenhouse gases in the atmosphere. This could end up having a greater adverse impact on wildlife and biodiversity than all other malevolent forces combined. According to the Intergovernmental Panel on Climate Change, increasing greenhouse gas emissions are primarily a function of rising populations and increased affluence.

As a consultant to the U.S. Fish and Wildlife Service, I have been privileged to help prepare comprehensive conservation plans (CCPs) on more than forty national wildlife refuges from the Caribbean to Alaska. In diverse ecosystems, I have witnessed the myriad impacts of voracious human demands on wildlife. Yet I have also seen the uplifting results of dedicated, indefatigable efforts to save biodiversity.

In the U.S. Virgin Islands, managers and concerned citizens are working hard to protect Sandy Point, Green Cay, and Buck Island National Wildlife Refuges. Population growth is not their friend. As noted in a recent CCP that I helped draft:

[The Caribbean] ecosystem is home to 78 threatened and endangered species (29 animals and 49 plants), including species of birds, reptiles, and amphibians, as well as unique and diverse habitats ranging from coral reefs, sandy beaches, and mangrove forests to limestone hills and forested mountains. . . .

Since the end of the Second World War, human population has increased dramatically on almost every island. . . . Negative ecological trends have all accelerated as a result of the demands explosive human growth has placed on the environment.

Within the U.S. Virgin Islands, the demands for space and land created by a rapidly growing human population of over 100,000 have resulted in extensive loss and degradation of natural ecosystems, especially on densely populated St. Thomas.[13]

Two of the many "critters" inhabiting this ecosystem—one tiny and one humongous—exemplify the threats and hopes facing wildlife in this brave, new, human-dominated world.

Three decades ago, the world's surviving population of the tiny, critically endangered St. Croix ground lizard (*Ameiva polops*) on fourteen-acre Green Cay National Wildlife Refuge could probably have fit in two buckets. The disappearance of this six-inch, inconspicuous reptile from the nearby island of St. Croix is believed due to human disturbance, land development, and the introduced Indian mongoose, which preyed on it. Determined recovery efforts at the refuge and nearby Buck Island Reef National Monument, to which it has been translocated, offer hope that *A. polops* will yet endure. Even so, it is disconcerting that the entire existing population still probably weighs less than a single adult human. Globally, the ratio of our species' aggregate biomass to theirs exceeds seven billion to one (7,000,000,000:1).

The lumbering leatherback turtle (*Dermochelys coriacea*) nests on beaches at Sandy Point National Wildlife Refuge not far from the ground lizard but outweighs it 10,000:1. At up to eight feet in length and a ton in weight, the leatherback is the largest, deepest diving, and widest ranging of all sea turtles. It is endangered because of overexploitation by people for its eggs and meat, incidental take by commercial fisheries, disorientation of hatchlings by beachfront lighting, and excessive nest predation.[14] Yet at Sandy Point National Wildlife Refuge, nesting leatherbacks have increased from fewer than twenty in 1982 to more than a hundred in recent years; average hatchling production has quintupled. This success is due to the tireless efforts of refuge manager Mike Evans and biologist Claudia Lombard and their supporting cast of conservationists and volunteers.

Limiting human numbers both locally and globally is crucial to saving these two endangered reptiles. Locally, population stabilization would reduce pressure to poach sea turtle eggs and trample the ground lizard's habitat; it would give them vital breathing room. Globally, it would help curb the unremitting buildup of greenhouse gases that threatens the Caribbean with more frequent and ferocious hurricanes, sea level rise, coral-killing warmer waters, and coral-dissolving acidifying waters. Humanely stabilizing and reducing the human population is a necessary condition for saving biodiversity and halting the brutal wave of extinction breaking over the Earth.

Lest We Forget

The late Donella Meadows, lead author of *The Limits to Growth*, used to write a syndicated newspaper column called *The Global Citizen*. In 1999, on the fiftieth anniversary of the conservation classic *A Sand County Almanac*, Meadows wrote a tribute to its author, Aldo Leopold, pioneering wildlife scientist and wilderness advocate.

Meadows noted that while environmentalists are sometimes accused of disliking people, Leopold wasn't like that at all. Rather, he saw us humans as the only creatures endowed with the ability to love nature and understand our connection to it. Leopold once spoke at the dedication of a monument to the passenger pigeon, and his remarks are telling: "For one species to mourn the death of another is a new thing under the sun. The Cro-Magnon who slew the last mammoth thought only of steaks. The sportsman who shot the last pigeon thought only of his prowess. The sailor who clubbed the last auk thought of nothing at all. But we, who have lost our pigeons, mourn the loss. Had the funeral been ours, the pigeons would hardly have mourned us."[15]

Our species is unique, because here and now only we have the ability to destroy, or to save, biodiversity. Only we have the ability to care one way or the other. The destiny of all wild living things is in our hands. Will we crush them or let them be wild and free? Limiting human population will not guarantee success, but not doing so means certain failure.

NOTES

1. M. A. Carrasco, A. D. Barnosky, and R. W. Graham, "Quantifying the Extent of North American Mammal Extinction Relative to the Pre-Anthropogenic Baseline," *PLoS ONE* 4, no. 12 (2009), DOI:10.1371/journal.pone.0008331.

2. J. A. Alroy, "A Multispecies Overkill Simulation of the End-Pleistocene Megafaunal Mass Extinction," *Science* 292 (2001): 1893–1896.

3. David Bunn, Andrea Mummert, Marc Hoshovsky, Kirsten Gilardi, and Sandra Shanks, *California Wildlife: Conservation Challenges: California's Wildlife Action Plan* (Sacramento: California Department of Fish and Game, 2009).

4. Stuart H. M. Butchart, Matt Walpole, Ben Collen, Arco van Strien, Jörn P. W. Scharlemann, Rosamunde E. A. Almond, Jonathan E. M. Baillie, et al., "Global Biodiversity: Indicators of Recent Declines," *Science* 328 (2010): 1164–1168.

5. James Sanderson and Michael Moulton, *Wildlife Issues in Our Changing World*, 2nd ed. (Boca Raton, FL: CRC, 1998).

6. David Biello, "Population Bomb Author's Fix for Next Extinction: Educate Women," *Scientific American*, August 2008.

7. Edward Wilson, "The Bottleneck," *Scientific American*, February 2002.

8. Conservation Measures Partnership, "Threats Taxonomy" (2010), www.conservation measures.org/initiatives/threats-actions-taxonomies/threats-taxonomy.

9. B. Ewing, S. Goldfinger, A. Oursler, A. Reed, D. Moore, and M. Wackernagel, *The Ecological Footprint Atlas* (Oakland, CA: Global Footprint Network, 2009), www.footprintnetwork.org.

10. William R. Catton, Jr., *Overshoot: The Ecological Basis of Revolutionary Change* (Urbana: University of Illinois Press, 1980).

11. Ewing et al., *The Ecological Footprint Atlas*.

12. The Jevons paradox is named for the English economist William Stanley Jevons, who first commented on this phenomenon in 1865. Jevons had observed that technological improvements in the efficiency of coal use in various industries actually increased overall coal consumption, rather than reducing it, as one might expect. The Jevons paradox is a particular example of what modern economists call the "rebound effect." Higher efficiency reduces the relative cost of using energy and resources, thereby stimulating demand and thus offsetting potential savings. Moreover, increased efficiency encourages economic growth, further increasing demand for energy/resources. Since the Jevons paradox applies to technological improvements, it may be possible to circumvent by means of green taxes or conservation standards such as nationwide caps on carbon emissions.

13. U.S. Fish and Wildlife Service (USFWS), Department of the Interior, "Leatherback Sea Turtle Fact Sheet" (Jacksonville, FL: USFWS North Florida Ecological Services Office, 2009), www .fws.gov/northlforida/SeaTurtles/Turtle%20Factsheets/leatherback-sea-turtle.htm.

14. U.S. Fish and Wildlife Service (USFWS), Department of the Interior, "Sandy Point, Green Cay, and Buck Island National Wildlife Refuges: Draft Comprehensive Conservation Plan and Environmental Assessment" (Atlanta, GA: USFWS Southeast Region, 2009), www .fws-gov/southeast/planning/PDFdocuments/VirginIslandsDraftCCP/Edited%20Draft%20 CCP%20Virgin%20Islands%20Refuges.pdf.o.

15. Donella Meadows, "Sand Country Almanac Fifty Years Later," Donella Meadows Archive, Sustainability Institute (1999), http://www.sustainer.org/dhm_archive/index.php?display _article=vn783leopolded.

Reading Questions and Further Readings

Reading Questions

1. The readings in this part span a period of rapid human population growth, as well as rapid technological advancement. How and why have earlier projections of natural resource limits proved inaccurate? Give examples from the readings, and show how these led to innovations in the concepts and models that are most widely employed to determine population and consumption impacts today.
2. McDonough/Braungart and Chertow outline analytical and active agendas that answer the population and consumption dilemma with what we might call more "sustainable" approaches. What are these approaches, and how do they take the problem of uneven access to resources—that is, patterns of poverty and broader definitions of equity—into account?
3. Compare Malthus's idea of a human carrying capacity to those implied by Kolankiewicz and Daily and Ehrlich. Is it possible to answer Kolankiewicz's call to conserve biodiversity without sacrificing principles of equity and environmental justice? How does the content of each text indicate how each author might respond?
4. According to Maniates, in what ways is it possible to revise our individual and collective sense of material needs? Give examples from the text.
5. In a world where material wealth is unevenly distributed, what role do the various authors propose that poverty plays in environmental degradation? How might each author define the importance of environmental justice to determining a balance between ecological vitality and human population? Substantiate your answer with examples from the texts.

Further Readings

Braungart, Michael, and William McDonough. *Cradle to Cradle: Remaking the Way We Make Things*. New York: North Point, 2002.

Cafaro, Philip, and Eileen Crist, *Life on the Brink: Environmentalists Confront Overpopulation*. Athens: University of Georgia Press, 2012.

Catton, William. *Overshoot: The Ecological Basis of Revolutionary Change*. Urbana: University of Illinois Press, 1982.

McKibben, Bill. *Deep Economy*. New York: Henry Holt, 2007.

Princen, Thomas, Michael Maniates, and Ken Conca, eds. *Confronting Consumption*. Cambridge, MA: MIT Press, 2002.

PART IV

Public Goods and Collective Action

Many environmental and natural resource problems ultimately reflect a lack of adequate human cooperation. Individuals' decisions can be collectively detrimental, leading to outcomes that are not beneficial to society as a whole. The collapse of fish populations due to overfishing, forest and land degradation, and the irreversible effects of climate change are all examples where people, acting independently and in their own interests, deplete a shared resource even when the outcomes are not in any individual's long-term interest. Economic theory has addressed the dilemma around shared, or "common-pool," resources for a long time, but the insights that can be drawn from conventional models based on assumptions of perfectly rational, egoistic, and atomistic agents are often limited. Many communities have demonstrated that individuals can and do work together successfully to manage common resources without degrading them, while in other settings resources are rapidly overused and ecosystems degraded. The texts in this part explore the role of markets, institutions, individual psychology, and ethics in collective resource management problems.

Garrett Hardin was a human ecologist whose 1968 article "The Tragedy of the Commons" brought broad-scale attention to the classic commons dilemma. Hardin's highly influential piece presented reasoning for how and why individuals, acting in their own self-interest, are destined to overuse resources from a social welfare perspective. He presents a canonical example based on common grazing land in medieval Europe, where each herder has the incentive to put an additional cow onto the land even though it is in each individual's interest to preserve the quality of the grazing land for the future. The problem is that each herder receives all the benefits from his cow's grazing, while the damage to the grazing land is shared by the entire population of herders; each herder's seemingly rational economic decision will cause the land to be depleted or even destroyed, to the detriment of all. This social dilemma is widely referred to as the "commons problem" or "commons dilemma." Hardin argues that common-pool resources will inevitably lead to resource degradation and ecosystem collapse without strict management by external authorities, namely by "enclosing," or privatizing, common-pool resources so that users have a direct private incentive for long-term preservation.

The commons dilemma is often seen as a primary underlying factor in environmental issues. However, since Hardin's seminal piece was published, researchers have conducted a wide range of studies to analyze more complex factors involved in common-pool resource use and to better understand the nuances around successful collective management. In "Revisiting the Commons: Local Lessons, Global Challenges," Elinor Ostrom and her coauthors discuss advances in this research and present a countervailing viewpoint to Hardin's characterization of commons problems and the proposed solution of establishing property rights. Indeed, many diverse studies have documented how various communities around the world have successfully managed common resources such as grazing lands, forests, irrigation waters, and fisheries, equitably and sustainably, and over long time horizons, without relying on external regulation. In "Rationality and Solidarities: The Social Organization of Common Property Resources in the Imdrhas Valley of Morocco," Peggy Petrzelka and Michael Bell present one such analysis, in which they compare two communities in southern Morocco where, they contend, differences in embedded social ties lead to starkly different outcomes in the health of their common-pool resources.

Successful cooperation has been shown to depend, for example, on how local institutions are structured, group size and diversity, and other factors related to the economic, political, and social setting. According to Ostrom's central thesis, for which she was awarded a Nobel Prize in Economic Sciences in 2009, social mechanisms can regulate the use of the commons. The Ostrom et al. piece also discusses challenges of "scaling up" the principles for sustainable resource management that are successful in local and regional contexts, particularly when facing global challenges such as climate change.

Ostrom's scholarship challenged conventional models of common-pool resource problems and thus paved the way for more contextualized understanding of collective resource management. Since her seminal work, other scholars have gone on to further challenge economic frameworks of common-pool resources. Petrzelka and Bell analyze social solidarities as a complementary model to rational choice frameworks, including Ostrom's, that assume that individuals are motivated narrowly by their own economic self-interest.

In "Averting the Tragedy of the Commons," Mark Van Vugt discusses theory and evidence that individuals care directly about how their actions affect others and the natural environment, aside from any implications for their own economic self-interest. He presents an alternative and complementary framework of resource-use decisions, centering on individuals' core motives, and offers mechanisms, based on social psychology, for averting commons-based tragedies.

Understanding commons and collective management problems in a holistic way requires not only the study of practical considerations involved in coordinating human behavior but also a consideration of the values and ethics surrounding these dilemmas. From an ethical perspective, the question of whether there exists an individual obligation to act in the collective best interest is central to understanding how individual action contributes to sustainable resource management. In "Climate, Collective Action and Individual Ethical Obligation,"

Marion Hourdequin explores the ethical complexity of personal responsibility for collective, global problems. In doing so, Hourdequin argues that the framework of the commons dilemma itself is grounded in an overly stark separation between what is individual and what is collective. She argues that a combination of individual moral change and the policy change suggested by Hardin are necessary to solve collective action problems such as climate change.

Another contemporary perspective, known as "free market environmentalism," challenges the traditional economic framework embodied by Hardin from a different angle, by focusing on the embedded costs of coordinating individual actions in a centralized manner. In the article "About Free-Market Environmentalism," Jonathan H. Adler posits that the costs of accessing information, contracting, and other inherent costs of transacting are significant enough to outweigh benefits of regulatory solutions. His perspective relies instead on decentralized markets to adequately reflect the societal values of resource preservation and to promote innovation that will foster outcomes beneficial to societal welfare over the long term.

The selected readings in this part provide economic, sociological, anthropological, psychological, and philosophical responses to the fundamental question of how human beings can coordinate their individual choices in a way that is beneficial to long-term societal welfare. The world's most pressing environmental problems often stem from this fundamental social dilemma, and the authors and perspectives represented here offer visions of the role of individuals and institutions in managing and valuing these collective action problems.

22

Excerpts from "The Tragedy of the Commons"

GARRETT HARDIN

The population problem has no technical solution; it requires a fundamental extension in morality.

At the end of a thoughtful article on the future of nuclear war, Wiesner and York[1] concluded that: "Both sides in the arms race are . . . confronted by the dilemma of steadily increasing military power and steadily decreasing national security. *It is our considered professional judgment that this dilemma has no technical solution.* If the great powers continue to look for solutions in the area of science and technology only, the result will be to worsen the situation."

I would like to focus your attention not on the subject of the article (national security in a nuclear world) but on the kind of conclusion they reached, namely that there is no technical solution to the problem. An implicit and almost universal assumption of discussions published in professional and semipopular scientific journals is that the problem under discussion has a technical solution. A technical solution may be defined as one that requires a change only in the techniques of the natural sciences, demanding little or nothing in the way of change in human values or ideas of morality.

In our day (though not in earlier times) technical solutions are always welcome. Because of previous failures in prophecy, it takes courage to assert that a desired technical solution is not possible. Wiesner and York exhibited this courage; publishing in a science journal, they insisted that the solution to the problem was not to be found in the natural sciences. They cautiously qualified their statement with the phrase, "It is our considered professional judgment. . . ." Whether they were right or not is not the concern of the present article. Rather, the concern here is with the important concept of a class of human problems which can be called "no technical solution problems," and, more specifically, with the identification and discussion of one of these.

It is easy to show that the class is not a null class. Recall the game of tick-tack-toe. Consider the problem, "How can I win the game of tick-tack-toe?" It is well known that I cannot, if I assume (in keeping with the conventions of game theory) that my opponent understands the game perfectly. Put another way, there is no "technical solution" to the problem. I can win only by giving a radical meaning to the word "win." I can hit my opponent over the head; or I can drug him; or I can falsify the records. Every way in which I "win" involves, in some sense, an abandonment of the game, as we intuitively understand it. (I can also, of course, openly abandon the game—refuse to play it. This is what most adults do.)

The class of "No technical solution problems" has members. My thesis is that the "population problem," as conventionally conceived, is a member of this class. How it is conventionally conceived needs some comment. It is fair to say that most people who anguish over the population problem are trying to find a way to avoid the evils of overpopulation without relinquishing any of the privileges they now enjoy. They think that farming the seas or developing new strains of wheat will solve the problem—technologically. I try to show here that the solution they seek cannot be found. The population problem cannot be solved in a technical way, any more than can the problem of winning the game of tick-tack-toe.

What Shall We Maximize?

Population, as Malthus said, naturally tends to grow "geometrically," or, as we would now say, exponentially. In a finite world this means that the per capita share of the world's goods must steadily decrease. Is ours a finite world?

A fair defense can be put forward for the view that the world is infinite; or that we do not know that it is not. But, in terms of the practical problems that we must face in the next few generations with the foreseeable technology, it is clear that we will greatly increase human misery if we do not, during the immediate future, assume that the world available to the terrestrial human population is finite. "Space" is no escape.[2]

A finite world can support only a finite population; therefore, population growth must eventually equal zero. (The case of perpetual wide fluctuations above and below zero is a trivial variant that need not be discussed.) When this condition is met, what will be the situation of mankind? Specifically, can Bentham's goal of "the greatest good for the greatest number" be realized?

No—for two reasons, each sufficient by itself. The first is a theoretical one. It is not mathematically possible to maximize for two (or more) variables at the same time. This was clearly stated by von Neumann and Morgenstern,[3] but the principle is implicit in the theory of partial differential equations, dating back at least to D'Alembert (1717–1783).

The second reason springs directly from biological facts. To live, any organism must have a source of energy (for example, food). This energy is utilized for two purposes: mere maintenance and work. For man, maintenance of life requires about 1600 kilocalories a day ("maintenance calories"). Anything that he does over and above merely staying alive will be defined as work, and is supported by "work calories" which he takes in. Work calories are used not only for what we call work in common speech; they are also required for all forms of enjoyment, from swimming and automobile racing to playing music and writing poetry. If our goal is to maximize population it is obvious what we must do: We must make the work calories per person approach as close to zero as possible. No gourmet meals, no vacations, no sports, no music, no literature, no art. . . . I think that everyone will grant, without argument or proof, that maximizing population does not maximize goods. Bentham's goal is impossible.

In reaching this conclusion I have made the usual assumption that it is the acquisition of energy that is the problem. The appearance of atomic energy has led some to question this assumption. However, given an infinite source of energy, population growth still produces an inescapable problem. The problem of the acquisition of energy is replaced by the problem of its dissipation, as J. H. Fremlin has so wittily shown.[4] The arithmetic signs in the analysis are, as it were, reversed; but Bentham's goal is still unobtainable.

The optimum population is, then, less than the maximum. The difficulty of defining the optimum is enormous; so far as I know, no one has seriously tackled this problem. Reaching an acceptable and stable solution will surely require more than one generation of hard analytical work—and much persuasion.

We want the maximum good per person; but what is good? To one person it is wilderness, to another it is ski lodges for thousands. To one it is estuaries to nourish ducks for hunters to shoot; to another it is factory land. Comparing one good with another is, we usually say, impossible because goods are incommensurable. Incommensurables cannot be compared.

Theoretically this may be true; but in real life incommensurables are commensurable. Only a criterion of judgment and a system of weighting are needed. In nature the criterion is survival. Is it better for a species to be small and hideable, or large and powerful? Natural selection commensurates the incommensurables. The compromise achieved depends on a natural weighting of the values of the variables.

Man must imitate this process. There is no doubt that in fact he already does, but unconsciously. It is when the hidden decisions are made explicit that the arguments begin. The problem for the years ahead is to work out an acceptable theory of weighting. Synergistic effects, nonlinear variation, and difficulties in discounting the future make the intellectual problem difficult, but not (in principle) insoluble.

Has any cultural group solved this practical problem at the present time, even on an intuitive level? One simple fact proves that none has: there is no prosperous population in the world today that has, and has had for some time, a growth rate of zero. Any people that has intuitively identified its optimum point will soon reach it, after which its growth rate becomes and remains zero.

Of course, a positive growth rate might be taken as evidence that a population is below its optimum. However, by any reasonable standards, the most rapidly growing populations on earth today are (in general) the most miserable. This association (which need not be invariable) casts doubt on the optimistic assumption that the positive growth rate of a population is evidence that it has yet to reach its optimum.

We can make little progress in working toward optimum population size until we explicitly exorcize the spirit of Adam Smith in the field of practical demography. In economic affairs, *The Wealth of Nations* (1776) popularized the "invisible hand," the idea that an individual who "intends only his own gain," is, as it were, "led by an invisible hand to promote . . . the public interest."[5] Adam Smith did not assert that this was invariably true, and perhaps neither did any of his followers. But he contributed to a dominant tendency of thought that has ever since interfered with positive action based on rational analysis, namely, the tendency to assume that decisions reached individually will, in fact, be the best decisions for an entire society. If this assumption is correct it justifies the continuance of our present policy of laissez-faire in reproduction. If it is correct we can assume that men will control their individual fecundity so as to produce the optimum population. If the assumption is not correct, we need to reexamine our individual freedoms to see which ones are defensible.

Tragedy of Freedom in a Commons

The rebuttal to the invisible hand in population control is to be found in a scenario first sketched in a little-known pamphlet[6] in 1833 by a mathematical amateur named William Forster Lloyd (1794–1852). We may well call it "the tragedy of the commons," using the word "tragedy" as the philosopher Whitehead used it:[7] "The essence of dramatic tragedy is not unhappiness. It resides in the solemnity of the remorseless working of things." He then goes on to say, "This inevitableness of destiny can only be illustrated in terms of human life by incidents which in fact involve unhappiness. For it is only by them that the futility of escape can be made evident in the drama."

The tragedy of the commons develops in this way. Picture a pasture open to all. It is to be expected that each herdsman will try to keep as many cattle as possible on the commons. Such an arrangement may work reasonably satisfactorily for centuries because tribal wars, poaching, and disease keep the numbers of both man and beast well below the carrying capacity of the land. Finally, however, comes the day of reckoning, that is, the day when the long-desired goal of social stability becomes a reality. At this point, the inherent logic of the commons remorselessly generates tragedy.

As a rational being, each herdsman seeks to maximize his gain. Explicitly or implicitly, more or less consciously, he asks, "What is the utility *to me* of adding one more animal to my herd?" This utility has one negative and one positive component.

1) The positive component is a function of the increment of one animal. Since the herdsman receives all the proceeds from the sale of the additional animal, the positive utility is nearly +1.

2) The negative component is a function of the additional overgrazing created by one more animal. Since, however, the effects of overgrazing are shared by all the herdsmen, the negative utility for any particular decision-making herdsman is only a fraction of –1.

Adding together the component partial utilities, the rational herdsman concludes that the only sensible course for him to pursue is to add another animal to his herd. And another; and another. . . . But this is the conclusion reached by each and every rational herdsman sharing a commons. Therein is the tragedy. Each man is locked into a system that compels him to increase his herd without limit—in a world that is limited. Ruin is the destination toward which all men rush, each pursuing his own best interest in a society that believes in the freedom of the commons. Freedom in a commons brings ruin to all.

Some would say that this is a platitude. Would that it were! In a sense, it was learned thousands of years ago, but natural selection favors the forces of psychological denial.[8] The individual benefits as an individual from his ability to deny the truth even though society as a whole, of which he is a part, suffers. Education can counteract the natural tendency to do the wrong thing, but the inexorable succession of generations requires that the basis for this knowledge be constantly refreshed.

A simple incident that occurred a few years ago in Leominster, Massachusetts, shows how perishable the knowledge is. During the Christmas shopping season the parking meters downtown were covered with plastic bags that bore tags reading: "Do not open until after Christmas. Free parking courtesy of the mayor and city council." In other words, facing the prospect of an increased demand for already scarce space, the city fathers reinstituted the system of the commons. (Cynically, we suspect that they gained more votes than they lost by this retrogressive act.)

In an approximate way, the logic of the commons has been understood for a long time, perhaps since the discovery of agriculture or the invention of private property in real estate. But it is understood mostly only in special cases which are not sufficiently generalized. Even at this late date, cattlemen leasing national land on the western ranges demonstrate no more than an ambivalent understanding, in constantly pressuring federal authorities to increase the head count to the point where overgrazing produces erosion and weed-dominance. Likewise, the oceans of the world continue to suffer from the survival of the philosophy of the commons. Maritime nations still respond automatically to the shibboleth of the "freedom of the seas." Professing to believe in the "inexhaustible resources of the oceans," they bring species after species of fish and whales closer to extinction.[9]

The National Parks present another instance of the working out of the tragedy of the commons. At present, they are open to all, without limit. The parks themselves are limited in extent—there is only one Yosemite Valley—whereas population seems to grow without limit. The values that visitors seek in the parks are steadily eroded. Plainly, we must soon cease to treat the parks as commons or they will be of no value to anyone.

What shall we do? We have several options. We might sell them off as private property. We might keep them as public property, but allocate the right to enter them. The allocation might be on the basis of wealth, by the use of an auction system. It might be on the basis of merit, as defined by some agreed-upon standards. It might be by lottery. Or it might be on a first-come, first-served basis, administered to long queues. These, I think, are all the reasonable possibilities. They are all objectionable. But we must choose—or acquiesce in the destruction of the commons that we call our National Parks.

Pollution

In a reverse way, the tragedy of the commons reappears in problems of pollution. Here it is not a question of taking something out of the commons, but of putting something in—sewage, or chemical, radioactive, and heat wastes into water; noxious and dangerous fumes into the air; and distracting and unpleasant advertising signs into the line of sight. The calculations of utility are much the same as before. The rational man finds that his share of the cost of the wastes he discharges into the commons is less than the cost of purifying his wastes before releasing them. Since this is true for everyone, we are locked into a system of "fouling our own nest," so long as we behave only as independent, rational, free-enterprisers.

The tragedy of the commons as a food basket is averted by private property, or something formally like it. But the air and waters surrounding us cannot readily be fenced, and so the tragedy of the commons as a cesspool must be prevented by different means, by coercive laws or taxing devices that make it cheaper for the polluter to treat his pollutants than to discharge them untreated. We have not progressed as far with the solution of this problem as we have with the first. Indeed, our particular concept of private property, which deters us from exhausting the positive resources of the earth, favors pollution. The owner of a factory on the bank of a stream—whose property extends to the middle of the stream, often has difficulty seeing why it is not his natural right to muddy the waters flowing past his door. The law, always behind the times, requires elaborate stitching and fitting to adapt it to this newly perceived aspect of the commons.

The pollution problem is a consequence of population. It did not much matter how a lonely American frontiersman disposed of his waste. "Flowing water purifies itself every 10 miles," my grandfather used to say, and the myth was near enough to the truth when he was a boy, for there were not too many people. But as population became denser, the natural chemical and biological recycling processes became overloaded, calling for a redefinition of property rights.

How to Legislate Temperance?

Analysis of the pollution problem as a function of population density uncovers a not generally recognized principle of morality, namely: *the morality of an act is a function of the state of the system at the time it is performed.*[10] Using the commons as a cesspool does not harm the general public under frontier conditions, because there is no public; the same behavior in a metropolis is unbearable. A hundred and fifty years ago a plainsman could kill an American bison, cut out only the tongue for his dinner, and discard the rest of the animal. He was not in any important sense being wasteful. Today, with only a few thousand bison left, we would be appalled at such behavior.

In passing, it is worth noting that the morality of an act cannot be determined from a photograph. One does not know whether a man killing an elephant or setting fire to the grassland is harming others until one knows the total system in which his act appears. "One picture is worth a thousand words," said an ancient Chinese; but it may take 10,000 words to validate it. It is as tempting to ecologists as it is to reformers in general to try to persuade others by way of the photographic shortcut. But the essence of an argument cannot be photographed: it must be presented rationally—in words.

That morality is system-sensitive escaped the attention of most codifiers of ethics in the past. "Thou shalt not . . ." is the form of traditional ethical directives which make no allowance for particular circumstances. The laws of our society follow the pattern of ancient ethics, and therefore are poorly suited to governing a complex, crowded, changeable world. Our epicyclic solution is to augment statutory law with administrative law. Since it is practically impossible to spell out all the conditions under which it is safe to burn trash in the back yard or to run an automobile without smog-control, by law we delegate the details to bureaus. The result is administrative law, which is rightly feared for an ancient reason—*Quis custodiet ipsos custodes?*—"Who shall watch the watchers themselves?" John Adams said that we must have "a government of laws and not men." Bureau administrators, trying to evaluate the morality of acts in the total system, are singularly liable to corruption, producing a government by men, not laws.

Prohibition is easy to legislate (though not necessarily to enforce); but how do we legislate temperance? Experience indicates that it can be accomplished best through the mediation of administrative law. We limit possibilities unnecessarily if we suppose that the sentiment of *Quis custodiet* denies us the use of administrative law. We should rather retain the phrase as a perpetual reminder of fearful dangers we cannot avoid. The great challenge facing us now is to invent the corrective feedbacks that are needed to keep custodians honest. We must find ways to legitimate the needed authority of both the custodians and the corrective feedbacks. . . .

Mutual Coercion Mutually Agreed Upon

The social arrangements that produce responsibility are arrangements that create coercion, of some sort. Consider bank-robbing. The man who takes money from a bank acts as if the bank were a commons. How do we prevent such action? Certainly not by trying to control his behavior solely by a verbal appeal to his sense of responsibility. Rather than rely on propaganda we follow Frankel's lead and insist that a bank is not a commons; we seek the definite social arrangements that will keep it from becoming a commons. That we thereby infringe on the freedom of would-be robbers we neither deny nor regret.

The morality of bank-robbing is particularly easy to understand because we accept complete prohibition of this activity. We are willing to say "Thou shalt not rob banks," without providing for exceptions. But temperance also can be created by coercion. Taxing is a good coercive device. To keep downtown shoppers temperate in their use of parking space we introduce parking meters for short periods, and traffic fines for longer ones. We need not actually forbid a citizen to park as long as he wants to; we need merely make it increasingly expensive for him to do so. Not prohibition, but carefully biased options are what we offer him. A Madison Avenue man might call this persuasion; I prefer the greater candor of the word coercion.

Coercion is a dirty word to most liberals now, but it need not forever be so. As with the four-letter words, its dirtiness can be cleansed away by exposure to the light, by saying it over and over without apology or embarrassment. To many, the word coercion implies arbitrary decisions of distant and irresponsible bureaucrats; but this is not a necessary part of its meaning. The only kind of coercion I recommend is mutual coercion, mutually agreed upon by the majority of the people affected.

To say that we mutually agree to coercion is not to say that we are required to enjoy it, or even to pretend we enjoy it. Who enjoys taxes? We all grumble about them. But we accept compulsory taxes because we recognize that voluntary taxes would favor the conscienceless. We institute and (grumblingly) support taxes and other coercive devices to escape the horror of the commons.

An alternative to the commons need not be perfectly just to be preferable. With real estate and other material goods, the alternative we have chosen is the institution of private property coupled with legal inheritance. Is this system perfectly just? As a genetically trained biologist I deny that it is. It seems to me that, if there are to be differences in individual inheritance, legal possession should be perfectly correlated with biological inheritance—that those who are biologically more fit to be the custodians of property and power should legally inherit more. But genetic recombination continually makes a mockery of the doctrine of "like father, like son" implicit in our laws of legal inheritance. An idiot can inherit millions, and a trust fund can keep his estate intact. We must admit that our legal system of private property plus inheritance is unjust—but we put up with it because we are not convinced, at the moment, that anyone has invented a better system. The alternative of the commons is too horrifying to contemplate. Injustice is preferable to total ruin.

It is one of the peculiarities of the warfare between reform and the status quo that it is thoughtlessly governed by a double standard. Whenever a reform measure is proposed it is often defeated when its opponents triumphantly discover a flaw in it. As Kingsley Davis has pointed out,[11] worshippers of the status quo sometimes imply that no reform is possible without unanimous agreement, an implication contrary to historical fact. As nearly as I can make out, automatic rejection of proposed reforms is based on one of two unconscious assumptions: (i) that the status quo is perfect; or (ii) that the choice we face is between reform and no action; if the proposed reform is imperfect, we presumably should take no action at all, while we wait for a perfect proposal.

But we can never do nothing. That which we have done for thousands of years is also action. It also produces evils. Once we are aware that the status quo is action, we can then compare its discoverable advantages and disadvantages with the predicted advantages and disadvantages of the proposed reform, discounting as best we can for our lack of experience. On the basis of such a comparison, we can make a rational decision which will not involve the unworkable assumption that only perfect systems are tolerable.

Recognition of Necessity

Perhaps the simplest summary of this analysis of man's population problems is this: the commons, if justifiable at all, is justifiable only under conditions of low-population density. As the human population has increased, the commons has had to be abandoned in one aspect after another.

First we abandoned the commons in food gathering, enclosing farm land and restricting pastures and hunting and fishing areas. These restrictions are still not complete throughout the world.

Somewhat later we saw that the commons as a place for waste disposal would also have to be abandoned. Restrictions on the disposal of domestic sewage are widely accepted in the Western world; we are still struggling to close the commons to pollution by automobiles, factories, insecticide sprayers, fertilizing operations, and atomic energy installations.

In a still more embryonic state is our recognition of the evils of the commons in matters of pleasure. There is almost no restriction on the propagation of sound waves in the public medium. The shopping public is assaulted with mindless music, without its consent. Our government is paying out billions of dollars to create supersonic transport which will disturb 50,000 people for every one person who is whisked from coast to coast 3 hours faster. Advertisers muddy the airwaves of radio and television and pollute the view of travelers. We are a long way from outlawing the commons in matters of pleasure. Is this because our Puritan inheritance makes us view pleasure as something of a sin, and pain (that is, the pollution of advertising) as the sign of virtue?

Every new enclosure of the commons involves the infringement of somebody's personal liberty. Infringements made in the distant past are accepted because no contemporary complains of a loss. It is the newly proposed infringements that we vigorously oppose; cries of "rights" and "freedom" fill the air. But what does "freedom" mean? When men mutually agreed to pass laws against robbing, mankind became more free, not less so. Individuals locked into the logic of the commons are free only to bring on universal ruin; once they see the necessity of mutual coercion, they become free to pursue other goals. I believe it was Hegel who said, "Freedom is the recognition of necessity."

The most important aspect of necessity that we must now recognize, is the necessity of abandoning the commons in breeding. No technical solution can rescue us from the misery of overpopulation. Freedom to breed will bring ruin to all. At the moment, to avoid hard decisions many of us are tempted to propagandize for conscience and responsible parenthood. The temptation must be resisted, because an appeal to independently acting consciences selects for the disappearance of all conscience in the long run, and an increase in anxiety in the short.

The only way we can preserve and nurture other and more precious freedoms is by relinquishing the freedom to breed, and that very soon. "Freedom is the recognition of necessity"—and it is the role of education to reveal to all the necessity of abandoning the freedom to breed. Only so, can we put an end to this aspect of the tragedy of the commons.

NOTES

1. J. B. Wiesner and H. F. York, "National Security and the Nuclear-Test Ban," *Scientific American* 211, no. 4 (1964): 27.

2. G. Hardin, "Interstellar Migration and the Population Problem," *Journal of Heredity* 50, no. 68 (1959); S. Von Hoernor, "The General Limits of Space Travel: We May Never Visit Our Neighbors, but We Should Start Listening and Talking to Them," *Science* 137, no. 18 (1962).

3. J. Von Neumann and O. Morgenstern, *Theory of Games and Economic Behavior* (Princeton, NJ: Princeton University Press, 1947), 11.

4. J. H. Fremlin, "How Many People Can the World Support?," *New Scientist*, no. 415 (1964): 285.

5. A. Smith, *The Wealth of Nations* (New York: Modern Library, 1937), 423.

6. W. F. Lloyd, *Two Lectures on the Checks to Population* (Oxford: Oxford University Press, 1833), reprinted (in part) in G. Hardin, ed., *Population, Evolution, and Birth Control* (San Francisco: Freeman, 1964), 37.

7. A. N. Whitehead, *Science and the Modern World* (New York: Mentor, 1948), 17.

8. G. Hardin, ed., *Population, Evolution, and Birth Control* (San Francisco: Freeman, 1964), 56.

9. S. McVay, "The Last of the Great Whales," *Scientific American* 216, no. 8 (1966): 13.

10. J. Fletcher, *Situation Ethics* (Philadelphia: Westminster, 1966).

11. J. D. Roslansky, ed., *Genetics and the Future of Man* (New York: Appleton-Century-Crofts, 1966), 177.

23

Revisiting the Commons

Local Lessons, Global Challenges

ELINOR OSTROM, JOANNA BURGER, CHRISTOPHER B. FIELD,
RICHARD B. NORGAARD, AND DAVID POLICANSKY

In a seminal paper, Garrett Hardin argued in 1968 that users of a commons are caught in an inevitable process that leads to the destruction of the resources on which they depend. This article discusses new insights about such problems and the conditions most likely to favor sustainable uses of common-pool resources. Some of the most difficult challenges concern the management of large-scale resources that depend on international cooperation, such as fresh water in international basins or large marine ecosystems. Institutional diversity may be as important as biological diversity for our long-term survival.

Thirty years have passed since Garrett Hardin's influential article "The Tragedy of the Commons."[1] At first, many people agreed with Hardin's metaphor that the users of a commons are caught in an inevitable process that leads to the destruction of the very resource on which they depend. The "rational" user of a commons, Hardin argued, makes demands on a resource until the expected benefits of his or her actions equal the expected costs. Because each user ignores costs imposed on others, individual decisions cumulate to a tragic overuse and the potential destruction of an open-access commons. Hardin's proposed solution was "either socialism or the privatism of free enterprise."[2]

The starkness of Hardin's original statement has been used by many scholars and policy-makers to rationalize central government control of all common-pool resources[3] and to paint a disempowering, pessimistic vision of the human prospect.[4] Users are pictured as trapped in a situation they cannot change. Thus, it is argued that solutions must be imposed on users by external authorities. Although tragedies have undoubtedly occurred, it is also obvious that for thousands of years people have self-organized to manage common-pool resources, and users often do devise long-term, sustainable institutions for governing these resources.[5-7] It is time for a reassessment of the generality of the theory that has grown out of Hardin's original paper. Here, we describe the advances in understanding and managing commons problems that have been made since 1968. We also describe research challenges, especially those related to expanding our understanding of global commons problems.

An important lesson from the empirical studies of sustainable resources is that more solutions exist than Hardin proposed. Both government owner-ship and privatization are themselves subject to failure in some instances. For example, Sneath shows great differences in grassland degradation under a traditional, self-organized group-property regime versus central government management. A satellite image of northern China, Mongolia, and southern Siberia[8] shows marked degradation in the Russian part of the image, whereas the Mongolian half of the image shows much less degradation. In this instance, Mongolia has allowed pastoralists to continue their traditional group-property institutions, which involve large-scale movements between seasonal pastures, while both Russia and China have imposed state-owned agricultural collectives that involve permanent settlements. More recently, the Chinese solution has involved privatization by dividing the "pasture land into individual allocations for each herding household."[8] About three-quarters of the pasture land in the Russian section of this ecological zone has been degraded and more than one-third of the Chinese section has been degraded, while only one-tenth of the Mongolian section has suffered equivalent loss.[8,9] Here, socialism and privatiza-tion are both associated with more degradation than resulted from a traditional group-property regime.

Most of the theory and practice of successful management involves resources that are effectively managed by small to relatively large groups living within a single country, which involve nested institutions at varying scales. These resources continue to be important as sources of sustained biodiversity and human well-being. Some of the most difficult future problems, however, will involve resources that are difficult to manage at the scale of a village, a large watershed, or even a single country. Some of these resources—for example, fresh water in an international basin or large marine ecosystems—become effectively depletable only in an international context.[10] Management of these resources depends on the cooperation of appropriate international institutions and national, regional, and local institutions. Resources that are intrinsically dif-ficult to measure or that require measurement with advanced technology, such as stocks of ocean fishes or petroleum reserves, are difficult to manage no mat-ter what the scale of the resource. Others, for example global climate, are largely self-healing in response to a broad range of human actions, until these actions exceed some threshold.[11]

Although the number and importance of commons problems at local or regional scales will not decrease, the need for effective approaches to commons problems that are global in scale will certainly increase. Here, we examine this need in the context of an analysis of the nature of common-pool resources and the history of successful and unsuccessful institutions for ensuring fair access and sustained availability to them. Some experience from smaller systems transfers directly to global systems, but global commons introduce a range of new issues, due largely to extreme size and complexity.[12]

The Nature of Common-Pool Resources

To better understand common-pool resource problems, we must separate concepts related to resource systems and those concerning property rights. We use the term common-pool resources (CPRs) to refer to resource systems regardless of the property rights involved. CPRs include natural and human-constructed resources in which (i) exclusion of beneficiaries through physical and institutional means is especially costly, and (ii) exploitation by one user reduces resource availability for others.[13] These two characteristics—difficulty of exclusion and subtractability—create potential CPR dilemmas in which people following their own short-term interests produce outcomes that are not in anyone's long-term interest. When resource users interact without the benefit of effective rules limiting access and defining rights and duties, substantial free-riding in two forms is likely: overuse without concern for the negative effects on others, and a lack of contributed resources for maintaining and improving the CPR itself.

CPRs have traditionally included terrestrial and marine ecosystems that are simultaneously viewed as depletable and renewable. Characteristic of many resources is that use by one reduces the quantity or quality available to others, and that use by others adds negative attributes to a resource. CPRs include earth-system components (such as groundwater basins or the atmosphere) as well as products of civilization (such as irrigation systems or the World Wide Web).

Characteristics of CPRs affect the problems of devising governance regimes. These attributes include the size and carrying capacity of the resource system, the measurability of the resource, the temporal and spatial availability of resource flows, the amount of storage in the system, whether resources move (like water, wildlife, and most fish) or are stationary (like trees and medicinal plants), how fast resources regenerate, and how various harvesting technologies affect patterns of regeneration.[14] It is relatively easy to estimate the number and size of trees in a forest and allocate their use accordingly, but it is much more difficult to assess migratory fish stocks and available irrigation water in a system without storage capacity. Technology can help to inform decisions by improving the identification and monitoring of resources, but it is not a substitute for decision-making. On the other hand, major technological advances in assessing groundwater storage capacity, supply, and associated pollution have allowed more effective management of these resources.[15] Specific resource systems in particular locations often include several types of CPRs and public goods with different spatial and temporal scales, differing degrees of uncertainty, and complex interactions among them.[16]

Institutions for Governing and Managing Common-Pool Resources

Solving CPR problems involves two distinct elements: restricting access and creating incentives (usually by assigning individual rights to, or shares of, the resource) for users to invest in the resource instead of overexploiting it. Both changes are needed. For example, access to the north Pacific halibut fishery was not restricted before the recent introduction of individual transferable quotas and catch limits protected the resource for decades. But the enormous competition to catch a large share of the resource before others did resulted in economic waste, danger to the fishers, and reduced quality of fish to consumers. Limiting access alone can fail if the resource users compete for shares, and the resource can become depleted unless incentives or regulations prevent overexploitation.[17,18]

Four broad types of property rights have evolved or are designed in relation to CPRs (Table 23.1). When valuable CPRs are left to an open-access regime, degradation and potential destruction are the result. The proposition that resource users cannot themselves change from no property rights (open access) to group or individual property, however, can be strongly rejected on the basis of evidence: Resource users through the ages have done just that.[5-7,13,15,19] Both group-property and individual-property regimes are used to manage resources that grant individuals varying rights to access and use of a resource. The primary difference between group property and individual property is the ease with which individual owners can buy or sell a share of a resource. Government property involves ownership by a national, regional, or local public agency that can forbid or allow use by individuals. Empirical studies show that no single type of property regime works efficiently, fairly, and sustainably in relation to all CPRs. CPR problems continue to exist in many regulated settings.[17] It is possible, however, to identify design principles associated with robust institutions that have successfully governed CPRs for generations.[19]

The Evolution of Norms and Design of Rules

The prediction that resource users are led inevitably to destroy CPRs is based on a model that assumes all individuals are selfish, norm-free, and maximizers of short-run results. This model explains why market institutions facilitate an efficient allocation of private goods and services, and it is strongly supported by empirical data from open, competitive markets in industrial societies.[20] However, predictions based on this model are not supported in field research or in laboratory experiments in which individuals face a public good or CPR problem and are able to communicate, sanction one another, or make new rules.[21] Humans adopt a narrow, self-interested perspective in many settings, but can also use reciprocity to overcome social dilemmas.[22] Users of a CPR include (i) those who always behave in a narrow, self-interested way and

TABLE 23.1. Types of Property-Rights Systems Used to Regulate
Common-Pool Resources

Property rights	Characteristics
Open access	Absence of enforced property rights
Group property	Resource rights held by a group of users who can exclude others
Individual property	Resource rights held by individuals (or firms) who can exclude others
Government property	Resource rights held by a government that can regulate or subsidize use

Source: D. Feeny, F. Berkes, B. J. McCay, and J. M. Acheson, "The Tragedy of the Commons: Twenty-Two Years Later," *Hum. Ecol.* 18, no. 1 (1990): 1–19.

never cooperate in dilemma situations (free-riders); (ii) those who are unwilling to cooperate with others unless assured that they will not be exploited by free-riders; (iii) those who are willing to initiate reciprocal cooperation in the hopes that others will return their trust; and (iv) perhaps a few genuine altruists who always try to achieve higher returns for a group.

Whether norms to cope with CPR dilemmas evolve without extensive, self-conscious design depends on the relative proportion of these behavioral types in a particular setting. Reciprocal cooperation can be established, sustain itself, and even grow if the proportion of those who always act in a narrow, self-interested manner is initially not too high.[23] When interactions enable those who use reciprocity to gain a reputation for trustworthiness, others will be willing to cooperate with them to overcome CPR dilemmas, which leads to increased gains for themselves and their offspring.[24] Thus, groups of people who can identify one another are more likely than groups of strangers to draw on trust, reciprocity, and reputation to develop norms that limit use. In earlier times, this restricted the size of groups who relied primarily upon evolved and shared norms. Citizen-band radios, tracking devices, the Internet, geographic information systems, and other aspects of modern technology and the news media now enable large groups to monitor one another's behavior and coordinate activities in order to solve CPR problems.

Evolved norms, however, are not always sufficient to prevent overexploitation. Participants or external authorities must deliberately devise (and then monitor and enforce) rules that limit who can use a CPR, specify how much and when that use will be allowed, create and finance formal monitoring arrangements, and establish sanctions for nonconformance. Whether the users themselves are able to overcome the higher level dilemmas they face in bearing the cost of designing, testing, and modifying governance systems depends on the benefits they perceive to result from a change as well as the expected costs of negotiating, monitoring, and enforcing these rules.[25] Perceived benefits are greater when the resource reliably generates valuable products for the users. Users need some autonomy to make and enforce their own rules, and they must highly value the future sustainability of the resource. Perceived costs are higher

TABLE 23.2. Relationship of Governance Structures and Cropping Intensities

Parameter	Farmer-owned systems (N = 97)	Government-owned systems (N = 21)	F	P
Head-end crop intensities	246%	208%	10.51	0.002
Tail-end crop intensities	237%	182%	20.33	0.004

Source: W. F. Lam, Governing Irrigation Systems in Nepal: Institutions, Infrastructure, and Collective Action (Oakland, CA: ICS, 1998).
A crop intensity of 100% means that all land in an irrigation system is put to full use for one season or partial use over multiple seasons, amounting to the same coverage. Similarly, a crop intensity of 200% is full use of all land for two seasons; 300% is full use for three seasons.

when the resource is large and complex, users lack a common understanding of resource dynamics, and users have substantially diverse interests.[26]

The farmer-managed irrigation systems of Nepal are examples of well-managed CPRs that rely on strong, locally crafted rules as well as evolved norms.[27] Because the rules and norms that make an irrigation system operate well are not visible to external observers, efforts by well-meaning donors to replace primitive, farmer-constructed systems with newly constructed, government-owned systems have reduced rather than improved performance.[28] Government-owned systems are built with concrete and steel headworks, in contrast to the simple mud, stone, and trees used by the farmers. However, the cropping intensity achieved by farmer-managed systems is significantly higher than on government systems (Table 23.2). In a regression model of system performance, controlling for the size of the system, the slope of the terrain, variation in farmer income, and the presence of alternative sources of water, both government ownership and the presence of modern headworks have a negative impact on water delivered to the tail end of a system, hence a negative impact on overall system productivity.[27]

Imposing strong limits on resource use raises the question of which community of users is initially defined as having use rights and who is excluded from access to a CPR. The very process of devising methods of exclusion has substantial distributional consequences.[29] In some instances, those who have long exercised stewardship over a resource can be excluded. A substantial distributional issue will occur, for example, as regulators identify who will receive rights to emit carbon into the atmosphere. Typically, such rights are assigned to those who have exercised a consistent pattern of use over time. Thus, those who need to use the resource later may be excluded entirely or may have to pay a very large entry cost.

The counterpoint to exclusion is too rapid inclusion of users. When any user group grows rapidly, the resource can be stressed. For example, in the last 10 years the annual sales of personal watercraft (PWCs) have risen in the United States from about 50,000 to more than 150,000 a year. This has placed a burden on the use of surface water and created conflicts with homeowners, other boaters, fishermen, and naturalists. The rapid rise of PWCs has created a burden on the use of shorelines, contributed to a disproportionate increase in accidents and injuries, and caused disturbances to aquatic natural resources.[30] Traditional users of the water surface feel threatened by the invasion of their space by a new, faster, and louder boat that reduces the value of surface waters. In many other settings, when new users arrive through migration, they do not share a similar understanding of how a resource works and what rules and norms are shared by others. Members of the initial community feel threatened and may fail to enforce their own self-restraint, or they may even join the race to use up the resource.[31]

Given the substantial differences among CPRs, it is difficult to find effective rules that both match the complex interactions and dynamics of a resource and are perceived by users as legitimate, fair, and effective. At times, disagreements about resource assessment may be strategically used to propose policies that disproportionately benefit some at a cost to others.[4] In highly complex systems, finding optimal rules is extremely challenging, if not impossible. But despite such problems, many users have devised their own rules and have sustained resources over long periods of time. Allowing parallel self-organized governance regimes to engage in extensive trial-and-error learning does not reduce the probability of error for any one resource, but greatly reduces the probability of disastrous errors for all resources in a region.

Lessons from Local and Regional Common-Pool Resources

The empirical and theoretical research stimulated over the past 30 years by Garrett Hardin's article has shown that tragedies of the commons are real, but not inevitable. Solving the dilemmas of sustainable use is neither easy nor error-free even for local resources. But a scholarly consensus is emerging regarding the conditions most likely to stimulate successful self-organized processes for local and regional CPRs.[6,26,32] Attributes of resource systems and their users affect the benefits and costs that users perceive. For users to see major benefits, resource conditions must not have deteriorated to such an extent that the resource is useless, nor can the resource be so little used that few advantages result from organizing. Benefits are easier to assess when users have accurate knowledge of external boundaries and internal microenvironments and have reliable and valid indicators of resource conditions. When the flow of resources is relatively predictable, it is also easier to assess how diverse management regimes will affect long-term benefits and costs.

Users who depend on a resource for a major portion of their livelihood, and who have some autonomy to make their own access and harvesting rules, are more likely than others to perceive benefits from their own restrictions, but they need to share an image of how the resource system operates and how their actions affect each other and the resource. Further, users must be interested in the sustainability of the particular resource so that expected joint benefits will outweigh current costs. If users have some initial trust in others to keep promises, low-cost methods of monitoring and sanctioning can be devised. Previous organizational experience and local leadership reduces the users' costs of coming to agreement and finding effective solutions for a particular environment. In all cases, individuals must overcome their tendency to evaluate their own benefits and costs more intensely than the total benefits and costs for a group. Collective-choice rules affect who is involved in deciding about future rules and how preferences will be aggregated. Thus, these rules affect the breadth of interests represented and involved in making institutional changes, and they affect decisions about which policy instruments are adopted.[33]

The Broader Social Setting

Whether people are able to self-organize and manage CPRs also depends on the broader social setting within which they work. National governments can help or hinder local self-organization. "Higher" levels of government can facilitate the assembly of users of a CPR in organizational meetings, provide information that helps identify the problem and possible solutions, and legitimize and help enforce agreements reached by local users. National governments can at times, however, hinder local self-organization by defending rights that lead to overuse or maintaining that the state has ultimate control over resources without actually monitoring and enforcing existing regulations.

Participants are more likely to adopt effective rules in macro-regimes that facilitate their efforts than in regimes that ignore resource problems entirely or that presume that central authorities must make all decisions. If local authority is not formally recognized by larger regimes, it is difficult for users to establish enforceable rules. On the other hand, if rules are imposed by outsiders without consulting local participants, local users may engage in a game of "cops and robbers" with outside authorities. In many countries, two centuries of colonization followed by state-run development policy that affected some CPRs has produced great resistance to externally imposed institutions.

The broader economic setting also affects the level and distribution of gains and costs of organizing the management of CPRs. Expectations of rising resource prices encourage better management, whereas falling, unstable, or uncertain resource prices reduce the incentive to organize and assure future availability.[34] National policy also affects factors such as human migration rates, the flow of capital, technology policy, and hence the range of conditions local institutions must address to work effectively. Finally, local institutions are only rarely able to cope with the ramifications of civil or international war.

Challenges of Global Commons

The lessons from local and regional CPRs are encouraging, yet humanity now faces new challenges to establish global institutions to manage biodiversity, climate change, and other ecosystem services.[35] These new challenges will be especially difficult for at least the following reasons.

Scaling-up problem. Having larger numbers of participants in a CPR increases the difficulty of organizing, agreeing on rules, and enforcing rules. Global environmental resources now involve 6 billion inhabitants of the globe. Organization at national and local levels can help, but it can also get in the way of finding solutions.

Cultural diversity challenge. Along with economic globalization, we are in a period of reculturalization. Increasing cultural diversification offers increased hope that the diversity of ways in which people have organized locally around CPRs will not be quickly lost, and that diverse new ways will continue to evolve at the local level. However, cultural diversity can decrease the likelihood of finding shared interests and understandings. The problem of cultural diversity is exacerbated by "north-south" conflicts stemming from economic differences between industrialized and less-industrialized countries.

Complications of interlinked CPRs. Although the links between grassland and forest management are complex, they are not so complex as those between maintaining biodiversity and ameliorating climate change. As we address global issues, we face greater interactions between global systems. Similarly, with increased specialization, people have become more interdependent. Thus, we all share one another's common interests, but in more complex ways than the users of a forest or grassland. While we have become more complexly interrelated, we have also become more "distant" from each other and our environmental problems. From our increasingly specialized understandings and particular points on the globe, it is difficult to comprehend the significance of global CPRs and how we need to work together to govern these resources successfully. And given these complexities, finding fair solutions is even more challenging.

Accelerating rates of change. Previous generations complained that change occurred faster and faster, and the acceleration continues. Population growth, economic development, capital and labor mobility, and technological change push us past environmental thresholds before we know it. "Learning by doing" is increasingly difficult, as past lessons are less and less applicable to current problems.

Requirement of unanimous agreement as a collective-choice rule. The basic collective-choice rule for global resource management is voluntary assent to negotiated treaties.[36] This allows some national governments to hold out for special privileges before they join others in order to achieve regulation, thus strongly affecting the kinds of resource management policies that can be adopted at this level.

We have only one globe with which to experiment. Historically, people could migrate to other resources if they made a major error in managing a local CPR. Today, we have less leeway for mistakes at the local level, while at the global level there is no place to move.

These new challenges clearly erode the confidence with which we can build from past and current examples of successful management to tackle the CPR problems of the future. Still, the lessons from successful examples of CPR management provide starting points for addressing future challenges. Some of these will be institutional, such as multilevel institutions that build on and complement local and regional institutions to focus on truly global problems. Others will build from improved technology. For example, more accurate long-range weather forecasts could facilitate improvements in irrigation management, or advances in fish tracking could allow more accurate population estimates and harvest management. And broad dissemination of widely believed data could be a major contributor to the trust that is so central to effective CPR management.

In the end, building from the lessons of past successes will require forms of communication, information, and trust that are broad and deep beyond precedent, but not beyond possibility. Protecting institutional diversity related to how diverse peoples cope with CPRs may be as important for our long-run survival as the protection of biological diversity. There is much to learn from successful efforts as well as from failures.

REFERENCES AND NOTES

1. G. Hardin, "The Tragedy of the Commons," *Science* 162 (1968): 1243.

2. G. Hardin, "Extensions of 'The Tragedy of the Commons,'" *Science* 280, no. 5364 (1998): 682.

3. J. E. M. Arnold, *Managing Forests as Common Property* (Rome: FAO Forestry Paper 136, 1998); D. Feeny, S. Hanna, and A. F. McEvoy, "Questioning the Assumptions of 'The Tragedy of the Commons' Model of Fisheries," *Land Econ.* 72, no. 2 (1996): 187–205; F. Berkes and C. Folke, eds., *Linking Social and Ecological Systems: Management Practices and Social Mechanisms for Building Resilience* (New York: Cambridge University Press, 1998); A. C. Finlayson and B. J. McCay, "Crossing the Threshold of Ecosystem Resilience: The Commercial Extinction of Northern Cod," ibid., 311–337; R. Repetto, *Skimming the Water: Rent-Seeking and the Performance of Public Irrigation Systems* (Washington, DC: World Resources Institute Research Report 4, 1986).

4. D. Ludwig, R. Hilborn, and C. Walters, "Uncertainty, Resource Exploitation, and Conservatism: Lessons from History," *Science* 260, no. 17 (1993): 17.

5. B. J. McCay and J. M. Acheson, *The Question of the Commons: The Culture and Ecology of Communal Resources* (Tucson: University of Arizona Press, 1987); F. Berkes, D. Feeny, B. J. McCay, and J. M. Acheson, "The Benefits of the Commons," *Nature* 340 (1989): 91; F. Berkes, *Common Property Resources: Ecology and Community-Based Sustainable Development* (London: Belhaven, 1989); D. W. Bromley et al., *Making the Commons Work: Theory, Practice, and Policy* (San Francisco: ICS, 1992); S. Y. Tang, *Institutions and Collective Action: Self-Governance in Irrigation* (San Francisco: ICS, 1992); E. Pinkerton, ed., *Co-operative*

Management of Local Fisheries: New Directions for Improved Management and Community Development (Vancouver: University of British Columbia Press, 1989); C. Hess, *Common-Pool Resources and Collective Action: A Bibliography*, vol. 3, and *Forest Resources and Institutions: A Bibliography* (Bloomington, IN: Workshop in Political Theory and Policy Analysis, 1996, www.indiana.edu/~workshop/wsl/wsl.html.

6. R. Wade, *Village Republics: Economic Conditions for Collective Action in South India* (San Francisco: ICS, 1994).

7. D. Feeny, F. Berkes, B. J. McCay, and J. M. Acheson, "The Tragedy of the Commons: Twenty-Two Years Later," *Hum. Ecol.* 18, no. 1 (1990): 1–19.

8. D. Sneath, "State Policy and Pasture Degradation in Inner Asia," *Science* 281, no. 5380 (1998): 1147.

9. C. Humphrey and D. Sneath, eds., *Culture and Environment in Inner Asia*, vol. 1 (Cambridge, UK: White Horse, 1996).

10. R. Costanza, F. Andrade, P. Antunes, M. van den Belt, D. F. Boesch, F. Catarino, S. Hanna, et al., "Principles for Sustainable Governance of the Oceans," *Science* 281 (1998): 198.

11. W. S. Broecker, "Thermohaline Circulation, the Achilles Heel of Our Climate System," *Science* 278 (1997): 1582.

12. M. McGinnis and E. Ostrom, "Design Principles for Local and Global Commons," in *The International Political Economy and International Institutions*, vol. 2, ed. O. R. Young (Cheltenham, UK: Elgar, 1996), 465–493; R. O. Keohane and E. Ostrom, eds., *Local Commons and Global Interdependence: Heterogeneity and Cooperation in Two Domains* (London: Sage, 1995); S. Buck, *The Global Commons: An Introduction* (Washington, DC: Island, 1998).

13. E. Ostrom, R. Gardner, and J. Walker, *Rules, Games, and Common-Pool Resources* (Ann Arbor: University of Michigan Press, 1994).

14. E. Schlager, W. Blomquist, and S. Y. Tang, "Mobile Flows, Storage and Self-Organized Institutions for Governing Common Pool Resources," *Land Econ.* 70, no. 3 (1994): 294–317.

15. W. Blomquist, *Dividing the Waters: Governing Groundwater in Southern California* (San Francisco: ICS, 1992).

16. R. Norgaard, "Intergenerational Commons, Globalization, Economism and Unsustainable Development," *Adv. Hum. Ecol.* 4 (1995): 141; C. Gibson, *Politicians and Poachers: The Political Economy of Wildlife Policy in Africa* (New York: Cambridge University Press, 1999); A. Agrawal, *Greener Pastures: Politics, Markets, and Community among a Migrant Pastoral People* (Durham, NC: Duke University Press, 1999).

17. Organisation for Economic Co-operation and Development (OECD), *Towards Sustainable Fisheries: Economic Aspects of the Management of Living Marine Resources* (Paris: OECD, 1997); National Research Council, *Sustaining Marine Fisheries* (Washington, DC: National Academy Press, 1999).

18. H. S. Gordon, "The Economic Theory of a Common-Property Resource: The Fishery," *J. Pol. Econ.* 62, no. 2 (1954): 124; B. J. McCay, "Social and Ecological Implications of ITQs: An Overview," *Coastal Ocean Management* 3 (1995): 3.

19. E. Ostrom, *Governing the Commons: The Evolution of Institutions for Collective Action* (New York: Cambridge University Press, 1990).

20. C. R. Plott, "Laboratory Experiments in Economics: The Implications of Posted-Price Institutions," *Science* 232 (1986): 732; K. A. McCabe, S. J. Rassenti, and V. L. Smith, "Smart Computer-Assisted Markets," *Science* 254 (1991): 534.

21. See S. Bowles, R. Boyd, E. Fehr, and H. Gintis, *Homo reciprocans: A Research Initiative on the Origins, Dimensions, and Policy Implications of Reciprocal Fairness* (working paper, University of Massachusetts, 1997); E. Ostrom and J. M. Walker, "Neither Markets nor States: Linking Transformation Processes in Collective Action Arenas," in *Perspectives on Public*

Choice: A Handbook, ed. D. C. Mueller (New York: Cambridge University Press, 1997), 35–72; J. M. Orbell, A. J. C. van de Kragt, and R. M. Dawes, "Explaining Discussion-Induced Cooperation," *J. Personality & Soc. Psychol.* 54, no. 5 (1988): 811; E. Ostrom, "A Behavioral Approach to the Rational Choice Theory of Collective Action," *Am. Pol. Sci. Rev.* 92, no. 1 (1998): 1–22. In these experiments, the formal structure of a dilemma is converted into a set of decisions made by subjects who are financially rewarded as a result of their own and others' decisions. See also J. H. Kagel and A. E. Roth, eds., *The Handbook of Experimental Economics* (Princeton, NJ: Princeton University Press, 1995). The model is also not as robust in explaining exchange behavior in traditional societies where evolved norms still strongly affect behavior.

22. L. Cosmides and J. Tooby, "Cognitive Adaptations for Social Exchange," in *The Adapted Mind: Evolutionary Psychology and the Generation of Culture*, ed. J. H. Barkow, L. Cosmides, and J. Tooby (New York: Oxford University Press, 1992), 163–228; L. Cosmides and J. Tooby, "Better than Rational: Evolutionary Psychology and the Invisible Hand," *Am. Econ. Rev.* 84, no. 2 (1994): 327; E. Hoffman, K. McCabe, and V. Smith, "Social Distance and Other-Regarding Behavior in Dictator Games," *Am. Econ. Rev.* 86, no. 3 (1996): 653.

23. R. Axelrod, *The Evolution of Cooperation* (New York: Basic Books, 1984); R. Axelrod, "An Evolutionary Approach to Norms," *Am. Pol. Sci. Rev.* 80, no. 4 (1986): 1095.

24. M. A. Nowak and K. Sigmund, "Tit for Tat in Heterogeneous Populations," *Nature* 355 (1992): 250; D. M. Kreps, P. Milgrom, J. Roberts, and R. Wilson, "Rational Cooperation in the Finitely Repeated Prisoners' Dilemma," *J. Econ. Theory* 27 (1982): 245.

25. H. Demsetz, "Toward a Theory of Property Rights," *Am. Econ. Rev.* 57, no. 2 (1967): 347; D. North, "Economic Performance through Time," *Am. Econ. Rev.* 84 (1994): 359; C. M. Rose, *Property & Persuasion: Essays on the History, Theory, and Rhetoric of Ownership* (Boulder, CO: Westview, 1994); J. E. Krier, "The Tragedy of the Commons, Part Two," *Harv. J. L. Pub. Pol'y* 15 (1992): 325; F. Michelman, "Tutelary Jurisprudence and Constitutional Property," in *Liberty, Property, and the Future of Constitutional Development*, ed. E. F. Paul and H. Dickman (Albany: State University of New York Press, 1990), 127–171; V. Ostrom, "Courts and Collectives," *BYU L. Rev.* 3 (1990): 857.

26. E. Ostrom, "Reformulating the Commons," in *Protecting the Commons: A Framework for Resource Management in the Americas*, ed. J. Burger, E. Ostrom, R. B. Norgaard, D. Policansky, and B. D. Goldstein (Washington, DC: Island, 2001).

27. W. F. Lam, *Governing Irrigation Systems in Nepal: Institutions, Infrastructure, and Collective Action* (Oakland, CA: ICS, 1998).

28. W. F. Lam, "Improving the Performance of Small-Scale Irrigation Systems: The Effects of Technological Investments and Governance Structure on Irrigation Performance in Nepal," *World Dev.* 24, no. 8 (1996): 1301.

29. G. D. Libecap, "Distributional Issues in Contracting for Property Rights," *J. Instl. Theor. Econ.* 145, no. 1 (1989): 6.

30. J. Burger, "Effects of Motorboats and Personal Watercraft on Flight Behavior over a Colony of Common Terns," *Condor* 100 (1998): 528; L. Whiteman, "Making Waves," *National Parks* 71 (1997): 22.

31. F. G. Speck and W. S. Hadlock, "A Report on the Tribal Boundaries and Hunting Areas of the Malecite Indian of New Brunswick," *Am. Anthropol.* 48, no. 3 (1946): 355; C. Safina, "Where Have All the Fishes Gone?," *Issues Sci. Technol.* 10, no. 3 (1994): 37.

32. J. M. Baland and J. P. Platteau, *Halting Degradation of Natural Resources: Is There a Role for Rural Communities?* (Oxford, UK: Clarendon, 1996); M. A. McKean, "Success on the Commons," *J. Theor. Pol.* 4, no. 3 (1992): 247.

33. J. Buchanan and G. Tullock, *The Calculus of Consent* (Ann Arbor: University of Michigan Press, 1962); J. B. Wiener, "Global Environmental Regulation: Instrument Choice in Legal Context," *Yale L.J.* 108, no. 4 (1999): 677.

34. C. W. Clark and G. R. Munro, "Renewable Resources as Natural Capital: The Fishery," in *Investing in Natural Capital: The Ecological Economics Approach to Sustainability*, ed. A. M. Jansson, M. Hammer, C. Folke, and R. Costanza (Washington, DC: Island, 1994), 343–361.

35. See O. Young, ed., *Science Plan for Institutional Dimensions of Global Environmental Change* (Bonn, Germany: International Human Dimensions Programme on Global Environmental Change, 1999); O. Young, ed., *Global Governance: Drawing Insights from the Environmental Experience* (Cambridge, MA: MIT Press, 1997); P. Haas, R. Keohane, and M. Levy, *Institutions for the Earth: Sources of Effective Environmental Protection* (Cambridge, MA: MIT Press, 1993).

36. J. B. Wiener, "On the Political Economy of Global Environmental Regulation," *Georgetown L.J.* 87 (1998/1999): 749–794.

24

Excerpts from "Rationality and Solidarities: The Social
Organization of Common Property Resources in the
Imdrhas Valley of Morocco"

PEGGY PETRZELKA AND MICHAEL M. BELL

Introduction

Despite some two centuries of the development and spread of capitalist institutions, much of importance in the world is still not privately owned. Grazing
lands, woodlands, hunting grounds, fisheries, irrigation and drinking waters,
roads—all these and more frequently remain held in common. Yet social scientists long neglected the importance of common property. And when they did
turn their gaze toward common property, it was often with an eyebrow lifted in
skepticism and even scorn toward what was seen as a backward impediment to
industrial and industrializing societies.

   The dominant theory of what has come to be called common property
resources and common property resource systems (CPRs) draws principally
from the rational choice tradition—the idea that individuals act in their own
self-interest. In addition, probably the two most influential works in what might
be termed the classical literature on common property (both of which were written with that lifted eyebrow) were based on rational choice orientations: Mancur
Olson's 1965 book, *The Logic of Collective Action: Public Goods and the Theory of
Groups*, and Garrett Hardin's 1968 article in *Science*, "The Tragedy of the Commons." Much of the leading work since that time has retained these orientations,
as, for example, when Ostrom,[1] Stevenson,[2] and Sandler[3] argue that the functioning and malfunctioning of CPRs can be understood as the aggregate product of the private decision making of individuals acting in the rational pursuit of
their particular interests.

   There is good reason for impatience with the rational choice perspective,
however. First and most fundamentally, CPRs are *social* phenomena, as the
word "common" in the acronym's root indicates. Involvement in a CPR is not
an isolated activity, despite the rational choice focus on the individual. Second,
it is tautological and reductionist to ascribe all human motivation to the self-
interested decisions of rational actors—even "broadly rational" ones—as many
critics of rational choice have observed.[4] Humans are moved by considerations

of the interests of the self and by considerations of the interests of others, both by what might be termed "interests" proper and by what might be termed "sentiments."[5] Third, a rational choice perspective encourages a radical decontextualization of choice and actions, excluding from view many of the factors that define a particular action as "rational" in a CPR. In sum, a CPR is not an entity complete and entire unto itself; it cannot be analyzed apart from the overall social system of which it is necessarily a part.

In this article, we examine two Imazighen (Berber)[6] communities in southern Morocco and the social organization of their CPRs. Based on field research in these two villages, we argue that while both communities have the same established rules for managing their CPRs, there are distinct differences in what has occurred, and continues to occur, on them. These differences are due to more than rational actors' private calculations of personal gain. They can be equally attributed to the character of social ties in each community as a whole. Our point, then, is not to reject rational choice out of hand; our plea is for an understanding of rationality in its *full* social context.

Where Common Property Literature Has Been

Despite the widespread evidence that CPRs have persisted over the centuries[7] the classical roots of the CPR literature present a rather bleak portrayal of their possibilities. Olson ominously concluded that: "unless there is coercion or some other special device to make individuals act in their common interest, *rational, self-interested individuals will not act to achieve their common or group interests*" (emphasis in the original).[8] Likewise Hardin prophesied that because herders on a commons will seek to increase their herd size as much as possible until overgrazing sets in, and similarly for any public resource managed as a commons, "the inherent logic of the commons remorselessly generates tragedy. . . . Ruin is the destination toward which all men rush, each pursuing his own best interest in a society that believes in the freedom of the commons."[9]

Since the 1970s a large literature critical of Hardin and Olson has emerged. Ostrom perhaps represents the most significant work in what might be termed the second phase of CPR research. Responding to Olson and Hardin, Ostrom[10] points out that individuals *can* work together collectively and that successful CPRs *do* exist, a point Goldman[11] amplifies. The question Ostrom directs attention to is "how it is possible that some individuals organize themselves to govern and manage CPRs and others do not?"[12] She then goes on to document a set of design principles on which successful CPRs are based. These principles include an ownership arrangement within which management rules are developed, boundaries are clearly defined and known to all, group size is known and enforced, sanctions work to ensure compliance, and mechanisms exist for resolving conflict.

Ostrom's work represents an important advance in CPR research. However, it remains wholly within the confines of a strict—albeit better informed—rational choice perspective. For Ostrom, the actor is still self-interested and still a "broadly rational actor." But her work complemented the emergence of a third phase of CPR research that finds the traditional rational choice perspective too limiting.

Transitional to this third phase was the work of Bromley and Cernea[13] and Niamir.[14] Bromley and Cernea argued that the breakdown of many traditional CPRs in the 20th century represented not the remorseless tragedy of self-interested actors on a commons but rather the effects of externally imposed social change due to colonialism, nationalization, marketization, and modernization. They note that colonization brought with it the taking of lands and the implementation of non-CPR management forms, and they identify this exploitative period as the precipitating factor leading to the breakdown of many common property resource systems in poor countries. Following colonialism, state intervention through nationalization and still later through marketization of common property contributed greatly to the breakup of many systems of common property management. This sequence of changes led to increased stratification in CPR communities, lessening the commitment to "follow the rules" of a CPR.[15] Thus, Bromley and Cernea and others brought a previously missing perspective—power—to our understanding of CPRs.

Niamir noted that changes in a community's demographics and in the local physical environment could also undermine CPRs. Droughts have put a tremendous strain on the productivity of rangeland resources in many regions, reducing their carrying capacity, making them more vulnerable to overuse, and resulting in greater concentration of livestock. Increases in population have brought more demands on the land, resulting in pressures to divide CPRs into private holdings in an attempt to grow more food for household consumption and for the market.

But neither Niamir nor Bromley and Cernea questioned the grounding of a rational choice perspective on CPRs; rather, they argued that other factors, such as environmental change and the legacy of colonialism, were more significant in the recent breakdown of many CPRs. It is to others that we must turn for this foundational critique of rational choice.

Where Common Property Literature May Be Heading

The third phase's critique of self-interest stems from a new focus on the character of social ties within CPR communities. Granovetter[16] and Portes and Sensenbrenner[17] argued for examining economic behavior and social institutions in terms of their "embeddedness" within networks of social relations.

Other researchers have echoed this with specific reference to CPRs. Fisher was one of the first to stress "the importance of looking at embedded social relationships in understanding the commons."[18]

Other CPR researchers have made much the same point without drawing explicitly on the concept of embeddedness. Ireson[19] and Wade[20] noted that CPRs that have persisted typically exhibit strong social bonds, respect for local leadership, repeated social interaction among members, shared cultural norms, and cooperative social institutions. They also note some of the social changes that may lead to the breakdown of these characteristics. For example, in his discussion of CPRs in a Laotian village, Ireson notes difficulties in maintaining collective activity as social and economic differences increase in the village. Degradation of the CPR occurs because wealthier villagers "may become willing to risk village displeasure" as "they no longer need to depend on village assistance."[21]

We should applaud this recognition of the importance of embeddedness of social relationships. But these analyses are still incomplete. The picture we are left with is still of a rational actor, albeit one who has to pay attention to the embeddedness of a CPR. Rationality itself is not yet fully contextualized in this model.

Moreover, neither is a CPR fully contextualized in most of this research. What is needed is a more holistic conception that looks beyond the CPR itself, and beyond those who are directly involved in it, to the overall social system within which a CPR is only a part. To our knowledge, Mearns's[22] case study of common grazing in Mongolia is the only work to attempt such a conception. In her analysis of herders and their involvement in other agricultural communal activities, she emphasizes the need to look at commons within the social context of their other activities. We agree.

Just as a CPR is one social structure within a larger social system, so too do the participants in a CPR have multiple forms of involvement in a larger social group—in a word, a wider *community*. And to understand why this social structure is working, or not, we need to examine how it is influenced by this larger system, this wider community. The experience we gain from our involvement in one social setting does not disappear when we are involved in a different one. Therefore, examining a CPR in isolation, as well as only the individuals directly involved in it, will not provide an accurate representation of all the social dynamics at play.

The Dialogue of Solidarities

There is still something missing from all these analyses: a recognition of the social dynamism between solidarities based on *interests* and solidarities based on *sentiments*. Bell[23] argues that any collective action that persists over space and time must not only be conceived as a solidarity of interests, it must equally depend upon a solidarity of sentiments—upon the social coordination of reciprocal or complementary concerns for the interests of others and upon the affective and normative ties that lead to such coordination. As Bell notes,

> in reciprocal [and complementary] action, there is almost always a time delay involved. Sometimes I'll have to wait my turn. But how do I know that you, my partner, will come through when it is my turn—when it is your turn to wait? Because of my sense that we also have a solidarity of sentiments. We have affection for each other . . . and a sense of common commitment to certain norms of behavior.[24]

Thus, the existence of a solidarity of sentiments builds the trust necessary to cross the reaches of time and space inherent in a solidarity of interests. Moreover, a solidarity of interests builds the trust necessary to maintain a solidarity of sentiments. As Bell also notes,

> If you do not come through, if you violate my trust or if I violate yours, chances are my affection for you and your affection for me will soon disappear—as well as our sense of a common normative commitment.[25]

Sustained collective action is an interactive process in which solidarities of interests and solidarities of sentiments mutually constitute and reconstitute each other, a process Bell terms a *dialogue of solidarities* (Figure 24.1).

The cynic's retort to the idea of a dialogue of solidarities is that sentiments are an indefensible theoretical construct. Any affective or normative commitment,

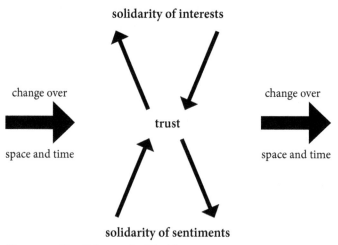

Figure 24.1. The dialogue of solidarities

says the cynic, is based ultimately on our interests in attracting the affection of others and our need to constrain our activities to fit social norms if we are to achieve our interests. Indeed, in recognition of these objectives and needs of interests, a number of rational choice theorists have tried to incorporate a theory of norms within a rational framework.[26] The result is to reduce sentiments to interests.

It is important to recognize that we often experience norms as a kind of constraint. We often decide not to violate norms because, frankly, we think we might not get away with it. The theory of the dialogue of solidarities does not deny the importance of normative constraint. Rather, we argue that it is necessary to distinguish between *normative constraint* and *normative commitment*, with the former a result of the interaction of interests and norms and the latter a result of the interaction of sentiments and norms. (And here we elaborate the model of the dialogue of solidarities outlined in Bell.)[27] Thus, it is important not to associate the following of norms only with the sentiments' side of the dialogue of solidarities. We often follow norms for which we feel very little sympathy. But nor should following norms be associated only with the interests side of the dialogue. In contrast to normative constraint, normative commitment represents a generalized concern for others that is not experienced as merely an imposition. (The generalized character of this concern, we note, is what distinguishes normative sentiments from the more specific concerns of affective sentiments.)

It is also important to recognize that sentiments are not merely sugar and roses. Hate, anger, and vindictiveness are equally sentimental orientations toward others. Indeed, the common negative affections of some toward others has served as a basis for a solidarity of sentiments equal in social power to sentimental solidarities built upon positive affections.

In other words, a dialogue of solidarities does not always result in collective action that would generally be described as socially "good" or "useful." But all collective action that persists over time and space is far more likely when people act on both ties of interests *and* ties of sentiments. Therefore, the dialogue of solidarities seems an important point of analytic entry into the dynamics of CPRs.

These dynamics often depend upon the interaction of ties of interests and sentiments seemingly far removed from a particular CPR. Without some preexisting basis in sentimental solidarity, it is difficult to form the basis of trust necessary even to begin to coordinate interests. Moreover, one of the surest signs that social actors commonly look for to verify that sentimental ties really exist is the actions of others in other settings. Our point is that a dialogue of solidarities needs to be understood within the context of a larger social system—within the often chaotic, shifting, and overlapping context of the *dialogues* of solidarities that constitute real social life. We now turn to applying the dialogue of solidarities to the case study.

Methods and Setting

... The Imazighen of the High Atlas are primarily farming people and therefore directly dependent upon the local natural resource base for their economic survival. The common property in the Imdrhas Valley consists mainly of pastures, rangelands, and almous (spring-fed areas where vegetation grows). Together, these areas provide a source of forage for livestock and are the main source of fuelwood, collected by girls and women. Some common property areas are restricted solely to sheep and goat grazing; others are strictly for fuelwood and forage collection.[28]

To sustain these activities, Imazighen have traditionally managed their common property with an ancient system of pasture and rangeland regulation, called the *agdal* system. *Agdal* is Tamazight for "to hold in reserve."[29] The term is used to refer both to the system of common property management and to the actual areas that are under *agdal* protection.

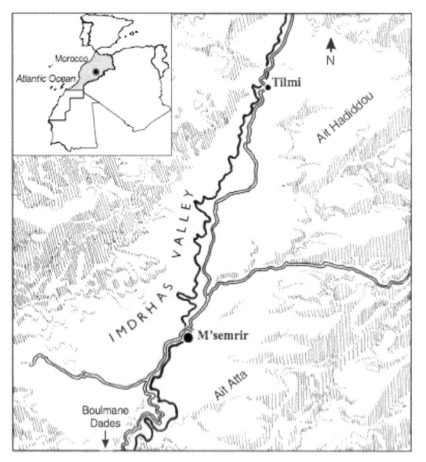

Figure 24.2. Imdrhas Valley, Morocco

In Tilmi and M'semrir, *agdal* use is organized through a representative council, known as the *jemaa*.[30] The *jemaa* supervises *agdal* use and establishes schedules for collecting forage and fuelwood. Community members are informed about the dates through word of mouth spread at the local mosque or at the weekly *suq* (market). The *jemaa* keeps the common lands closed if too much degradation has occurred, or if there has not been adequate rainfall, to allow regeneration of vegetation.

Borders of the various *igudlan* (the plural of *agdal*) are generally known to all in an area and members of certain tribal segments have the right to graze livestock and collect fuelwood in specific *igudlan*. The system is guarded by *nuadar*, two men in each village who are selected annually by the *jemaa*. The *nuadar* patrol the collective pasturelands to ensure that the number of livestock is kept in compliance, that girls and women are not in "shepherd's land," that shepherds are not in "women's land," and that those not of the tribal segment (as well as those from neighboring tribes) have not entered the land. *Nuadar* also patrol the common property to ensure that the opening and closing dates of the collective lands are respected. If violators are caught, they must pay a penalty, referred to as *izmas*.

These characteristics of the *igudlan* have all the design principles of successful CPRs presented by Ostrom:[31] an ownership arrangement in which management rules are developed, an established structure provides equitable distribution of benefits, distinct territorial boundaries are widely known, group size and rules are also known, and incentives and sanctions exist for co-owners to follow the accepted institutional arrangements. And for centuries the *agdal* system worked well in the High Atlas.

Yet in the Ait Atta village of M'semrir, the system is disintegrating, while it remains more stable in the Ait Hadiddou village of Tilmi. This disintegration in M'semrir is manifested in overstocking, in the common presence of livestock herds in women's fuelwood collection areas, in specific members of the village not being informed of opening dates of the common property, and in the recent inequitable redivision and privatization of portions of M'semrir's *agdal*. Some of this breakdown seems easy to ascribe to the socially erosive forces of colonialism and persistent drought that undermined the collective working of the *igudlan*.[32] Colonization by the French in the early 1930s and the region's long periods of drought are in line with the arguments of Bromley and Cernea[33] and Niamir[34] concerning the external factors that can impact CPRs. But M'semrir and Tilmi, as closely neighboring villages in the same valley, experienced these forces in much the same way. Yet current conditions of *igudlan* in the villages are sharply different. Moreover, they have experienced something else in quite a different way—the quality of the social ties within the two communities and the activities that maintain them. It is to the quality of social ties and their implications for the *igudlan* that we now turn.

Social Organization in the Imdrhas Valley

M'semrir is the first of the two communities one reaches along the dirt road into the Imdrhas Valley. The *Caid* (mayor-like figure) of the region resides there, as do the *Gendarme* (state police), doctor, and agricultural workers. Some of these officials are Imazighen, and some Arab, but all are from outside of the village. Government bureaus have been established in the village since the 1960s. Therefore, M'semrir in the last 30 years has become more than the village of the Ait Atta alone.

M'semrir is also where the large suq for the surrounding villages takes place. And it is the point tourists stop to drink mint tea before either heading back down the valley or continuing on the loop of a popular gorge trip. As a result of this tourist influx, several cafes and hotels have recently been established in the village.

Introduction of cash crops (apples and potatoes) in the last two decades by the Ministry of Agriculture has also brought changes. Upon returning from the fields one day, Nejla pointed out former communal property that is now privatized. When asked why they split up the land, she replied, "Everyone wanted their own land, so they could grow whatever crops they choose." The introduction of these crops has encouraged the push for further privatization of the communal land.

A greater market orientation in cattle grazing has also promoted privatizing M'semrir's *igudlan*. Some 90 percent of the imported cows in the Imdrhas Valley are in M'semrir.[35] The cattle are kept in individual compounds where they are carefully fed so farmers can better recoup their investments. Meanwhile, much of the area of the *agdal* that was once communal pasturage has been divided into private fields.

The Imazighen notion that "wealth is on the hoof"[36] still holds true—but with a greater emphasis on the wealth part of the equation than ever before. The money gained from marketing cash crops and selling imported cattle has also been used by many local farmers to increase their herds of sheep—an animal that is regarded as a "bank"—and goats that they graze on the remaining communal lands. One area extension agent estimated stocking rates of sheep and goats in the *igudlan* of M'semrir are 50 to 100 percent higher than the level established for the area. This, then, directly affects overall carrying capacity and increases pressure and cheating on the lands still held in common.

The prevalence of cheating was made evident one spring morning when walking with Tooda, a local woman, to collect fuelwood high in the mountains. Tooda pointed out sheep manure and noted that it should not be there since this was "women's" land. Upon returning to the village, Tooda pointed out shepherd's huts that had been newly erected on the communal property. "They build these at night," she said, "so that in the morning they can say, 'I have a house here, so this is my land.'"

Moreover, 25 percent of the males have migrated from M'semrir since the 1960s,[37] sending money back from their work in Holland, France, and larger cities of Morocco. This money has been used to establish cafes, purchase televisions, and buy imported cattle. The flow of money from outmigrants is leading to marked socioeconomic and cultural differences within the community. As one walks around M'semrir, many visual cues to these differences stand out in the landscape and in the passing populace: television antennas, cement houses instead of the indigenous mud and straw types, women and girls donning cotton shawls instead of the indigenous wool shawls—all indicators of new wealth and a changing culture.

The local political structure has also changed in M'semrir. Traditionally, the villagers elected the leader of the *jemaa*, the Amghrar. Today, the Amghrar is still elected by the people, but he (it is always a he) is only installed in that position if desired by the Caid. The Caid is now appointed from the ranks of state government—not by the local Imazighen tribal leaders as in the first decades of this century.[38] And the people can no longer vote the Amghrar out, unless the Caid agrees. Consequently, the actions of the Amghrar have shifted from representing and enacting the wishes of the villagers to legitimating the authority structure of the central state and the king.

Disagreement over M'semrir's current Amghrar has deeply divided the community, and songs have been made up about his actions. One song refers to his ill division of the communal land, where he allegedly gave the land only to those he liked. The villagers laugh as they sing the song, but their bitterness is evident as they discuss the inequitable distribution. The Amghrar has also neglected to inform the entire community of the opening dates for the communal grazing land. For example, in one specific area of the common land, villagers are given one day to collect forage before livestock is taken there to graze. Villagers and government officials alike said the area would not be opened in the summer of 1993, as there had not been enough rain to adequately restore the vegetation. One day in July, while Amina was talking about this area she said, "They didn't open it this year. There's no rain and no irrigation canal there."

Loho, her daughter-in-law who was preparing tea, interrupted, "Abisha told me people are going."

Amina responded, "They'll have to pay izmas if they are caught."

"No," said Loho, "it's open. They just didn't tell everyone."

Apparently, the Amghrar had allowed those who were "on his side" to enter the common property area. Those not "on his side" were told the area was closed. A Ministry of Agriculture official confirmed this, saying, "It is true. And the people cannot do anything, because the Amghrar is a friend of the Caid."

Amina echoed this. Thus, collective land in M'semrir is now being used as an instrument of exclusion by the Amghrar. And since the Caid now effectively controls the Amghrar, those villagers who do not like his actions feel helpless, but they are unable to vote him out.

Thirteen kilometers up the mountain valley the people of Ait Hadiddou live in their village of Tilmi. The road becomes increasingly worse as one ascends the valley. The small market in Tilmi does not encourage many merchants into the village, and there are no cafes or hotels to entice the few tourists who do venture this far. The Caid comes to the village once a week (if that) on market day, as does the doctor.

Only 4 percent of males have migrated to larger cities.[39] Less than 10 percent of the cattle here are imported, and the cash crops grown in M'semrir do not fare as well in the agronomic conditions of Tilmi, where there are many springs and wet conditions do not encourage cultivation of these crops.

Due to the small number of tourists, difficulties in growing cash crops, and the minimal number of men working outside of the village, there are fewer opportunities for capital accumulation in Tilmi. Purchase of new technologies is kept at a minimum, and the status symbols seen in M'semrir are not so prevalent. There are few cement houses, few television antennas, and few women and girls who wear cotton shawls and gold. Consequently, within Tilmi there appears to be little variation in economic status, resulting in less social stratification.

Tilmi has also maintained the more traditional forms of local democracy. Disputes still occur here, but villagers did not appear to have a problem telling the Amghrar when they are not happy with him. One day a fellow villager, Said, came to discuss irrigation issues with the Amghrar. Over tea, Said explained the situation. Unhappy with the Amghrar's response, Said shook his head and said, "The Amghrar is no good." While Said didn't go away happy that day and perhaps is not pleased with the Amghrar's effectiveness as a local leader, he continues to visit the Amghrar, and still considers him somewhat of a friend.

When the Amghrar is away, Abo, his wife, takes over. One day Abo stood in her courtyard, talking with one villager while another waited outside for his turn. It became obvious Abo was resolving a dispute. After the two men left Abo commented, "I'm a *Tamghrart*," making a play on words. Tamghrart could mean "female amghrar" (of which there are none) but more commonly means "old woman." The point is that Tilmi is a place where people have less fear of speaking their minds to those in power, where in special circumstances those in power are women, and where women sometimes feel able to contest gender relations in the village, at least relatively empowered women like Abo do.

Traditionally, when work of mutual benefit needed to get done, Imazighen communities practiced *touiza*, volunteer labor that occurs both at the community and the family level. When asked about touiza villagers in M'semrir repeatedly said, "How long ago." In M'semrir, each family now harvests its own crop, or helps and is helped by one or two other households, and community work like fixing the road is now done by paid laborers. But the role of touiza is still an active one in Tilmi. When the harvesting of cereals begins, groups of approximately 15 girls and women will gather together and work consecutively on each other's fields, singing as they work. Touiza also extends into nonagricultural activities, such as clearing the road of snow. When such work needs to be done, the Amghrar's assistant will go through the village, hollering for workers to come help. A group of available males will work together, chanting and singing, in much the same manner as the females when they are in the fields or gathering fuelwood. In Tilmi, generalized reciprocity (exchange without expectation of direct equal return) has remained as touiza, while in M'semrir, touiza has become direct reciprocity (expectation of equal return).[40]

In addition to touiza, dancing also remains a common feature of everyday life in Tilmi. When a major life event occurs, such as circumcisions or marriages, the dancing continues for days. Moreover, the Ait Hadiddou organize these events as a community. With circumcision, the Ait Hadiddou perform a community-wide ceremony followed by three days of celebration, with dancing every evening and the talib reading from the Koran during the day. But in M'semrir, only one or two households at a time get together for circumcision ceremonies. And while weddings occur throughout the summer in M'semrir, among the Ait Hadiddou most weddings take place during a single week in October. This is for practical purposes (to prevent conflict with harvest time and so guests traveling a distance do so only once a year), but also to celebrate the event at a community level. Villagers are invited by the throwing of *amzeet* (almonds, dates, and figs) off the roofs of the grooms' homes to the crowd waiting below. On the first day of the ceremony, all the brides proceed into the village together. After dinner, they gather outside, each holding a light or candle, and stand there while the community dances until 2:00 or 3:00 in the morning. The next day, the dancing begins again in the afternoon.

The Ait Hadiddou of Tilmi appear to *like* each other. And they continually show it, simultaneously demonstrating and rebuilding their social ties. They work together. They celebrate together. They sing together. They dance together.

Perhaps most striking is the fact that when the weather permits, some of the people of Tilmi gather outside for singing and dancing to celebrate the sunset and the end of another day. As much as this activity fits every romantic cliché about traditional life before the age of globalization, marketization, modernization, and rationalization, it is a remarkable empirical reality in Tilmi.

Most significant for our topic here, the Ait Hadiddou still manage their land together. While some former communal land has been converted to private plots, the *igudlan* surrounding Tilmi are for the most part still intact. For example, during the months of June, July, and August, everyone who owns cattle and wishes to participate brings them to a public gathering place. The herd is collected and taken en masse to the communal land, where it is guarded and cared for by each participating family in turn. Rotational grazing is practiced, forage from the mountains and foothills is conserved, and the stocking rates remain closer to sustainable levels.[41] The issues heard in M'semrir regarding the communal property, such as cheating and inequitable decisions, were not witnessed in Tilmi.

We believe that patterns of interests and sentiments and the solidarities built upon them interact in these two Imazighen communities. In Tilmi, broad-based solidarities of interests are closely associated with equally broad-based solidarities of sentiments. The Ait Hadiddou retain patterns of local political participation, touiza, and *igudlan* that take into account the interests of others in the community. As well, they continue to enact patterns of open ritual, open political criticism, and open informal interaction that demonstrate their positive sympathies for each other. In M'semrir, however, there is a conspicuous lack of community-wide solidarities of interests. The relatively private orientation of ritual and social interaction indicate that suspicion and distrust have riven the web of sentimental ties in M'semrir as well.

To be sure, it is empirically quite difficult to document a direct connection between singing and dancing together in the evening as the sun sets and the collective decision to honor the *jemaa*'s proscriptions about where to collect fodder and when to enter the communal land. But that lack of a direct connection is part of the social power of sentimental solidarities. If the connection was immediately clear, it would reduce sentimental acts to mere interest, and if interest is the sole motivating factor, there would be no need for sentimental acts to begin with. Local interaction would amount to no more than bargaining, direct exchanges of labor and information, and the assertion of power.

Our point is not that Tilmi is all self-sacrifice for the good of the whole, but here solidarities of interests interweave and overlap with solidarities of sentiments, which seems to be what is now missing in M'semrir. Unfortunately, we have no direct historical evidence on the quality of social ties in M'semrir in the years before the tourists and the imported cows came to the village and before the Caid became the agent of the King. But local people suggest that M'semrir was once far more like Tilmi is today, at least in some aspects.[42]

All of which is not to say that sentimental ties no longer play an important role in M'semrir. Some people in M'semrir *are* told when and where the communal pastures are opened up for forage collection, and likely the families of those people would turn to each other to arrange a circumcision celebration. Ties of interests and sentiments still dialogically interact in M'semrir. What has changed is that those interactions have fractured along zones of exclusion, hierarchy, and hostility.

Conclusion

The long-term maintenance of the solidarity of interests created by a CPR depends equally upon a corresponding solidarity of sentiments—on an interweaving and interacting system of ties of interest and sentiment. Ties of interest are socially embedded, as others have written. Yet we need to recognize that they are embedded not only in other ties of interest but also sentimental ties of affection and normative commitment. . . .

Life is not ideal in Tilmi. People are poor. Women have far lower social power and harder lives than men. Moreover, many of the communal rituals of collective sentiment are on the wane. Weddings have begun to take place at other times of the year. Often residents remark that the crowd at the evening dance used to be much larger; at one time, nearly everyone from the village would attend. But when attempting to understand what enables a CPR to retain its vitality, examination of the larger social system and its dialogic ties seems essential. These are the crucial points that the Ait Hadiddou make to an increasingly rationalistic world: that people who like each other generally get along better, and that singing and dancing together after the sun has set and partaking in the collective celebration of important events helps keep the grass green for everyone.

NOTES

1. Elinor Ostrom, *Governing the Commons: The Evolution of Institutions for Collective Action* (Cambridge: Cambridge University Press, 1992); Elinor Ostrom "Community and the Endogenous Solution of Common Problems," *Journal of Theoretical Politics* 4 (1990): 343–351.

2. Glenn G. Stevenson, *Common Property Economics: A General Theory and Land Use Applications* (Cambridge: Cambridge University Press, 1991).

3. Todd Sandler, *Collective Action: Theory and Applications* (Ann Arbor: University of Michigan Press, 1992).

4. Michael M. Bell, "The Dialogue of Solidarities, or, Why the Lion Spared Androcles," *Sociological Focus* 31 (1998): 189–199; Michael M. Bell *An Invitation to Environmental Sociology* (Thousand Oaks, CA: Pine Forge, 1998); Amartya Sen, "Rational Fools: A Critique of the Behaviourial Foundations of Economic Theory," *Philosophy and Public Affairs* 6 (1977): 317–344; Neil Smelser, "The Rational Choice Perspective: A Theoretical Assessment," *Rationality and Society* 4 (1992): 381–410.

5. Bell, "Dialogue of Solidarities"; Bell, *Invitation to Environmental Sociology*.

6. The indigenous people of Morocco. Literally meaning "free man," the term "Berber," while commonly used, was imposed on the Imazighen and is not their own term.

7. Michael Goldman, *Privatizing Nature* (New Brunswick, NJ: Rutgers University Press, 1998).

8. Mancur Olson, *The Logic of Collective Action: Public Goods and the Theory of Groups* (Cambridge, MA: Harvard University Press, 1965), 2.

9. Garrett Hardin, "The Tragedy of the Commons," *Science* 162 (1968): 1244.

10. Ostrom, "Community and the Endogenous Solution."

11. Goldman, *Privatizing Nature*.

12. Ostrom, "Community and the Endogenous Solution," 27.

13. David W. Bromley and Michael M. Cernea, "The Management of Common Property Natural Resources," World Bank Discussion Paper 57 (Washington, DC: World Bank, 1989).

14. Maryam Niamir, *Community Forestry: Herders' Decision-Making in Natural Resources Management in Arid and Semi-Arid Africa* (Rome: FAO, 1990).

15. Bromley and Cernea, "Management"; Niamir, *Community Forestry*.

16. Mark Granovetter, "Economic Action and Social Structure: The Problem of Embeddedness," *American Journal of Sociology* 91 (1985): 481–510.

17. Alejandro Portes and Julia Sensenbrenner, "Embeddedness and Immigration: Notes on the Social Determinants of Economic Action," *American Journal of Sociology* 98 (1993): 1320–1350.

18. Robert J. Fisher, "Indigenous Forest Management in Nepal: Why Common Property Is Not a Problem," in Michael Allen, ed., *Anthropology of Nepal: Peoples, Problems and Processes* (Kathmandu: Mandala Book Point, 1994), 71.

19. Randall W. Ireson, "Village Irrigation in Laos: Traditional Patterns of Common Property Resource Management," paper for the annual meeting of the Rural Sociological Society, Columbus, Ohio, August 19–21, 1991.

20. Robert Wade, *Village Republics: Economic Conditions for Collective Action in South India* (Cambridge: Cambridge University Press, 1988).

21. Ireson, "Village Irrigation," 12.

22. Robin Mearns, "Community, Collective Action and Common Grazing: The Case of Post-Socialist Mongolia," *Journal of Development Studies* 32 (1996): 297–339.

23. Bell, "Dialogue of Solidarities"; Bell, *Invitation to Environmental Sociology*.

24. Bell, "Dialogue of Solidarities," 183.

25. Ibid.

26. Ostrom, "Community and the Endogenous Solution"; Jon Elster, *The Cement of Society: A Study of Social Order* (Cambridge: Cambridge University Press, 1989).

27. Bell, "Dialogue of Solidarities"; Bell, *Invitation to Environmental Sociology*.

28. Common property can be located within village boundaries, surrounding the village, or in high-altitude pasturelands. Our discussion focuses on that within and surrounding the village.

29. Claude Lefebure, "Acces aux Ressources Collectives et Structure Sociale: L'estivage chez les Ayt Atta (Maroc)," in *Production Pastoral et Societe* (Paris: Editions de la MSH / Cambridge University Press, 1979).

30. The *jemaa* is a village council of male elders. After King Hassan II's ascent to the throne in 1961, a strong effort to change the political and social structure of Morocco began. In an effort to break the political and military power of the Imazighen, the government attempted to disband the *jemaa* and created the Rural Commune, an organization that overlooks a specific region rather than a particular village. M'semrir and Tilmi are in different rural communes, but under the same Caid jurisdiction. While in some areas of Morocco the power of the *jemaa* has been severely weakened, villagers consistently noted it was the jemaa that dealt with issues regarding *igudlan* (plural of *agdal*).

31. Ostrom, "Community and the Endogenous Solution."

32. Jeanne Chiche, "Les Pratiques de l'Usage des Ressources Communes," in Alain Bourbouze and Roberto Rubino, eds., *Terres Collectives en Mediterranee* (Rome: FAO, 1992), 41–56.

33. Bromley and Cernea, "Management."

34. Niamir, *Community Forestry*.

35. Ofice Regional de Mise en Valeur Agricole (ORMVAO), *Projet Integre De Developpement Agricole Du Sous Bassin Du Dades* (Ouarzazate, Morocco: ORMVAO, 1992).

36. David M. Hart, *Dadda 'Atta and His Forty Grandsons* (Cambridge, UK: Middle East and North African Studies Press, 1981), 93.

37. ORMVAO, *Projet Integre*.

38. Negib Bouderbala, "La Terre Collective au Maroc: Droite et Fait," in Alain Bourbouze and Roberto Rubino, eds., *Terres Collectives en Mediterranee* (Rome: FAO, 1992), 28–41.

39. ORMVAO, *Projet Integre*.

40. Marshall Sahlins, *Stone Age Economics* (Chicago: Aldine-Atherton, 1972).

41. ORMVAO, *Projet Integre*.

42. There has always been a historical difference between the tribes in religion and related ideas of proper behavior. Islam is less infused among the Aid Hadiddou, while the Ait Atta of M'semrir view the dancing of the Ait Hadiddou as "shameful" and not proper Muslim behavior. As Hart notes, the "devotion and piety of the Ait Atta is beyond question. They also see themselves as very good Muslims." David M. Hart, *The Ait 'Atta of Southern Morocco: Daily Life and Recent History* (Cambridge, UK: Middle East and North African Studies Press, 1984), 98.

25

Averting the Tragedy of the Commons

MARK VAN VUGT

Within a short (evolutionary) time frame, *Homo sapiens* has become a global force dominating the natural world. Currently the human population worldwide amounts to 6.6 billion, and it is expected to rise to almost 9 billion by 2050. It is doubtful whether the Earth's ecosystems can sustain such large numbers, particularly at the current standard of living. Human activities are responsible for depleting natural resources, polluting the environment, and reducing biodiversity. Human-made environmental problems create economic and social conflicts with potentially devastating consequences for the health and well-being of ourselves and future generations. This is nothing new. Our species has had a long history of causing ecological destruction; yet due to a rise in population and technological know-how, these effects are now felt globally.

It is widely accepted that we need to move toward greater environmental sustainability. Yet making the necessary changes has proved very difficult, in part because there are conflicting interests between relevant parties.[1] As the World Commission on Environment and Development recognized a while ago: "The Earth is one, but the world is not."[2]

The Tragedy of the Commons

The social dynamics underlying many environmental challenges are famously captured by Garrett Hardin in an article in *Science* titled "The Tragedy of the Commons," one of the most frequently cited works in the social sciences.[3] The essay tells the story of how the management of a communal pasturage by a group of herdsmen turns into ecological disaster when each individual, upon realizing that adding extra cattle benefits him personally, increases his herd, thereby unintentionally causing the destruction of the commons.

The tragedy of the commons has become central to our understanding of many local and global ecological problems. As an evolutionary biologist, Hardin argued that nature favors individuals who exploit common resources at the expense of the more restrained users. He also argued that voluntary contributions to create institutions for managing the commons often fall short because of (the fear of) free-riders. To save the commons, Hardin therefore recommended "mutual coercion, mutually agreed upon by the majority of the people affected."[4]

The tragedy of the commons has generated much research activity in the behavioral sciences, from psychology to political science and from economics to biology. But despite its compelling logic, it has been criticized for two main reasons. First, scientists studying real-world environmental problems have found many instances of successful community-resource-management projects around the world, such as the maintenance of common agricultural land, irrigation systems, and lake and shore fisheries.[5] Rather than a "free for all," these commons are strictly regulated in terms of access and intensity of use. A second more fundamental criticism concerns the validity of the assumption that commons users are driven exclusively by narrow (economic) self-interest. Although this is clearly an important motive, recent theoretical and empirical developments in social psychology, evolutionary biology, anthropology, and experimental economics suggest that individuals are not indifferent to the welfare of others, their group, or the natural environment. Using experimental game paradigms, such as the prisoner's dilemma, the public goods dilemma, or the commons dilemma (the latter is also known as the resource dilemma or common pool resource game), researchers have discovered myriad motives beyond self-interest that influence decision making in commons dilemmas.[6]

Key Strategies for Protecting the Environment

In combination with field data, the experimental games literature suggests four key components of strategies for successful resource management: information, identity, institutions, and incentives. These four I's correspond, by and large, to four core motives for decision making in social dilemmas: understanding, belonging, trusting, and self-enhancing (for an overview, see Table 25.1). These motives are fundamental psychological processes—likely shaped by the evolutionary selection pressures—that influence our thinking, feeling, and behaving in social interactions.[7]

Information

People have a fundamental need to understand their environment to predict what will happen in case of uncertainties. Environmental uncertainty tends to promote overuse because most users are optimistic about the future and underestimate the damage they are doing to the environment.[8] Managing environmental resources therefore depends first and foremost on reliable information about the use and availability of resources like, for instance, drinking water, fossil fuels, and fish stocks. Science plays a vital role in reducing environmental uncertainty. Gathering reliable information is much easier when resources have clearly defined boundaries (e.g., land is easier to control than water or air).

TABLE 25.1. Four I's: Core Motives and Foci of Interventions of Successful Commons Resource Management and Potential Constraints

Focus of intervention	Core motive	Description	Aim of intervention	Potential constraint
Information	Understanding	The need to understand the physical and social environment	Reducing environmental and social uncertainty	Global environmental problems are inherently uncertain
Identity	Belonging	The need for positive social identity	Improving and broadening one's sense of community	Resource competition between communities increases overuse
Institutions	Trusting	The need to build trusting relationships	Increasing acceptance of commons rules and institutions	Authorities are not always seen as legitimate and fair
Incentives	Self-enhancing	The need to improve oneself and increase one's resources	Punishing overuse and rewarding responsible use	Economic incentives undermine intrinsic motivation to conserve

Global environmental trends are highly complex and uncertain, which undermines effective behavioral change. In contrast, information about local environmental destruction is generally more persuasive, in part because the contingencies between actions and outcomes are easier to understand. A perceptible local resource threat such as an acute food or water shortage is an example. My colleagues and I conducted a survey among 120 households during the 1997 water shortage in the United Kingdom and found that the perceived severity of the shortage was positively associated with households' efforts to conserve water.[9] People's attributions of the causes underlying the water shortage made a difference. When people believed the shortage was caused by other households, they consumed more (and conserved less) water than when they believed it was caused by the weather. In addition, people made more efforts to conserve when they believed their own contribution made a difference in alleviating the crisis (cf. self-efficacy).

It appears that, when crafting messages to raise public awareness about environmental matters, simple information is often most effective—particularly when decision makers are already contemplating changing their behavior. For instance, labels with comparative information about energy use and emissions of household appliances work best when consumers are already thinking "green" but lack specific technical knowledge. Environmental and social scientists must work more closely together to enhance people's understanding of environmental problems and to design public campaigns providing accurate information.[10]

Identity

As a group-living species, humans have a deep sense of belonging to social groups. Research suggests that people easily identify with and form attachments with other individuals in sometimes very large groups.[11] The strength of their social identity affects how much people are willing to help their group or community, for instance in protecting the environment.[12] High-identifying group members sometimes even compensate for the resource overuse of fellow group members.[13]

There are several ways in which people's identity and belongingness needs could be mobilized to foster proenvironmental action. First, people identify strongest with primary groups such as friends and family, and therefore an appeal to the interests of those groups will generally be more persuasive (e.g., "think of your children's future"). In addition, when people identify with a group, they are more likely to share costly environmental information.[14] For instance, in comparing lobster-fishing communities in Maine it was found that fishermen in communities with dense social networks exchanged catch information more frequently than did those in more loosely connected communities, resulting in more sustainable fishing.[15]

Third, when people identify with a social group they are more concerned about their in-group reputation, and this can promote proenvironmental action.[16] Asking households to make a public commitment, for instance, reduces energy use by 20%.[17] Providing households with normative social feedback—sticking a "smiley" or "frowney" face on their home energy bill when their energy use is less or more than the neighborhood average—leads to similar reductions.[18] Finally, environmental pressure groups routinely, and with some success, apply reputation tactics in so-called "naming and shaming" campaigns to force polluting companies to change their policies.

Human belongingness needs are embedded within a marked in-group/out-group psychology. Many studies show that our social identities are boosted through inducing competition, either real or symbolic, between groups.[19] Yet creating intergroup competition in environmental dilemmas can be a double-edged sword. Resources that are shared between several communities, such as river irrigation systems or sea fisheries, are generally at greater risk of depletion.[20] In such cases, it would be helpful to promote a superordinate social identity—for instance by promoting trade between the communities or by emphasizing a common threat such as the collapse of the local economy.

Institutions

A third condition for successful resource management is the presence of legitimate commons institutions. Authorities play a key role in governing local and global environmental resources, but who is prepared to trust and empower them? Institutions are essentially public goods that are in danger of being undermined by free-riders, individuals who profit from their existence but don't contribute to their upkeep. One way out of this dilemma is to appoint a leader or authority to regulate resource access (the Hardin solution). Yet this creates a second-order free-rider problem also known as the "who guards the guards" paradox: How can authorities be trusted to look after the common good?[21]

Trust is a core motive in social relationships.[22] Having confidence in the benevolence of other individuals and institutions lies at the heart of any collective effort to protect the environment. Commons users generally trust others to exercise voluntary restraint, but if institutional changes are necessary (e.g., during a resource crisis) they want leaders and authorities that can be trusted to look after the common good.

To get trust, authorities must employ fair decision-making rules and procedures. Regardless of whether people receive bad or good outcomes, they want to be treated fairly and respectfully. A study on the 1991 California water shortage[23] showed that Californians only cooperated with local water authorities in implementing drastic water-saving measures if they believed the authorities made efforts to listen to their concerns and provide accurate, unbiased information about the shortage. Moreover, procedural concerns were particularly important for residents with a strong sense of community identity. A survey on the 1994 British railway privatization found that train users who did not trust private companies to look after this public good were more likely to take a car instead.[24] Thus, trust in institutions plays a crucial role in managing urgent and complex environmental challenges.

Incentives

There is no denying that many proenvironmental actions are driven by self-enhancing motives, notably the desire to seek rewards and avoid punishments. Monetary incentive schemes in the form of subsidies appear effective in fostering the adoption of home saving devices such as solar panels, water meters, and roof insulation. Financial incentives also promote sustainable practice within industry. In the United States, market-based systems of tradable environmental allowances have become quite popular in recent years. This scheme permits companies to buy and sell "pollution" credits, and this system is believed to have contributed to a decline in acid rain.[25] Furthermore, in applying penalties for environmental damage, it seems better to start with a modest punishment and then gradually increase it after repeated violation, such as with catch quotas in fisheries.[26]

The core-motives approach provides various novel insights into why particular incentive schemes might work better than others and why some might

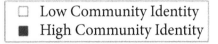

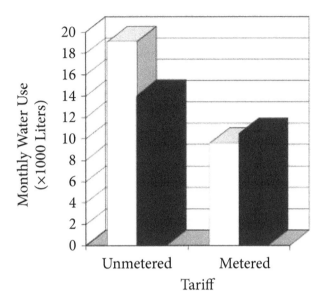

Figure 25.1. Average monthly water use among a sample of 593 households in the United Kingdom in 1997 (data are corrected for income, household size, and pre-meter use). Water use varied between households as a function of tariff (metered vs. unmetered) and level of community identity (high vs. low). Adapted from M. Van Vugt, "Community Identification Moderating the Impact of Financial Incentives in a Natural Social Dilemma: A Water Shortage," *Personality and Social Psychology Bulletin* 27 (2001): 1440–1449. Copyright 2001, Society for Personality and Social Psychology. Adapted with permission.

not work at all. First, not everyone is equally motivated by economic self-interest.[27] In a water-conservation study, I asked households to complete a short community-identity scale,[28] with statements such as "I feel strongly attached to this community" and "There are many people in my community whom I think of as good friends" (1 = strongly disagree, 5 = strongly agree). Water records (corrected for various demographic variables and previous use) showed that households that identified strongly with their community did not need a financial incentive (through a metered tariff) to consume less water but those that weakly identified with their community did (see Fig. 25.1). This implies that economic incentives work better when other core needs are unfulfilled.

Second, interventions that fulfill various core motives simultaneously are likely to be most successful. The Van Vugt and Samuelson[29] study showed that, during a shortage, conservation efforts were highest among households with a water meter, because having a meter both gave them a financial incentive to conserve (thus furthering their self-enhancement) and enhanced their knowledge about appropriate water-saving measures (thus furthering their understanding). Thus, metered households were better able to adjust their behavior during the shortage.

Third, incentive schemes might be counterproductive if they undermine other core needs. Handing out small fines for littering might signal that the problem is more widespread than it actually is (undermining the need for trust) or transform it from an ethical-environmental issue into an economic issue (altering the understanding of the problem).[30] Particular incentive schemes might also create mistrust in authorities. When the Dutch government built a special lane for carpoolers in 1993 along one of the busiest highways in the Netherlands, it cut travel times substantially for car sharers. Yet single drivers reacted strongly against the lane, and after widespread protest and a legal challenge the lane closed within a year. Survey data suggested that many drivers did not trust the intentions of the authorities, and whereas some single drivers showed open resistance against the lane (by going to court to get access to it), others showed more subtle attitudinal shifts in favor of driving their car alone.[31]

Conclusions and Implications

More research is needed to establish the interplay between these core motives in shaping environmental decisions. For instance, do people with strong community ties also have better knowledge of local environmental problems? Do sanctioning schemes enhance or undermine people's trust in commons institutions and in other users? Individual differences in core motives may also matter. We already know that environmental appeals are more persuasive among car drivers with cooperative dispositions.[32] Similarly, I suspect that people with high belongingness needs will be influenced more by community-based incentive schemes, whereas individuals with low belongingness needs might respond better to individual financial incentives. Also, we know very little about how these core motives change across the lifespan: Do people's belongingness needs become weaker or stronger as they grow older, and how do their social networks change? Finally, are there other central motives shaping people's decision making in commons dilemmas, such as autonomy or caring needs? We know that humans evolved on the savannah in Africa, and living in this environment may have endowed us with "biophilia,"[33] an innate tendency to enjoy and care for the natural world. Across cultures, people are attracted to the same savannah-type landscapes, and in both Europe and the United States, zoos attract more visitors annually than all professional sports events combined. Exposing children to enjoyable social outdoor experiences such as camping, trekking, or scouting may promote their lifelong environmental commitment. To develop these and other interventions to protect our environment and avert a commons tragedy requires a good understanding of human nature, which social psychology can provide.

NOTES

1. T. Dietz, E. Ostrom, and P. C. Stern, "The Struggle to Govern the Commons," *Science* 302 (2003): 1907–1912.
2. World Commission on Environment and Development, *Our Common Future* (Oxford: Oxford University Press, 1987), 27.
3. G. Hardin, "The Tragedy of the Commons," *Science* 162 (1968): 1243–1248.
4. Ibid., 1247.
5. E. Ostrom, *Governing the Commons: The Evolution of Institutions for Collective Action* (Cambridge: Cambridge University Press, 1990).
6. M. Weber, S. Kopelman, and D. M. Messick, "A Conceptual Review of Social Dilemmas: Applying a Logic of Appropriateness," *Personality and Social Psychology Review* 8 (2004): 281–307.
7. S. T. Fiske, *Social Beings: A Core Motives Approach to Social Psychology* (Hoboken, NJ: Wiley, 2004).
8. S. Opotow and L. Weiss, "New Ways of Thinking about Environmentalism: Denial and the Process of Moral Exclusion in Environmental Conflict," *Journal of Social Issues* 56 (2000): 475–490.
9. M. Van Vugt and C. D. Samuelson, "The Impact of Metering in a Natural Resource Crisis: A Social Dilemma Analysis," *Personality and Social Psychology Bulletin* 25 (1999): 731–745.
10. Dietz et al., "Struggle."
11. R. F. Baumeister and M. R. Leary, "The Need to Belong: Desire for Interpersonal Attachment as a Fundamental Human Motivation," *Psychological Bulletin* 117 (1995): 497–529.
12. M. Van Vugt, "Community Identification Moderating the Impact of Financial Incentives in a Natural Social Dilemma: A Water Shortage," *Personality and Social Psychology Bulletin* 27 (2001): 1440–1449.
13. M. B. Brewer and R. M. Kramer, "Choice Behavior in Social Dilemmas: Effects of Social Identity, Group Size and Decision Framing," *Journal of Personality and Social Psychology* 3 (1986): 543–549.
14. D. De Cremer and M. Van Vugt, "Social Identification Effects in Social Dilemmas: A Transformation of Motives," *European Journal of Social Psychology* 29 (1999): 871–893.
15. D. J. Penn, "The Evolutionary Roots of Our Environmental Problems: Toward a Darwinian Ecology," *Quarterly Review of Biology* 78 (2003): 275–301.
16. C. Hardy and M. Van Vugt, "Nice Guys Finish First: The Competitive Altruism Hypothesis," *Personality and Social Psychology Bulletin* 32 (2006): 1402–1413; M. Milinski, D. Semmann, H. Krambeck, and J. Marotzke, "Stabilizing the Earth's Climate Is Not a Losing Game: Supporting Evidence from Public Goods Experiments," *Proceedings of the National Academy of Sciences* 103 (2006): 3994–3998.
17. Penn, "Evolutionary Roots."
18. P. W. Schultz, J. M. Nolan, R. B. Cialdini, N. J. Goldstein, and V. Griskevicius, "The Constructive, Destructive, and Reconstructive Power of Social Norms," *Psychological Science* 18 (2007): 429–434.
19. De Cremer and Van Vugt, "Social Identification Effects."
20. Ostrom, *Governing the Commons*.

21. R. O. O'Gorman, J. Henrich, and M. Van Vugt, "Constraining Free-Riding in Public Goods Games: Designated Solitary Punishers Can Sustain Human Cooperation," *Proceedings of Royal Society-B* 276 (2008): 323–329.

22. Fiske, *Social Beings*.

23. T. R. Tyler and P. Degoey, "Collective Restraint in Social Dilemmas: Procedural Justice and Social Identification Effects on Support for Authorities," *Journal of Personality and Social Psychology* 69 (1995): 482–497.

24. M. Van Vugt, P. A. M. Van Lange, R. M. Meertens, and J. A. Joireman, "How a Structural Solution to a Real-World Social Dilemma Failed: A Field Experiment on the First Carpool Lane in Europe," *Social Psychology Quarterly* 59 (1996): 364–374.

25. Dietz et al., "Struggle."

26. Ostrom, *Governing the Commons*.

27. P. A. M. Van Lange, E. De Bruin, W. Otten, and J. A. Joireman, "Development of Prosocial, Individualistic, and Competitive Orientations: Theory and Preliminary Evidence," *Journal of Personality and Social Psychology* 73 (1997): 733–746.

28. Van Vugt et al., "How a Structural Solution."

29. Van Vugt and Samuelson, "Impact of Metering."

30. A. E. Tenbrunsel and D. M. Messick, "Sanctioning Systems, Decision Frames, and Cooperation," *Administrative Science Quarterly* 44 (1999): 684–707.

31. Cf. cognitive dissonance reduction; Van Vugt et al., "How a Structural Solution."

32. Van Vugt et al., "How a Structural Solution."

33. E. O. Wilson, *The Creation: An Appeal to Save Life on Earth* (New York: Norton, 2006).

26

Excerpts from "Climate, Collective Action and Individual Ethical Obligations"

MARION HOURDEQUIN

In recent papers, Walter Sinnott-Armstrong[1] and Baylor Johnson[2] have argued that under current circumstances, individuals do not have obligations to reduce their personal contributions to greenhouse gas (GHG) emissions. Johnson argues that climate change has the structure of a tragedy of the commons, and that there is no unilateral obligation to reduce emissions in a commons. Rather, one's moral obligation is to work toward a collective agreement that solves the problem. . . .

This paper challenges the conclusions of Johnson and Sinnott-Armstrong, arguing that although we have moral obligations to work toward collective agreements that will slow global climate change and mitigate its impacts, it is also true that individuals have obligations to reduce their personal contributions to the problem. . . .

A relational conception of persons provides an alternative framework in which it is possible to re-conceptualize collective action problems in a way that dissolves the stark contrast between the individually and the collectively rational. The relational perspective, which I develop by drawing on ideas in Confucian philosophy, emphasizes the role of self-cultivation and individual moral development as the basis for social change. Although this view initially appears subject to the kinds of objections that Johnson and Sinnott-Armstrong raise— that individual action in the absence of collective action achieves nothing—a subtle characterization of the view shows that Confucian self-cultivation is essentially social in nature. Individual moral development thus involves the support and instantiation of social institutions that make possible social transformation. I argue that the relational perspective developed in Confucian thought can fruitfully inform our approach to global climate change and help reconcile individual and political action to mitigate it. . . .

Recall, first, the logic of collective action problems that leads Johnson to conclude that unilateral action to reduce one's emissions in a tragedy of the commons will be fruitless, and even self-defeating:

> The only incentive players have is to maximize benefits from [their individual] use of the commons.

> The only way players can communicate is by increasing or reducing use of the
> commons.
>
> Use of the commons is shared, [however not all costs and benefits associated with
> use are shared.][3]

Commons problems presuppose that individuals are rational economic actors
who seek to take personal advantage of the commons to the greatest degree possible and who do not influence one another's thinking or decision making in
morally salient ways. A single individual's restraint will be exploited by others.
Thus, unilateral restraint is not only irrational, but morally impotent: unilateral
action will do nothing to save the commons from overexploitation.

A Confucian perspective on morality challenges this view. First, and perhaps
most importantly, Confucian philosophy does not understand the individual as
an isolated, rational actor. Instead, the Confucian self is defined relationally.[4]
Persons are constituted by and through their relations with others. According to
Confucianism, we learn how to be persons—how to be moral and how to live in
a community with others—first in the family. There, children witness generosity and care and learn the virtues of respect and gratitude. Children also learn
to understand themselves as an integral part of a human community, where
their actions not only have material consequences, but also symbolic meaning.
Though it may not matter, functionally, what kind of vessel one uses to pour
water or what kind of material is used to make one's garments, such choices
may have significant symbolic importance within the culture, such that one may
express respect by making one choice and disrespect by making another.

The Confucian model is, further, one in which individuals look to one
another as examples, learning from one another what constitutes virtuous
behavior.[5] Confucius believes that moral models have magnetic power, and
virtuous individuals can effect moral reform through their actions by inspiring
others to change themselves.[6]

Whether virtuous individuals have the moral powers Confucius attributes
to them may be controversial. In contrast to the Confucian view, the rational actor model suggests that altruistic individuals simply provide opportunities of which others take advantage. However, even if the Confucian optimism
about a single individual's transformative powers is overstated, the conception
of the self that figures in Confucian ethics provides an important counterpoint to the model of the rational economic actor: it represents not only an
alterative possibility for the construction of human identities, but a possibility
that many people actually embrace—at least in certain contexts. This possibility may, in tum, help provide an important way out of the seemingly inexorable logic of collective action problems, whose seriousness and intractability
arises in part from the framing of the problems themselves. In particular, I
argue below that when persons conceive of themselves relationally, as opposed

to atomistically, individual "unilateral" actions can both catalyze and support emerging collective agreements.

Contrasting Garrett Hardin's approach to commons problems with a Confucian one may help illustrate more clearly these points. Hardin argues, with respect to his canonical example of the sheep pasture, that in the absence of top-down measures to limit grazing on the commons, individuals will fail to show the restraint necessary to sustain the commons as a resource for all. The solution to such problems, according to Hardin, requires coercion, or as he puts it "mutual coercion, mutually agreed upon by the majority of the people affected."[7] Hardin is deeply skeptical of the prospects for "conscience" to play a role in the solution to collective action problems; instead, he recommends privatization of resources, taxation and legislative prohibitions of certain behaviors. Hardin's recommendations find echoes in the views of Johnson, for if Hardin is right (as Johnson seems to accept), then it is not individual action, but only large-scale social policy that can resolve problems such as climate change.

Yet Hardin's view diverges significantly from a Confucian perspective, which explicitly rejects coercion as a route to genuine social reform.[8] On the Confucian model, although coercion may keep people out of trouble, it cannot accomplish thoroughgoing social change, involving the transformation of minds as well as actions. Thus, reform achieved primarily through coercion will be both shallow and unstable. In order to solve a collective action problem, it is not just incentives for individually rational agents that need to change. Policy is not enough: what is also crucial is moral change in individuals. Changing institutions without changing people will not resolve tensions between the individual and the collective good.

In the contemporary context, the Confucian point is this: while one can provide both carrots, in the form of economic incentives, and sticks, in the form of laws and regulations, to foster reductions in greenhouse gas emissions, such top-down measures comprise only part of the solution. If people do not recognize and affirm the need to control their greenhouse gas emissions, the effectiveness of such efforts may be limited or unstable.[9] Aldo Leopold made a similar point long ago when he complained about farmers who were willing to enact conservation on their lands only when such measures were paid for by others, and argued for an environmentalism based on a broadened sense of responsibility, embodied in an ecological conscience.[10]

I do not want to argue so much for the ineffectiveness of policy-level changes as I do for the efficacy of individual action to reduce one's own emissions. In this regard, the Confucian model is instructive because it asks us to recognize the possibility that persons need not—and many do not—see themselves as rational economic actors, making decisions based only on a preference structure that stands independent of social consequences or others' values and decisions. If persons are constituted relationally, as Confucians suggest, then one individual's actions cannot be treated as independent of others', and one's personal actions cannot be understood in isolation from their social meaning. Whether one chooses to drive a hybrid electric vehicle and to minimize one's miles driven, or to drive a gas-guzzling SUV with no thought about the number of car trips one takes, one communicates not merely information about how much of the atmospheric commons one is using up, but also sends a message regarding one's concern (or lack thereof) for the commons. When people see themselves as connected members of a moral community, they react to such messages in moral ways: by admiring, and in some cases at least, emulating those whose actions protect the commons, and by criticizing, and in some cases, openly reprimanding those whose actions do not. Such responses can assist in the responsible management of common resources. What seem unlikely to foster such responses, however, are conceptual frameworks that treat individuals as atomistic economic actors whose personal efforts to reduce damage to the commons are viewed as irrational and of little or no moral value. As research by Robert Frank et al. strongly suggests,[11] people see the *Homo economicus* framework not only as descriptive, but as normative, such that the application of models of narrow economic rationality to environmental problems may encourage people to conceptualize themselves and act in accordance with the assumptions of these models.

Yet despite the prevalence of such models in our contemporary society (which no doubt gives support to those who choose not to do anything to reduce their personal emissions), there is good evidence that many people do not understand themselves or their decisions exclusively or even primarily through this lens. For example, people seem to view and judge others' personal choices about what kind of car to drive as moral choices that reflect their overall commitment (or lack thereof) to solving the "collective action problem" of climate change. In recent research, Thomas Turrentine and Kenneth Kurani found in interviews with car owners that people who drove hybrids expressed anger toward purchasers of SUVs, presumably based on their environmental impact.[12] Furthermore, individuals who purchased hybrid electric vehicles often cited other hybrid owners as models for their purchasing decisions, and Turrentine and Kurani found that people rarely purchase hybrid vehicles based on the kind of decision-making process described by the rational economic actor model. People rarely know how much they spend on fuel, for example, or how many miles per gallon their cars achieve, and when considering a hybrid purchase, they rarely calculate how long it would take for the fuel efficiency payback to compensate for the premium in the purchase price.[13] Instead, people choose hybrid cars to make a statement, to express their commitment to the environment, and to discuss with others their choice.[14]

The traditional framework of collective action suggests that such individual actions can have no positive effect on the development of a large-scale solution to the problem. But if the data from hybrid purchasing decisions are any indication, then it seems that one individual's environmentally conscious decision can spur another's, and decisions about what kind of car to buy, how much to drive, and so on, are viewed by many as falling within the moral sphere. Such decisions are the subject of others' moral judgments and can be the basis for social approbation and disapprobation—and we know from models in evolutionary game theory that moralistic punishment of selfish (or otherwise socially disapproved) actions can lead to the stabilization of altruistic (or socially approved) behaviors within a population.[15]

The critical point is that individual rationality is not simply a matter of preference satisfaction independent of the effects of one's actions on others. If individuals do not see themselves as rational economic actors in the narrow sense described by the assumptions of a tragedy of the commons framework—and evidence suggests that they do not—then there are ways in which so-called "unilateral" actions by individuals can influence other individuals not to take advantage of the "excess resources" remaining in the commons, but to see the restraint of others as a model for their own exercise of restraint.[16] Furthermore, if there is a sufficient critical mass of individuals who are committed to such restraint, these individuals can exercise moral suasion over the more obdurate members of the community, and they are better positioned to form a bloc whose commitment to protecting the commons can be voiced effectively through legislative or other channels.[17]

Just as individuals are not atomistic, narrowly self-interested actors, isolated from one another in their decisions and values, actions at the individual level are not divorced from those in the political sphere. A commitment at the personal level may in fact spur greater awareness and more careful consideration of the kinds of political changes that may be most effective. After installing solar panels on one's house, for example, one may recognize more clearly the importance of net metering legislation and feel more inclined to lobby for it than one would based on the abstract recognition that such legislation would be a good thing.

If we take seriously a Confucian conception of persons as both descriptive and normative, then we ought not assume—as the logic of collective action problems does—that the motives of individuals who partake of the shared resources of a commons are narrowly self-interested. Confucian morality counsels against such an individualistic self-conception, and the work of Turrentine and Kurani suggests that the presuppositions of the collective action framework are not necessarily borne out, even in a highly individualistic culture such as that of the United States.[18] Since philosophical and economic characterizations of persons have both descriptive and normative functions, such that these characterizations may produce the very kinds of persons they describe, we should consider carefully the heuristic value of thinking about our climate change obligations in a traditional collective action context. Although the collective action framework may be useful for certain purposes, it is not clear that it provides an adequate justification for the view that in the absence of a collective agreement, individuals have no obligation to reduce their personal greenhouse gas emissions. Collective agreements can emerge in a variety of ways, and the hybrid vehicle example above suggests that individual consumer decisions, personal conversations about such decisions, and similar small-scale, local actions may turn out to be important catalysts for emerging collective agreements, and may support and reinforce agreements and policies at larger scales. Hence, the distinction between acting unilaterally and acting to fulfill one's responsibilities as part of a collective agreement is not sharp, but rather a matter of degree. As such, a strong distinction between one's obligations under a collective agreement and one's obligations in the absence of such an agreement is unjustified.

Part of what makes personal choices effective and morally important is that personal choices have a communicative and social function. On the Confucian view, individual actions gain their moral value in a social context. Thus, although Confucius emphasizes the importance of virtuous action at the personal level, he also insists that one ought not to ignore one's political obligations in order to preserve personal moral purity, nor withdraw from society in order to live—in isolation—in accord with one's individual values.[19] With respect to climate change, a Confucian perspective therefore supports a personal moral obligation to reduce one's greenhouse gas emissions, but insists that one's obligations do not end there: one ought also to work for larger scale social reform, for regional, national, and international policies to reduce emissions and mitigate the effects of global warming.

Conclusion

In the case of climate change, an atomistic conception of persons can lead to two different kinds of problems. The first is one in which individualism supports the view that an individual's *only* responsibility is at the personal level; the second is one that supports the view that the individual's only responsibility in a "tragedy of the commons" is at the level of policy. The arguments of Johnson and Sinnott-Armstrong respond to the first problematic kind of individualism. It is this kind of individualism that can lead people to believe that so long as they "live lightly on the planet," they are not responsible for the depletion of resources and damage to the earth's climate and need not take political action to address these problems. This kind of individualism fails to recognize the individual's role in society and broad responsibility to promote good social decisions.

However, a second kind of individualism is equally problematic: this is the individualism that finds its expression in the view that one's only responsibility is to change society without changing oneself. This kind of individualism, which rests on the sort of assumptions that characterize collective action problems, fails to recognize the connections between the personal and the social, the expressive function of personal action, the importance of integrity, the role of individual action in constructing one's moral identity, and the effect of individual action on one's relations with others, and on *their* actions. I have tried, in this paper, to highlight the problems with this latter kind of individualism, and the promise of abandoning it.

Johnson and Sinnott-Armstrong hold that there is no personal moral obligation to reduce GHG emissions because personal reductions cannot mitigate the problem of global climate change. I have argued that personal reductions *can* make a contribution, and hence that if there is an obligation to ameliorate climate change, it includes a personal obligation to control one's own emissions. However, in order to see how personal reductions can contribute—and in order for such contributions to be most effective—we may need to conceptualize persons differently, and more relationally, than is the case in a tragedy of the commons model.

NOTES

1. W. Sinnott-Armstrong, "It's Not *My* Fault: Global Warming and Individual Moral Obligations," in W. Sinnott-Armstrong and R. Howarth, eds., *Perspectives on Climate Change: Science, Economics, Politics, Ethics* (Amsterdam: Elsevier, 2005), 285–307.

2. B. Johnson, "Ethical Obligations in a Tragedy of the Commons," *Environmental Values* 12 (2003): 271–287.

3. Ibid., 275.

4. D. Hall and R. Ames, *Thinking from the Han: Self, Truth, and Transcendence in Chinese and Western Culture* (Albany: SUNY Press, 1998); H. Rosemont, "Rights-Bearing Individuals and Role-Bearing Persons," in Mary I. Bockover, ed., *Rules, Rituals, and Responsibilities: Essays Dedicated to Herbert Fingarette* (LaSalle, IL: Open Court, 1991).

5. In the *Analects*, Confucius observes that there is no one from whom he cannot learn: "There is no man who does not have something of the way of Wen and Wu in him. Superior men have got hold of what is of major significance while inferior men have got hold of what is minor significance. From whom, then, does the Master not learn?" Confucius, *The Analects* (*Lun yu*), trans. D. C. Lau (Hong Kong: Chinese University Press, 1983), 19.22.

6. Ibid.

7. G. Hardin, "The Tragedy of the Commons," *Science* 162 (1968): 1247.

8. "Guide them by edicts, keep them in line with punishments, and the common people will stay out of trouble, but will have no sense of shame. Guide them by virtue, keep them in line with the rites, and they will, besides having a sense of shame, reform themselves" (Confucius, *Analects*, 2.3).

9. See Elinor Ostrom's *Governing the Commons: The Evolution of Institutions for Collective Action* (New York: Cambridge University Press, 1990) for detailed discussion of the ways in which the costs of monitoring and enforcement of collective action arrangements may impede success when the arrangement is imposed from outside (e.g., by a central government), whereas management of common resources may meet greater success where collective action arrangements are the result voluntary self-organization.

10. A. Leopold, "The Land Ethic," in *A Sand County Almanac, and Sketches Here and There* (New York: Oxford University Press, 1987).

11. R. H. Frank, T. Gilovich, and D. Regan, "Does Studying Economics Inhibit Cooperation?," *Journal of Economic Perspectives* 7 (1993): 159–171.

12. T. Turrentine and K. Kurani, "Car Buyers and Fuel Economy?," *Energy Policy* 35 (2007): 1213–1223.

13. Ibid.

14. Ibid., 1221.

15. R. Boyd and P. Richerson, "Punishment Allows the Evolution of Cooperation (or Anything Else) in Sizable Groups," *Ethology and Sociobiology* 13 (1992): 171–195.

16. D. Jamieson, "When Utilitarians Should Be Virtue Theorists," *Utilitas* 19, no. 2 (2007): 179, too notes the importance of the "example-setting and role-modeling" aspects of individual behavior.

17. See P. Rozin, "The Process of Moralization," *Psychological Science* 10 (1999): 218–221, for discussion of how moralization of certain behaviors can lead to social change. Rozin argues that moralization of behaviors leads to institutional prohibitions, education to discourage these behaviors, transmission of norms from parents to children, and censure of those practicing the negatively moralized behavior.

18. Although I have used a Confucian framework to illustrate the possibilities entailed by adopting a relational conception of persons, relational conceptions can be found in the Western tradition as well. For example. Piers Stephens, "Green Liberalisms: Nature, Agency and the Good," *Environmental Politics* (2001): 12, points to John Stuart Mill as implicitly advocating "the idea of the agent as a social and relational being." Similarly, many feminist ethical theories emphasize the relational dimensions of human life. E.g., N. Noddings, *Caring: A Feminine Approach to Ethics and Moral Education* (Berkeley: University of California Press, 1984).

19. M. Hourdequin, "Engagement, Withdrawal, and Social Reform: Confucian and Contemporary Perspectives," *Philosophy East and West* 60 (2010): 369–390.

27

Excerpts from "About Free-Market Environmentalism"

JONATHAN H. ADLER

The Theory of Free Market Environmentalism

Conventional environmental policymaking presupposes that only government action can improve environmental quality. In this view, environmental problems arise from "market failures" that produce "externalities." Government regulation is needed to correct environmental concerns that the market has "failed" to handle because they are "external" to the price signals that regulate marketplace transactions. The conventional paradigm of environmental policy justifies the regulation of economic activity because it assumes all activities—from purchasing clothing to driving a car to turning on a light bulb—have an impact on the environment that is not factored into the cost of the product or service. Economic central planning may be intellectually and historically discredited, but the "market failure" thesis justifies environmental central planning, an endeavor just as prone to ruin. In the words of Competitive Enterprise Institute president Fred L. Smith, Jr., "The disastrous road to serfdom can just as easily be paved with green bricks as with red ones."[1] Embracing the "market failure" rationale leads to policy failure.

Free market environmentalism (FME) rejects the "market failure" model. "Rather than viewing the world in terms of market failure, we should view the problem of externalities as a failure to permit markets and create markets where they do not yet—or no longer—exist," argues Smith.[2] Resources that are privately owned or managed and, therefore, are in the marketplace are typically well-maintained. Resources that are unowned or politically controlled, and therefore outside the market, are more apt to be inadequately managed. "At the heart of free market environmentalism is a system of well-specified property rights to natural resources," explain Terry Anderson and Donald Leal, authors of *Free Market Environmentalism*.[3] Adds Smith, "Rather than the silly slogan of some environmentalists, that 'trees should have standing,' our argument is that behind every tree should stand an owner who can act as its protector."[4]

FME owes an intellectual debt to ecologist Garrett Hardin's discussion of the "tragedy of the commons."[5] Hardin noted when a resource is unowned or owned in common, such as the grazing pasture in a medieval village, there is no incentive for any individual to protect it. In the medieval village it is in every cattleowner's self-interest to have his herd graze the pasture as much as possible and before any other herd. Every cattleowner who acquires additional cattle gains the benefits of a larger herd, while the cost of overusing the pasture is borne by all members of the village. Inevitably, the consequence is an overgrazed pasture, and everyone loses. Indeed, the cattleowner with foresight will anticipate that the pasture will become barren in the future, and this will give him additional incentive to overgraze. Refusing to add another cow to one's own herd does not change the incentive of every other cattleowner to do so.

The world's fisheries offer a contemporary example of the tragedy of the commons. Because oceans are unowned, each fishing fleet has no incentive to conserve or replenish the fish it takes and it has every incentive to take as many fish as possible lest the benefits of a larger catch go to someone else.[6] Private ownership overcomes the commons problem because owners can prevent overuse by controlling access to the resource. As Hardin noted, "The tragedy of the commons as a food basket is averted by private property, or something formally like it."[7]

Although environmental activists often disparage private ownership, the record of private owners in conserving resources is superior to that of government agencies. For instance, Terry Anderson observes that "well-established private rights to Great Lakes timber resulted in efficient markets rather than the 'rape and run' tactics alleged by conservationists."[8] As R. J. Smith explains,

> Wherever we have exclusive private ownership, whether it is organized around a profit-seeking or nonprofit undertaking, there are incentives for the private owners to preserve the resource. . . . [P]rivate ownership allows the owner to capture the full capital value of the resource, and self-interest and economic incentive drive the owner to maintain its long-term capital value.[9]

Unlike public officials, private owners directly benefit from sound management decisions and suffer from poor ones.

For incentives to work, the property right to a resource must be definable, defendable, and divestible. Owners must be free to transfer their property rights to others at will. Even someone indifferent or hostile to environmental protection has an incentive to take environmental concerns into account, because despoiling the resource may reduce its value in the eyes of potential buyers. The role of government is to protect property rights for environmental resources and secure the voluntary agreements property owners contract to carry out. Moreover, FME advocates insist on the application of common law liability rules to environmental harms, such as polluting a neighbor's property, to protect property rights and to provide additional incentives for good stewardship. To harm someone's property by polluting it is no more acceptable than vandalizing it.

The importance of private ownership to sound conservation is clear from America's environmental history. When environmental groups like the National Audubon Society and the Nature Conservancy act to protect habitat and ecologically sensitive areas by purchasing land and establishing sanctuaries, they act in the marketplace to advance environmental values. R. J. Smith explains:

> Private ownership includes not only hunting preserves, commercial bird breeders, parrot jungles, and safari parks, it also includes wildlife sanctuaries, Audubon Society refuges, World Wildlife Fund preserves, and a multitude of private, nonprofit conservation and preservation projects.[10]

These organizations raise money by soliciting contributions to acquire ownership in preferred lands. Were it not for the institution of private property, these ventures to protect the environment would be impossible.

Private efforts to support the reintroduction of wolves in Montana offers another example of how market transactions can advance environmental goals. The hostility of ranchers has been a major obstacle to reintroducing predators into the wild. In the 1970s, free market economists argued that ranchers' fear of livestock losses would be addressed if those who wanted to reintroduce wolves would agree to compensate ranchers who suffered economic loss due to predators.[11]

Defenders of Wildlife adopted the idea and established a Wolf Compensation Fund. "What we're trying to do is devise a system whereby all those people who care about endangered species restoration actually pay some of the bills," explained Defenders of Wildlife staffer Hank Fischer. "What this solution attempts to do is utilize economic forces—in other words, to make it desirable to have wolves."[12] After several years, the fund had paid ranchers approximately $12,000 for livestock losses. The program has flaws—some ranchers complain that compensation is not always paid, and federal regulations still prevent ranchers from killing wolves to protect livestock—but the Fund remains an example of how marketplace transactions can further environmental goals even when no goods are exchanged.

FME proponents would terminate government programs that cause environmental harm or inhibit private-sector solutions to environmental problems. Free market environmental policies would establish property rights, where possible, so as to internalize "externalities." In some cases, FME proponents would counsel more modest steps. For instance, in the difficult case of automobile air pollution, a "polluter pays" approach would replace regulations mandating specific emissions-control equipment and annual emissions testing. An owner would be assessed a fee proportional to the amount of pollution his auto generated. Since fees would vary in relation to pollution emission levels, owners would have incentives to have their automobiles repaired or replaced when they began to pollute significantly. Technologies currently exist to monitor emissions as autos move on the highway, providing potential enforcement mechanisms that will not inconvenience most owners of vehicles whose emission levels are negligible. This solution is not ideal because a genuine market is not created; but it is more market-oriented that current air pollution policies.[13]

The Environmental Establishment's Response

Most environmental activists reject free market environmentalism. Economist Thomas Michael Power and *Sierra* associate editor Paul Rauber write: "Markets are not neutral, technological devices. They are social institutions whose use has profound consequences. All societies purposely limit the extent of the market in order to protect basic values."[14] Wedded to the state as the instrument of reform, environmental activists cannot accept the idea that market forces will produce the results they desire even when it is apparent that government regulation will not.

Nonetheless, FME has changed the discussion of environmental issues. Many environmentalists now seem to understand why environmental policies should be examined in economic terms. Says Roberto Repetto, director of the economics program at World Resources Institute, "If we can enact policies that adjust prices so they more accurately reflect all the costs associated with producing a particular pollutant or using particular resources, then society will make better decisions."[15] Repetto advocates pollution taxes and other government interventions as ways to "internalize" externalities. Such policies are not FME, but they are evidence that the terms of debate are shifting.

Some environmentalists also see the strategic political benefit of market rhetoric and some free market policies. Ned Ford, energy chair of the Sierra Club's Ohio chapter, argues that "by forcing the marketplace to the lowest cost solution that really works, environmentalists gain credibility and enhance the opportunity for further reduction."[16] Even President Bill Clinton has acknowledged the importance of developing a "market-based environmental-protection strategy," noting that "Adam Smith's invisible hand can have a green thumb."[17] Too often, however, market rhetoric merely merchandises government regulatory policies. Environmentalist groups rarely adopt FME policies fully, opting instead to pick and choose free market precepts.

Attempts to use "market mechanisms" to reach predetermined environmental outcomes are the most common example of this tactic. The Environmental Defense Fund (EDF), for instance, advocates widespread use of "pollution credit trading" as a market-oriented policy. Setting an emission level as an environmental target, the EDF proposal allows companies the freedom to determine how best to reach it. Companies could buy and sell emission allotments among themselves to find the least-cost means to reach a goal set by government regulation. Explains EDF's Dan Dudek, "Who is better to know [what to do] than the people who own and operate" the facility causing pollution?[18]

FME advocates note that this approach will not necessarily produce sound environmental policy. The Clean Air Act Amendments of 1990, for instance, include an elaborate EDF-designed pollution-credit trading scheme for sulfur oxide emissions to control acid rain. Many companies favored the policy because, by allowing them to select the least-cost pollution reduction measures, they might save millions of dollars in compliance costs. But was a sulfur oxide emission reduction plan needed at all? The most extensive US study of acid rain to date suggests that acid rain was not a substantial threat to forests and streams, despite environmentalist claims to the contrary.

John Baden warns against market mechanisms that are used "simply as tools for the efficient delivery of environmental goals . . . [while] the goals themselves remain collectively determined."[19] CEI's Fred Smith calls such policies "market socialism," as they resemble the efforts in Communist countries to use market mechanisms to reach politically determined production quotas. EDF's emission trading scheme is structurally the equivalent of the tradeable wheat production quotas established in parts of Eastern Europe. Notes Smith, "the efficiency gains of market systems occur not only in production, but in allocation as well. This means that markets are as effective at determining what is to be done as they are at determining how it should be accomplished."[20]

The Road Ahead

Free market environmental policies will not soon be embraced by many environmental regulators, activists, or lobbyists. But the fact that "market" language—incentives, costs, trade-offs, markets, property rights, and so on—dominates current environmental discussions suggests that it presents a serious challenge to conventional environmental policy approaches. Even the most statist environmental activist organizations feel the need to embrace "market" perspectives on certain issues. As the editor of the Worldwatch Institute's book series seemed to lament in the introduction to a book on "market" mechanisms, "market economies will remain the dominant economic system for the foreseeable future."[21] Markets may well become the basis for the next generation of environmental protections as well.

Advancing the free market environmental agenda will certainly be a challenge. On top of the obvious political obstacles, there are serious implementation questions that need to be addressed. There are tremendous legal and cultural barriers to the extension of market institutions in many areas. The technical requirements of property rights definition and enforcement are also substantial. It is one thing to create rights in instream water flows, as is done in many states; it is quite another to contemplate property rights in the air or the deep seas.

The relevant question, however, is whether these obstacles are any greater than asking the federal government to plan our collective environmental future. The experience with environmental policy to date suggests not. Central planning has clearly failed. It is time to give market institutions a chance.

NOTES

1. Fred L. Smith, Jr., "The Market and Nature," *Freeman*, September 1993, 352.
2. Fred L. Smith, Jr., "Conclusion: Environmental Policy at the Crossroads," in Michael S. Greve and Fred L. Smith, Jr., eds., *Environmental Politics: Public Costs, Private Rewards* (New York: Praeger, 1992), 192.
3. Terry L. Anderson and Donald R. Leal, *Free Market Environmentalism* (San Francisco: Pacific Research Institute, 1991), 3.
4. Fred Smith, "Conclusion," 192.
5. See Garrett Hardin, "The Tragedy of the Commons," *Science* 162, no. 3859 (1968): 1243–1248.
6. See Kent Jeffreys, "Rescuing the Oceans," in Ron Bailey, ed., *The True State of the Planet* (New York: Free Press, 1995).
7. Hardin, "Tragedy of the Commons," 1243.
8. Terry L. Anderson, "The New Resource Economics: Old Ideas and New Applications," *American Journal of Agricultural Economics*, December 1982, 933.
9. R. J. Smith, "Resolving the Tragedy of the Commons by Creating Private Property Rights in Wildlife," *Cato Journal*, Fall 1981, 456–457.
10. Ibid., 456.
11. Ryan Amacher, Robert D. Tolison, and Thomas D. Willet, "The Economics of Fatal Mistakes: Fiscal Mechanisms for Preserving Endangered Predators," in Terry Anderson and P. J. Hill, eds., *Wildlife in the Marketplace* (Lanham, MD: Rowman and Littlefield, 1995), 43–60.
12. Hank Fischer, "Free Market Environmentalism's Bottom Line," *Morning Edition*, NPR, September 1, 1992.
13. See Jonathan Adler, *Reforming Arizona's Air Pollution Policy* (Phoenix: Goldwater Institute, 1993).
14. Thomas Michael Power and Paul Rauber, "The Price Is Everything," *Sierra*, November–December 1993, 94.
15. Joe Alper, "Protecting the Environment with the Power of the Market," *Science* 250, no. 5116 (1993): 1884–1885.
16. Power and Rauber, "The Price Is Everything," 89.
17. Ibid., 88.
18. Karen Riley, "Rewards for Friends of the Earth," *Washington Times*, November 22, 1992.
19. Power and Rauber, "The Price Is Everything," 92.
20. Fred L. Smith, Jr., *Europe, Energy and the Environment: The Case against Carbon Taxes* (Washington, DC: Competitive Enterprise Institute, 1992), 8.
21. Linda Starke, foreword to David Malin Roodman, *The Natural Wealth of Nations* (New York: Norton, 1998), 12.

Reading Questions and Further Readings

Reading Questions

1. What general solutions do Garrett Hardin propose to get us out of a tragedy of the commons?
2. Describe Elinor Ostrom et al.'s critique of Garrett Hardin. In her and her coauthors' view, why do the "commons" not necessarily spell tragedy?
3. How does the classical assumption of perfectly rational and self-interested individuals limit our understanding of collective action problems?
4. In contrast to Hardin, Hourdequin argues that we integrate both personal and political commitments when trying to understand the ethics of climate change. Describe the reasoning for this, and how it might change how we think of solutions to collective action problems.
5. What is "free-market environmentalism," and what does this framework imply about the specific role of governments in managing of natural resources?

Further Readings

Meadows, Donella, Jorgen Randers, and Dennis Meadows. *Limits to Growth: The 30-Year Update*. White River Junction, VT: Chelsea Green, 2004.

Nowak, Martin, and Roger Highfield. *SuperCooperators: The Mathematics of Evolution, Altruism and Human Behavior*. Edinburgh, UK: Canongate Books, 2011.

Ostrom, Elinor. *Governing the Commons: The Evolution of Institutions for Collective Action*. New York: Cambridge University Press, 1990.

Tomasello, Michael. *Why We Cooperate*. Boston: Boston Review Books, 2009.

PART V

Values and Justice

What is it to have environmental values, and how should environmental harms and benefits be distributed? These questions are fundamental to understanding the interface of environment and society, especially as we wrestle with challenges like climate change that are multifaceted and global in scope. In *Sand County Almanac*, Aldo Leopold advocated for the historical expansion of values to include the biotic community. Contemporary questions posed by global climate change involve the value of nonsentient nature and of future generations. Environmental Studies has concerned itself with such questions of value throughout its history, including the challenges that arise when the scope of consideration expands.

This part features a partially chronological description of the expansion and extension of environmental values. Henry David Thoreau celebrates the quality of wildness, both found in nature and in domesticated settings, and its contribution to individual, spiritual fulfillment. He laments society's detachment from "this vast, savage, howling Mother of ours, Nature lying all around, with such beauty, and such affection for her children, as the leopard," and argues that "in Wildness is the preservation of the world." Thoreau's influence is clear in contemporary examples of connecting to nature like "living off the land" and the increased interest in producing one's own food.

Significant theoretical questions still remain about what is valuable and how strong these values are. We can extend existing values, for instance, by internalizing environmental costs into economics (e.g., taxing coal in order pay for the pollution it causes), or construct new values that account for the uniqueness of the environment. Are there other reasons to value and preserve nature, or to have a constructive relationship with nature, in addition to the benefits that nature provides, such as clean water and air? Robert E. Goodin gives us one answer in his defense of "deep green" values such as naturalness. Using analogies to forgeries to distinguish between nature that has been faked or restored and nature that has not, he spells out a basic "green theory of value." This theory sees nature as a context that facilitates ordering and meaning in human life and is an alternative (or complement) to "shallow" theories, such as those typically presupposed by economics, that give a less central role to nature.

Even if we accept a "deeper" value for nature, another wrinkle concerns how long the environment should be preserved. Do we have responsibilities to those who are not yet born, who might find meaning in nature as well? John Passmore spells out the case for conservation for future generations, unpacking the complex issue of valuing future persons. He surveys basic conceptions of our moral relations to posterity (based on duty, utility, Christianity, contracts, and economics), describing the tension between present gains and conservation, both of which can benefit posterity. He also distinguishes the strengths and weaknesses of valuing the environment directly (i.e., finding a value for the environment in and of itself) from valuing the environment indirectly (i.e., for the services it provides).

Environmental values are often contested, and different people have different ideas about how much they should count. For example, I might find the beauty of a mountain range a reason to protect it, but others might find that reason insufficient. Even if we agree that the mountain range should be protected, the question arises, how long and at what cost? Alan Holland gives us one proposal as he describes the rise of such notions as sustainability, sustainable development, and other conceptions of value that extend into the future. Holland spells out strong and weak versions of sustainability and details the challenges of creating a "new ethic." As the scope of values expands, conceptual challenges for justice arise, especially in the global context.

In addition to the valuing of environmental goods, the distribution of environmental benefits and harms and participation in decision-making processes raise important questions of value. During and after Hurricane Katrina, for instance, people with less wealth and social capital suffered disproportionately, a disconcertingly common trend in natural disasters as well as in cases of environmental pollution. The field that studies disproportionate environmental impact is called environmental justice (also discussed in part 2). Global environmental change, including climate change and persistent pollutants, forces us to face new questions about intergovernmental responsibility, intergenerational damages, and duties to the far off. What are our responsibilities to people in far-off countries who might be impacted by the pollution from industrialized countries? David Schlosberg details the expansion of the environmental justice discourse, discussing conceptions of community and the inclusion of the nonhuman world. Our perception and assessment of risk is deeply value laden and instantiated in policy in multiple forms (e.g., cost-benefit analysis and the precautionary principle). Even when such questions are clarified and resolved, we are still faced with the task of weighing and negotiating between values that conflict (e.g., human, nonhuman animal, and natural values).

Once again, the scale of the challenge has expanded, and along with this expansion come new values and challenges. These readings tell a story of the broadening circle of environmental and human values, including the strengths and weaknesses of a more inclusive set of values.

28

Excerpts from "Walking"

HENRY DAVID THOREAU

I wish to speak a word for Nature, for absolute freedom and wildness, as contrasted with a freedom and culture merely civil—to regard man as an inhabitant, or a part and parcel of Nature, rather than a member of society. I wish to make an extreme statement, if so I may make an emphatic one, for there are enough champions of civilization: the minister and the school committee and every one of you will take care of that.

I have met with but one or two persons in the course of my life who understood the art of Walking, that is, of taking walks—who had a genius, so to speak, for *sauntering*, which word is beautifully derived "from idle people who roved about the country, in the Middle Ages, and asked charity, under pretense of going *à la Sainte Terre*," to the Holy Land, till the children exclaimed, "There goes a *Sainte-Terrer*," a Saunterer, a Holy-Lander. They who never go to the Holy Land in their walks, as they pretend, are indeed mere idlers and vagabonds; but they who do go there are saunterers in the good sense, such as I mean. Some, however, would derive the word from *sans terre*, without land or a home, which, therefore, in the good sense, will mean, having no particular home, but equally at home everywhere. For this is the secret of successful sauntering. He who sits still in a house all the time may be the greatest vagrant of all; but the saunterer, in the good sense, is no more vagrant than the meandering river, which is all the while sedulously seeking the shortest course to the sea. But I prefer the first, which, indeed, is the most probable derivation. For every walk is a sort of crusade, preached by some Peter the Hermit in us, to go forth and reconquer this Holy Land from the hands of the Infidels.

It is true, we are but faint-hearted crusaders, even the walkers, nowadays, who undertake no persevering, never-ending enterprises. Our expeditions are but tours, and come round again at evening to the old hearthside from which we set out. Half the walk is but retracing our steps. We should go forth on the shortest walk, perchance, in the spirit of undying adventure, never to return, prepared to send back our embalmed hearts only as relics to our desolate kingdoms. If you are ready to leave father and mother, and brother and sister, and wife and child and friends, and never see them again—if you have paid your debts, and made your will, and settled all your affairs, and are a free man—then you are ready for a walk.

To come down to my own experience, my companion and I, for I sometimes have a companion, take pleasure in fancying ourselves knights of a new, or rather an old, order—not Equestrians or Chevaliers, not Ritters or Riders, but Walkers, a still more ancient and honorable class, I trust. The Chivalric and heroic spirit which once belonged to the Rider seems now to reside in, or perchance to have subsided into, the Walker—not the Knight, but Walker, Errant. He is a sort of fourth estate, outside of Church and State and People. . . .

I think that I cannot preserve my health and spirits, unless I spend four hours a day at least—and it is commonly more than that—sauntering through the woods and over the hills and fields, absolutely free from all worldly engagements. You may safely say, A penny for your thoughts, or a thousand pounds. When sometimes I am reminded that the mechanics and shopkeepers stay in their shops not only all the forenoon, but all the afternoon too, sitting with crossed legs, so many of them—as if the legs were made to sit upon, and not to stand or walk upon—I think that they deserve some credit for not having all committed suicide long ago.

I, who cannot stay in my chamber for a single day without acquiring some rust, and when sometimes I have stolen forth for a walk at the eleventh hour, or four o'clock in the afternoon, too late to redeem the day, when the shades of night were already beginning to be mingled with the daylight, have felt as if I had committed some sin to be atoned for—I confess that I am astonished at the power of endurance, to say nothing of the moral insensibility, of my neighbors who confine themselves to shops and offices the whole day for weeks and months, aye, and years almost together. I know not what manner of stuff they are of, sitting there now at three o'clock in the afternoon, as if it were three o'clock in the morning. Bonaparte may talk of the three-o'clock-in-the-morning courage, but it is nothing to the courage which can sit down cheerfully at this hour in the afternoon over against one's self whom you have known all the morning, to starve out a garrison to whom you are bound by such strong ties of sympathy. I wonder that about this time, or say between four and five o'clock in the afternoon, too late for the morning papers and too early for the evening ones, there is not a general explosion heard up and down the street, scattering a legion of antiquated and house-bred notions and whims to the four winds for an airing—and so the evil cure itself.

How womankind, who are confined to the house still more than men, stand it I do not know; but I have ground to suspect that most of them do not *stand* it at all. When, early in a summer afternoon, we have been shaking the dust of the village from the skirts of our garments, making haste past those houses with purely Doric or Gothic fronts, which have such an air of repose about them, my companion whispers that probably about these times their occupants are all gone to bed. Then it is that I appreciate the beauty and the glory of architecture, which itself never turns in, but forever stands out and erect, keeping watch over the slumberers.

No doubt temperament, and, above all, age, have a good deal to do with it. As a man grows older, his ability to sit still and follow indoor occupations increases. He grows vespertinal in his habits as the evening of life approaches, till at last he comes forth only just before sundown, and gets all the walk that he requires in half an hour.

But the walking of which I speak has nothing in it akin to taking exercise, as it is called, as the sick take medicine at stated hours—as the Swinging of dumbbells or chairs; but is itself the enterprise and adventure of the day. If you would get exercise, go in search of the springs of life. Think of a man's swinging dumbbells for his health, when those springs are bubbling up in far-off pastures unsought by him!

Moreover, you must walk like a camel, which is said to be the only beast which ruminates when walking. When a traveler asked Wordsworth's servant to show him her master's study, she answered, "Here is his library, but his study is out of doors."

Living much out of doors, in the sun and wind, will no doubt produce a certain roughness of character—will cause a thicker cuticle to grow over some of the finer qualities of our nature, as on the face and hands, or as severe manual labor robs the hands of some of their delicacy of touch. So staying in the house, on the other hand, may produce a softness and smoothness, not to say thinness of skin, accompanied by an increased sensibility to certain impressions. Perhaps we should be more susceptible to some influences important to our intellectual and moral growth, if the sun had shone and the wind blown on us a little less; and no doubt it is a nice matter to proportion rightly the thick and thin skin. But methinks that is a scurf that will fall off fast enough—that the natural remedy is to be found in the proportion which the night bears to the day, the winter to the summer, thought to experience. There will be so much the more air and sunshine in our thoughts. The callous palms of the laborer are conversant with finer tissues of self-respect and heroism, whose touch thrills the heart, than the languid fingers of idleness. That is mere sentimentality that lies abed by day and thinks itself white, far from the tan and callus of experience.

When we walk, we naturally go to the fields and woods: what would become of us, if we walked only in a garden or a mall? Even some sects of philosophers have felt the necessity of importing the woods to themselves, since they did not go to the woods. "They planted groves and walks of Platanes," where they took *subdiales ambulationes* in porticos open to the air. Of course it is of no use to direct our steps to the woods, if they do not carry us thither. I am alarmed when it happens that I have walked a mile into the woods bodily, without getting there in spirit. In my afternoon walk I would fain forget all my morning occupations and my obligations to Society. But it sometimes happens that I cannot easily shake off the village. The thought of some work will run in my head and I am not where my body is—I am out of my senses. In my walks I would fain return to my senses. What business have I in the woods, if I am thinking of something out of the woods? I suspect myself, and cannot help a shudder when I find myself so implicated even in what are called good works—for this may sometimes happen.

My vicinity affords many good walks; and though for so many years I have walked almost every day, and sometimes for several days together, I have not yet exhausted them. An absolutely new prospect is a great happiness, and I can still get this any afternoon. Two or three hours' walking will carry me to as strange a country as I expect ever to see. A single farmhouse which I had not seen before is sometimes as good as the dominions of the King of Dahomey. There is in fact a sort of harmony discoverable between the capabilities of the landscape within a circle of ten miles' radius, or the limits of an afternoon walk, and the three-score years and ten of human life. It will never become quite familiar to you.

Nowadays almost all man's improvements, so called, as the building of houses and the cutting down of the forest and of all large trees, simply deform the land-scape, and make it more and more tame and cheap. A people who would begin by burning the fences and let the forest stand! I saw the fences half consumed, their ends lost in the middle of the prairie, and some worldly miser with a sur-veyor looking after his bounds, while heaven had taken place around him, and he did not see the angels going to and fro, but was looking for an old post-hole in the midst of paradise. I looked again, and saw him standing in the middle of a boggy Stygian fen, surrounded by devils, and he had found his bounds without a doubt, three little stones, where a stake had been driven, and looking nearer, I saw that the Prince of Darkness was his surveyor. . . .

What is it that makes it so hard sometimes to determine whither we will walk? I believe that there is a subtle magnetism in Nature, which, if we uncon-sciously yield to it, will direct us aright. It is not indifferent to us which way we walk. There is a right way; but we are very liable from heedlessness and stupid-ity to take the wrong one. We would fain take that walk, never yet taken by us through this actual world, which is perfectly symbolical of the path which we love to travel in the interior and ideal world; and sometimes, no doubt, we find it difficult to choose our direction, because it does not yet exist distinctly in our idea. . . .

Soon after, I went to see a panorama of the Mississippi, and as I worked my way up the river in the light of today, and saw the steamboats wooding up, counted the rising cities, gazed on the fresh ruins of Nauvoo, beheld the Indians moving west across the stream, and, as before I had looked up the Moselle, now looked up the Ohio and the Missouri and heard the legends of Dubuque and of Wenona's Cliff—still thinking more of the future than of the past or present—I saw that this was a Rhine stream of a different kind; that the foundations of castles were yet to be laid, and the famous bridges were yet to be thrown over the river; and I felt that this was the heroic age itself, though we know it not, for the hero is commonly the simplest and obscurest of men.

The West of which I speak is but another name for the Wild; and what I have been preparing to say is, that in Wildness is the preservation of the World. Every tree sends its fibers forth in search of the Wild. The cities import it at any price. Men plow and sail for it. From the forest and wilderness come the tonics and barks which brace mankind. Our ancestors were savages. The story of Romulus and Remus being suckled by a wolf is not a meaningless fable. The founders of every state which has risen to eminence have drawn their nourishment and vigor from a similar wild source. It was because the children of the Empire were not suckled by the wolf that they were conquered and displaced by the children of the northern forests who were.

I believe in the forest, and in the meadow, and in the night in which the corn grows. We require an infusion of hemlock, spruce or arbor vitae in our tea. There is a difference between eating and drinking for strength and from mere gluttony. The Hottentots eagerly devour the marrow of the koodoo and other antelopes raw, as a matter of course. Some of our northern Indians eat raw the marrow of the Arctic reindeer, as well as various other parts, including the summits of the antlers, as long as they are soft. And herein, perchance, they have stolen a march on the cooks of Paris. They get what usually goes to feed the fire. This is probably better than stall-fed beef and slaughterhouse pork to make a man of. Give me a wildness whose glance no civilization can endure—as if we lived on the marrow of koodoos devoured raw. . . .

Life consists with wildness. The most alive is the wildest. Not yet subdued to man, its presence refreshes him. One who pressed forward incessantly and never rested from his labors, who grew fast and made infinite demands on life, would always find himself in a new country or wilderness, and surrounded by the raw material of life. He would be climbing over the prostrate stems of primitive forest-trees. . . .

It is said to be the task of the American "to work the virgin soil," and that "agriculture here already assumes proportions unknown everywhere else." I think that the farmer displaces the Indian even because he redeems the meadow, and so makes himself stronger and in some respects more natural. I was surveying for a man the other day a single straight line one hundred and thirty-two rods long, through a swamp at whose entrance might have been written the words which Dante read over the entrance to the infernal regions, "Leave all hope, ye that enter"—that is, of ever getting out again; where at one time I saw my employer actually up to his neck and swimming for his life in his property, though it was still winter. He had another similar swamp which I could not survey at all, because it was completely under water, and nevertheless, with regard to a third swamp, which I did survey from a distance, he remarked to me, true to his instincts, that he would not part with it for any consideration, on account of the mud which it contained. And that man intends to put a girdling ditch round the whole in the course of forty months, and so redeem it by the magic of his spade. I refer to him only as the type of a class.

The weapons with which we have gained our most important victories, which should be handed down as heirlooms from father to son, are not the sword and the lance, but the bushwhack, the turf-cutter, the spade, and the bog hoe, rusted with the blood of many a meadow, and begrimed with the dust of many a hard-fought field. The very winds blew the Indian's cornfield into the meadow, and pointed out the way which he had not the skill to follow. He had no better implement with which to entrench himself in the land than a clamshell. But the farmer is armed with plow and spade.

In literature it is only the wild that attracts us. Dullness is but another name for tameness. It is the uncivilized free and wild thinking in Hamlet and the Iliad, in all the scriptures and mythologies, not learned in the schools, that delights us. As the wild duck is more swift and beautiful than the tame, so is the wild—the mallard—thought, which 'mid falling dews wings its way above the fens. A truly good book is something as natural, and as unexpectedly and unaccountably fair and perfect, as a wild-flower discovered on the prairies of the West or in the jungles of the East. Genius is a light which makes the darkness visible, like the lightning's flash, which perchance shatters the temple of knowledge itself—and not a taper lighted at the hearthstone of the race, which pales before the light of common day. . . .

I do not know of any poetry to quote which adequately expresses this yearning for the Wild. Approached from this side, the best poetry is tame. I do not know where to find in any literature, ancient or modern, any account which contents me of that Nature with which even I am acquainted. You will perceive that I demand something which no Augustan nor Elizabethan age, which no culture, in short, can give. Mythology comes nearer to it than anything. How much more fertile a Nature, at least, has Grecian mythology its root in than English literature! Mythology is the crop which the Old World bore before its soil was exhausted, before the fancy and imagination were affected with blight; and which it still bears, wherever its pristine vigor is unabated. All other literatures endure only as the elms which overshadow our houses; but this is like the great dragon-tree of the Western Isles, as old as mankind, and, whether that does or not, will endure as long; for the decay of other literatures makes the soil in which it thrives.

The West is preparing to add its fables to those of the East. The valleys of the Ganges, the Nile, and the Shine having yielded their crop, it remains to be seen what the valleys of the Amazon, the Plate, the Orinoco, the St. Lawrence, and the Mississippi will produce. Perchance, when, in the course of ages, American liberty has become a fiction of the past—as it is to some extent a fiction of the present—the poets of the world will be inspired by American mythology.

The wildest dreams of wild men, even, are not the less true, though they may not recommend themselves to the sense which is most common among Englishmen and Americans today. It is not every truth that recommends itself to the Common sense. Nature has a place for the wild Clematis as well as for the Cabbage. Some expressions of truth are reminiscent, others merely sensible, as the phrase is, others prophetic. Some forms of disease, even, may prophesy forms of health. The geologist has discovered that the figures of serpents, griffins, flying dragons, and other fanciful embellishments of heraldry, have their prototypes in the forms of fossil species which were extinct before man was created, and hence "indicate a faint and shadowy knowledge of a previous state of organic existence." The Hindus dreamed that the earth rested on an elephant, and the elephant on a tortoise, and the tortoise on a serpent; and though it may be an unimportant coincidence, it will not be out of place here to state, that a fossil tortoise has lately been discovered in Asia large enough to support an elephant. I confess that I am partial to these wild fancies, which transcend the order of time and development. They are the sublimest recreation of the intellect. The partridge loves peas, but not those that go with her into the pot.

In short, all good things are wild and free. There is something in a strain of music, whether produced by an instrument or by the human voice—take the sound of a bugle in a summer night, for instance—which by its wildness, to speak without satire, reminds me of the cries emitted by wild beasts in their native forests. It is so much of their wildness as I can understand. Give me for my friends and neighbors wild men, not tame ones. The wildness of the savage is but a faint symbol of the awful ferity with which good men and lovers meet.

I love even to see the domestic animals reassert their native rights—any evidence that they have not wholly lost their original wild habits and vigor; as when my neighbor's cow breaks out of her pasture early in the spring and boldly swims the river, a cold, gray tide, twenty-five or thirty rods wide, swollen by the melted snow. It is the buffalo crossing the Mississippi. This exploit confers some dignity on the herd in my eyes—already dignified. The seeds of instinct are preserved under the thick hides of cattle and horses, like seeds in the bowels of the earth, an indefinite period.

Any sportiveness in cattle is unexpected. I saw one day a herd of a dozen bullocks and cows running about and frisking in unwieldy sport, like huge rats, even like kittens. They shook their heads, raised their tails, and rushed up and down a hill, and I perceived by their horns, as well as by their activity, their relation to the deer tribe. But, alas! a sudden loud Whoa! would have damped their ardor at once, reduced them from venison to beef, and stiffened their sides and sinews like the locomotive. Who but the Evil One has cried "Whoa!" to mankind?

Indeed, the life of cattle, like that of many men, is but a sort of locomotiveness; they move a side at a time, and man, by his machinery, is meeting the horse and the ox halfway. Whatever part the whip has touched is thenceforth palsied. Who would ever think of a side of any of the supple cat tribe, as we speak of a side of beef? . . .

Here is this vast, savage, hovering mother of ours, Nature, lying all around, with such beauty, and such affection for her children, as the leopard; and yet we are so early weaned from her breast to society, to that culture which is exclusively an interaction of man on man—a sort of breeding in and in, which produces at most a merely English nobility, a civilization destined to have a speedy limit.

In society, in the best institutions of men, it is easy to detect a certain precocity. When we should still be growing children, we are already little men. Give me a culture which imports much muck from the meadows, and deepens the soil—not that which trusts to heating manures, and improved implements and modes of culture only!

Many a poor sore-eyed student that I have heard of would grow faster, both intellectually and physically, if, instead of sitting up so very late, he honestly slumbered a fool's allowance. . . .

I would not have every man nor every part of a man cultivated, any more than I would have every acre of earth cultivated: part will be tillage, but the greater part will be meadow and forest, not only serving an immediate use, but preparing a mould against a distant future, by the annual decay of the vegetation which it supports.

29

Excerpts from "Naturalness as a Source of Value"

ROBERT E. GOODIN

The question remains why we should attach special value to the *particular* sort of history which greens identify as a source of value. What is so especially valuable about something having processes? In the word of one memorable title, "What's wrong with plastic trees?"[1]

We might make a start on this question by assimilating it to the problem of fakes and forgeries.[2] The title just mentioned refers to events that are real, not merely feared or fantasized. When the city fathers discovered that real trees could no longer survive their polluted air surrounding the Los Angeles freeways, they tried planting plastic ones there instead. When they found that those plastic trees kept being chopped down, they professed genuine surprise. If so, I suspect that they were just about the only ones to be surprised.[3]

Most of us, confronting schemes involving the destruction of some especially unique bit of the natural landscape, would leap almost automatically to protest at its loss. But perhaps we are too quick to protest. Not so long ago, the Queensland government was proposing to allow the mining of sand on Fraser Island—the only purely sand island on the Great Barrier Reef. When the scheme met with fierce protests, the Deputy Prime Minister of Australia of the time ventured the opinion that such mining could be resumed just as soon as "the community becomes more informed and more enlightened as to what reclamation work is being carried out by mining companies." The same is often said by, or anyway on behalf of, a frequent despoiler of the American landscape, the US Army Corps of Engineers.[4]

Now, we might doubt those claims on any of several grounds. When confronting a particularly notorious mining company or the Army Corps of Engineers making such claims, we might well doubt their sincerity, or their capacity, or their willingness. But let us set all those doubts to one side.

Let us imagine, purely for the sake of argument, a best-case scenario. Let us suppose—*per impossible*, perhaps—that the developers in question offer an iron-clad, legally binding promise to recreate that landscape just as it was, once they have finished. Let us suppose that they provide detailed plans for how they would go about doing so. Let us suppose that we find those plans absolutely convincing. In short, let us suppose that we have no doubt that the landscape will indeed look just the same after they have finished as it did before they began.

Still, I think, we are inclined to object to the proposal.[5] Even if we are convinced that the landscape will look the same, still it will not really *be* the same. Previously it has been the work of nature; afterwards, it will have become the work of humanity. However talented as restorationists the developers' landscape architects might be, the one thing that they cannot possibly replicate is history. They might be able to restore something perfectly in every other respect, but the very act of restoration itself necessarily alters irrevocably a thing's history.

Just as a talented forger might replicate the *Mona Lisa* perfectly, so too might landscape architects replicate nature perfectly. But just as the one amounts to "faking" a painting, so too does the other amount to "faking" nature.[6] Fakes might look the same as the originals but they cannot possibly be the same, for they have different histories and different origins. And in so far as the historical origins of the original are what matter to us, neither can those fakes possibly have the same value for us.

The point of producing fakes, in general, is presumably always to substitute (and, indeed, to attempt to pass off) something less valuable for something more valuable. But whereas in the case of the faked *Mona Lisa* the fake might exist without the original being extinguished, in the case of the faked landscape here in the view of the fake—the restored landscape—would be in place of the original. It is not just a matter of creating in the world something extra, with less (but nonetheless real) value. It is a matter of extinguishing something more valuable and replacing it with something less valuable. To adapt the examples from aesthetics discussed earlier, it is as if we were to make a cast of the *Pietà* through some process which would destroy the original. Even if we could, using that cast, produce a perfect replica of the *Pietà*—even if we could produce such replicas in virtually unlimited quantities—I take it that few would be tempted by the prospect.

Putting the point in terms of "fakery" rather begs the question, of course. By pretending to be something that they are not, fakes necessarily concede their own inferiority. The things which they pretend to be must be somehow superior to that which they actually are, or else there simply would be no point in the pretence.

But it is not just the element of fraud that makes fakes, forgeries and counterfeits less valuable. The same is true (to a lesser but nonetheless significant extent) with restorations, reproductions and replicas, where the element of fraud is absent. It is the very act of copying something else which, in the end, crucially concedes the superiority of that which is copied. What is the point of copying, if you could have done better all by yourself, starting from scratch?

The developers in our hypothetical example can thus ill afford to accept "faking nature" as a description of what they propose doing. That much was probably obvious from the start. The more surprising implication of those observations just offered is that it is not merely the pejorative description which the developers must shun. If they are to avoid conceding the necessary inferiority of the landscape which they will leave, as compared to that which they had found, they must not even try to recreate the original. They must instead assert that they will create something different from and better than—or, anyway, not comparable to—the natural landscape which their handiwork will replace.

Landscape architects themselves may well suffer just such hubris. They may well regard that claim, however immodest, as being nonetheless perfectly true. Still, few firms proposing to despoil the natural environment would be willing to take any such stance publicly. That does not necessarily mean that they think that it is untrue, of course. They might merely regard it as impolitic. But even for such a claim to be impolitic, it is necessarily the case that it must be seen as untrue—indeed, untrue to the point of outrage—in the eyes of the public at large.

Why should that be so? There are many possible reasons, of very different characters. In practice, such attitudes may reflect no more than well-founded scepticism about the strictly limited capacities of restorationists. We may suppose that the technology is simply not (or, anyway, not yet) up to the task. That is undoubtedly true at present. As long as it remains so, that is a sufficiently compelling consideration in and of itself: there is hardly any pragmatic need to search for further arguments against any proposal involving destroying-then-recreating nature.

If that were all that was involved in objections to those proposals, though, then objections to restoring nature would be merely contingent rather than analytically necessary ones—as the parenthetical "anyway not yet" in the earlier formulation was meant to emphasize. Suppose technology someday makes sufficient strides to enable it, finally, to replicate nature perfectly.[7] Once a perfect replica really can be produced, we would have no basis for continuing to object—on these grounds, at least—to a strategy of destroying-then-recreating nature.

For larger philosophical purposes, then, we need to probe further. In the world as we know it, it may well be enough to say that the restorers just are not good enough. But for deeper theoretical purposes, we must try—however hard it may be—to contemplate a world where the restorer's art really has been mastered to perfection. Would we still have any objections, even if restorers could guarantee with complete certainty to restore the natural world to precisely its original state, after developers had despoiled it? And if so, how might we ground such objections?

For a clue, let us return to the problem of the value of the fake *Mona Lisa* as compared to that of the original. Intuitively, we want to say that the original is necessarily more valuable because of who painted it—Leonardo rather than some second-rate imitator. But if the agent responsible for the fake were so talented as to be able to reproduce the image *perfectly*, then she must be awfully good indeed. As sports commentators might say, "on the day" she was every bit as good as Leonardo himself. And if it is talent—as represented in the capacity to paint a picture like *Mona Lisa*—that makes us value paintings created by Leonardo, why not value equally paintings done by others who were, at least on the day the painting was done, every bit as good as Leonardo himself? After all, the painter of the perfect copy painted a *Mona Lisa*, too.

In the case of the fake *Mona Lisa*, the reason for rejecting that line of thought is clear enough, I suppose. The reason we value Leonardo's works especially highly has to do not only with their own merits, taken individually, but also with who their creator was and with *all else* that he did. We value the *Mona Lisa* in part as a beautiful painting, in its own right. But we also value it, in part, as a "manifestation of Leonardo's genius," as I said in introducing this example.

It might at first seem odd to suggest that, had Leonardo accomplished less, we would value the *Mona Lisa* less: after all, it would retain all its ethereal beauty, even as a one-off. In other ways, however, this theory is far from counterintuitive. We undoubtedly do tend to value lesser works from the hand of Leonardo more than on their purely aesthetic merits we ought to do, just because they are from his hand.[8] By the same token, we seem not to value paintings of considerable merit from those who produced only one such masterpiece in their lives as much as we do similar works from those who have produced a substantial corpus of such works—and, once again, this sort of analysis seems to help explain why.

Transpose that argument, now, from the realm of art into the realm of the natural environment. The analogy would suggest that one reason we might not be satisfied with restorers' recreation of nature is that they could only do it on a one-off basis. They might be able to reproduce any *particular* element in the natural world—just as the artistic forger might fortuitously manage to reproduce any particular image. But they would not be able to recreate nature as a whole, across the board, any more than the cleverest forger could manage to reproduce perfectly all of Leonardo's many works. And in so far as their talents are limited in this way, they deserve less respect than does the master craftsman whose talents are not similarly circumscribed.

This sort of argument carries us a considerable distance. But the philosopher's philosopher might insist—rightly, alas—that it still constitutes a merely contingent objection to a strategy of destroying-then-recreating nature. For it is a purely contingent matter that our capacity to create a perfect replica of nature as a whole is limited in this way. Although that is undeniably true, I am not at all sure that we need to be particularly embarrassed by that fact. After all, though a mere contingency, that contingency is virtually certain to be true for as long as any present policy-maker cares to contemplate. Still, for those who demand cast-iron arguments to show that nature's products are *necessarily* superior to humanity's, this argument will not (quite) suffice.

Much the same objection proves even more telling against another initially plausible analysis of this problem. Another reason for valuing Leonardo's original works more highly than the talented copyist's perfect reproductions has to do with their very *originality*. This analysis, too, would have us accord Leonardo's *Mona Lisa* extra value on account of who Leonardo was—but on this approach not merely, or even mainly, on account of what Leonardo actually did so much as on account of what else he *might* have done. Even if the copyist can reproduce perfectly each work that Leonardo actually executed, copyists by their nature require an original from which to copy. They may be able to paint a perfect *Mona Lisa* once they have seen Leonardo's, but they could never have come up with the image themselves in the first place. That, on one plausible account, is why the copyists' work must necessarily be disvalued compared to that of the original artists whose work they copy.[9]

Something very much like that might be said about the special value of naturally occurring properties. Much of the sense of wonder that people feel in confronting the natural world has to do with the endless variation and rich complexity that they find within it. And if that is any large part of what we value about natural, then simply being able to restore or recreate or replicate nature perfectly is not good enough. In supplanting nature, we have not just taken away what was actually there; we will also have taken away what all else might have been or might yet be there.

However plausible that may be as an argument in the aesthetics, it probably has to be regarded as fatally flawed in its environmental application. There, it is not just the philosopher's philosopher who would rightly worry about the possibility that artificial processes might be able someday to mimic natural ones perfectly.[10] As regards living organisms, at least, it seems that modern technology is—alas—every bit as capable of generating genetic variation, by inducing mutations, as is any naturally occurring process. Furthermore, through the further perfecting of recombinant DNA technology we might be able to meet or beat nature's capacity for "creativity." None of this is necessarily meant as praise for those modern technologies, mind you. But praise or criticism, those observations nonetheless undermine any analysis that traces the peculiar value of nature to its unique variability or creativity.[11]

There is a parallel problem with various other ways of characterizing what is valuable about nature in terms of "diversity," "complexity" or "fecundity." Certainly those are attributes of natural creation, and certainly they are valued ones. But of course genetic engineering can (in principle, and probably even in practice) generate just as much diversity, complexity and fecundity—yet it would not be the same. It is not diversity, complexity or fecundity *as such*, that we value. It is instead the diversity, complexity and fecundity of natural processes and their products. The value of such attributes derives from the fact that they result from natural processes, rather than the other way around. Pointing to such attributes cannot, then, explain why we value the natural processes that manifest them.

To explain that, we are, I think, driven to an argument along the following lines. (1) People want to see some sense and pattern to their lives. (2) That requires, in turn, that their lives be set in some larger context. (3) The products of natural processes, untouched as they are by human hands, provide precisely that desired context.[12] In what follows, I shall discuss each of these propositions in turn.

The first step in that argument is a familiar philosophical theme growing out of recent critiques of utilitarianism. In the now-familiar objection, utilitarianism as sometimes construed would seem to recommend that people be constantly investing the next ounce of their energies wherever it would do the most good for humanity at the margin. People following that advice would be spending twelve minutes raising money for starving Eritreans, two-and-a-half minutes feeding their own children, ten seconds signing a petition for world disarmament, eighteen minutes sweeping the floors of the local charity hospital, and so on. Such a life might be utility maximizing for the world at large. But from the inside, such a life would seem deeply unsatisfying. It would lack unity, coherence, purpose.[13] How telling that proposition is as a critique of utilitarianism is unclear.[14] But the truth of that basic proposition seems indisputable. What makes people's lives seem valuable to those who are living them is the unity and coherence of the projects comprising them.

The Green theory of value next adds a second thought closely related to that first one. Just as people want to see some unity and coherence among the various plans and projects that comprise their lives, so too do they want to see those plans and projects set in some larger context outside their own lives.[15] That is not necessarily to say that they want to "change the world" or even necessarily that they want to "make their mark on the world." It is just to say that, whatever the source of people's undeniable desire to see some coherence within their lives, the same thing would naturally lead them to want to see some continuity between their inner worlds and the external world. Whatever makes people strive for harmony within their own lives would also lead them to strive to lead their lives in harmony with the external world.[16]

The third and final step in the argument is just to say that natural processes provide just such a larger context. The products of a purely natural process are ones that are, by definition, not the product of deliberate human design. Things that are natural in that sense therefore provide a context of something outside of ourselves—a larger context within which we can set our own life plans and projects.[17]

NOTES

1. Martin H. Kriger, "What's Wrong with Plastic Trees?," *Science* 179 (1973): 446–455.

2. The analogy is strongest if we adopt a religious attitude towards nature, so in both cases we accord particular value to the unspoilt original (environment; work of art) as a manifestation of our respect for its creator. That line has obvious appeal for a "social ecologist" for whom the value of untouched nature is always necessarily derivative, in some way or another, from the values and interests of people (be they consumers or producers, present or future generations). But, as I shall argue below, we can derive value from certain attributes of a natural process of creation without talking about a Creator.

3. For interestingly varied reasons, legal philosophers who subsequently joined the debate inspired by Kriger's article were not. Lawrence H. Tribe, "Ways Not to Think about Plastic Trees: New Foundations for Environmental Law," *Yale Law Journal* 83 (1974): 1315–1348; Lawrence H. Tribe, "From Environmental Foundations to Constitutional Law: Learning from Nature's Future," *Yale Law Journal* 85 (1975): 545–556; Mark Sagoff, "On Preserving the Natural Environment," *Yale Law Journal* 84 (1974): 205–267.

4. Similarly, the US Army Corps of Engineers—a frequent despoiler of the American government—regularly asserts its capacity to restore whatever its activities might have damaged. For an excellent discussion both of those cases and of the larger issues they raise, see Robert Elliot, "Faking Nature," *Inquiry* 25 (1982): 81–93.

5. As indeed we would object even if the visual aspect of the landscape were never changed in the first place. Consider, in this connection, objections to the 1981 proposal to lease a National Trust property at Bradenham in Buckinghamshire to NATO for use as a bunker. The bunker would have been below ground, so (damage done by construction apart) the purely visual aspects of the site would have remained unchanged. Even granting that, though, protestors complained that their appreciation of the site would be lessened by the knowledge that there was a bunker beneath the ground. I am grateful to Albert Weale for this example.

6. Elliot, "Faking Nature."

7. And we have no reason, on this argument, to suppose that it cannot.

8. Dutton, in his editorial preface to *The Forger's Art*, reflects upon the all-too-standard case of "a musicologist who would wax ecstatic over some trifling bit of Mozart's juvenilia ("adumbrasions of the great *G Minor Symphony*") treats with barely disguised contempt a really outstanding and inventive quartet by the generally second-rate Luigi Cherubini." Denis Dutton, preface to Denis Dutton, ed., *The Forger's Art* (Berkeley: University of California Press, 1983), ix.

9. Thus, Otto Kurz says, "The reasoning 'if my work is mistaken for a Rembrandt, I am as good as Rembrandt' is illogical because it confuses original creation with imitation. . . ." Otto Kurz, *Fakes*, 2nd ed. (New York: Dover, 1967), 319. Of course, masters can serve as copyists, too. "A copy of a Lastman by Rembrandt may well be better"—more valuable—"than the original," Nelson Goodman says. Nelson Goodman, *Languages of Art* (Indianapolis, IN: Bobbs Merrill, 1968), ch. 3; Nelson Goodman, *Fact, Fiction and Forecast* (London: Athlone Press of University of London, 1983), 100. But if so, it can only be because it is a poor copy (different from but better than the original) or, if a perfect copy, because it is set in the context of a better painter's corpus (Rembrandt's, rather than Lastman's). In any case, as Rembrandt paintings go, it would definitely count among his lesser works, precisely because it embodied less of his own originality.

10. Landscapers might still be a slightly different matter: even master landscape gardeners such as Capability Brown still had to work from naturalistic models; and it is hard to imagine any

of them conceiving a design for something like the Grand Canyon or Yosemite, *ex ante* of anyone having described those canyons to them.

11. Notice, however, that that is once again an argument not so much for recreating nature as for being as creative as nature, perhaps in some other direction altogether. Supposing it is creativity that we value about nature, the implication is that if modern technology is to aspire to similar value as nature then it must not merely mimic nature but transcend it. On its face, that seems to be an argument for running roughshod over nature's creation. But, equally (in my view, preferably), it might be regarded as a *reductio ad absurdum* of that analysis of what it is that we value about nature.

12. "Untouched by human hands" is only a first approximation to the value at work here: for a more refined statement, see the discussion of "humanity as part of nature" in the next section.

13. Bernard Williams, "A Critique of Utilitarianism," in J. J. C. Smart and Bernard Williams, *Utilitarianism, For and Against* (Cambridge: Cambridge University Press, 1973); Bernard Williams, *Moral Luck* (Cambridge: Cambridge University Press, 1981). See also Richard Wollheim, *The Thread of Life* (Cambridge: Cambridge University Press, 1984); and Robert Nozick, *Philosophical Explanations* (Cambridge, MA: Harvard University Press, 1981), 403–405.

14. The utilitarian reply—trivial, but decisive—surely is simply to say that, assuming the psychological fact of the matter is as the critics claim, then utilitarian calculations should simply be revised accordingly.

15. A contingent preference along these lines would suffice for my purposes, just as long as that contingency was sufficiently common and the resulting preferences socially standard ones. A bolder version of this thesis might hold that people do not just happen to want to set their lives in some larger context but, rather, are morally obliged to do so. The sort of "humility" thus involved might be deemed a precondition for other human moral excellences—most especially, of to a capacity to appreciate the Good, as opposed merely to one's own good. Thomas Hill Jr., "Ideals of Human Excellence and Preserving the Natural Environment," *Environmental Ethics* 5 (1983): 211–224; R. Elliot, *Faking Nature*. Passmore's denigration of environmental vandalism suggests something of the same line of reasoning. John Passmore, *Man's Responsibility for Nature*, 2nd ed. (London: Duckworth, 1980), 28–42.

16. Tom Regan develops G. E. Moore's theses about "organic unity" in an ecological context. Tom Regan, *Bloomsbury's Prophet: G. E. Moore and the Development of His Moral Philosophy* (Philadelphia: Temple University Press, 1986), 199–202. Nozick develops similar themes in more theological directions (*Philosophical Explanations*).

17. There is an intermediate case consisting of those states of affairs which are the accidental, unintended consequences of intentional human actions which are directed at other ends. For those seeking to set their lives in the context of something outside themselves, that intermediate case is of intermediate value: less good (because less "outside of themselves") than nature, but better (because more "outside of themselves") than states which are the products of deliberate human design.

30

Excerpts from "Conservation"

JOHN PASSMORE

To conserve is to save, and the word "conservation" is sometimes so used as to include every form of saving, the saving of species from extinction or of wildernesses from land-developers as much as the saving of fossil fuels or metals for future use. Such organizations as the Australian Conservation Society, indeed, focus their attention on kangaroos and the Barrier Reef, not on Australia's reserves of oil and fuel. In accordance with what is coming to be the common practice, however, I shall use the word to cover only the saving of natural resources for later consumption. Where the saving is primarily a saving *from* rather than a saving *for*, the saving of species and wildernesses from damage or destruction, I shall speak, rather, of "preservation." My concern in the present chapter is solely with conservation, in the sense in which I have just defined it.

On a particular issue, conservationists and preservationists can no doubt join hands, as they did to prevent the destruction of forests on the West Coast of the United States. But their motives are quite different: the conserver of forests has his eye on the fact that prosperity, too, will need timber, the preserver hopes to keep large areas of forest forever untouched by human hands. They soon part company, therefore, and often with that special degree of hostility reserved for former allies.[1] So it is as well that they should be clearly distinguished from the outset.

The conservationist movement is now no novelty. It dates back in the United States to the latter half of the nineteenth century, although it suffered a marked decline in importance during the period 1920–60 when, under the influence of technological industrial advance, men were particularly disinclined to believe that natural resources were to any degree limited. Its spirit and ideals are summed up in the Declaration prepared in 1908 by a Conference of United States governors:

> We agree that the land should be so used that the erosion and soil wash shall cease; and that there should be reclamation of arid and semi-arid regions by means of irrigation, and of swamp and overflowed regions by means of drainage; that the waters should be so conserved and used as to promote navigation, to enable the arid regions to be reclaimed by irrigation, and to develop power in the interests of the people; that the forests which regulate our rivers, support our industries,

and promote the fertility and productiveness of the soil should be preserved and perpetuated; that the minerals found so abundantly beneath the surface should be so used as to prolong their utility; that the beauty, healthfulness, and habitability of our country should be preserved and increased; that sources of national wealth exist for the benefit of the people, and that monopoly thereof should not be tolerated.[2]

The conservationist, so much will be apparent, has no doubt that civilization ought to continue; he fully accepts the general principle that it is man's task to make of the world a better place for men to live in. Admittedly, the governors' declaration includes a side-reference to the need, in order to achieve that end, to preserve as well as to remake. But its main emphasis is on remaking, on the draining of swamps and the irrigating of deserts, the damming of rivers. The governors would have been astonished to hear the preservationist argue that deserts ought sometimes to be left unirrigated, rivers left undammed and swamps left undrained. What the conservationist opposes is not the harnessing of nature for man's economic purposes but carelessness and wastefulness in doing so. The many American conservationist acts of the late nineteenth century were largely directed against wasteful methods of oil-drilling or coal extraction. Conservation was identified with "careful husbandry."[3]

One might describe anti-pollution measures as a form of conservation. For air, water, lakes, rivers, the sea, are all of them economic resources. To pollute them is to use them wastefully; to cause them seriously to deteriorate is to make it impossible for posterity to continue to civilize the world—or perhaps, to continue to survive. There is, however, a fundamental difference between pollution and the exhaustion of resources, considered as ecological problems—a difference which profoundly affects our willingness to make sacrifices in order to solve them. Pollution is something we should like to get rid of in our own immediate interests; our doing so will benefit posterity, no doubt, but it will also benefit us, quite directly. Only in a few species cases—such as the storage of atomic wastes—does the campaigner against pollution call upon us to act solely for the benefit of posterity. There is no incompatibility between our enjoying and posterity's enjoying clean water, fresh air, open spaces, the company of birds and animals. The cleansing of the Pittsburgh air or the Willamette river are gains not only for posterity but for the present generation. In contrast, it is in the interest of posterity, not in our own interest, that we are called upon to diminish the rate at which we are depleting our natural resources.

So the conservationist program confronts us with a fundamental moral issue: ought we to pay any attention to the needs of posterity? To answer this question affirmatively is to make two assumptions: first, that posterity will suffer unless we do so; secondly, that if it will suffer, it is our duty so to act as to prevent or mitigate its sufferings. Both assumptions can be, and have been, denied. To accept them does not, of course, do anything to solve the problem of conservation, but to reject them is to deny that there is any such problem, to deny that our society would be a better one—morally better—if it were to halt the rate at which it is at present exhausting its resources. Or it is to deny this, at least, in so far as the arguments in favor of slowing-down are purely conservationist in character—ignoring for the moment, that is, such facts as that the lowering of the consumption rate is one way of reducing the incidence of pollution and that a high rate of consumption of metals and fossil fuels makes it impossible to preserve untouched the wilderness in which they are so often located.

To begin with the assumption that posterity will suffer unless we alter our ways, it is still often suggested that, on the contrary, posterity can safely be left to look after itself, provided only that science and technology continue to flourish. This optimistic interpretation of the situation comes especially from economists and from nuclear physicists. So Norman McRae, writing in the *Economist*, happily informs us that with the aid of the new sensors, carried by satellites, which are now available to geologists, vastly increased reserves of fossil fuels and minerals will soon be available for man's use. (They would certainly have to be vast: if our rate of consumption increases at the rate at which it has been increasing over the last decade even resources five times as great as those already known would be exhausted within the life-time of most children now less than ten years old.)[4] "My own guess," he writes of fossil fuel in particular, "is that we will have an embarrassingly large fuel surplus"—this by the year 2012, when so many calculations predict the exhaustion of petroleum. "And nuclear fission . . . should [by that time]," he continues, "give to mankind a virtually unlimited source of industrial power, with the oceans serving as boundless reservoir of fuel—even for a world population far larger and many times richer than today's." If there is a problem, on his view, it lies only in the disposal of wastes—a pollution rather than a conservation problem.[5] The nuclear physicist Alvin Weinberg is no less confident that "the problem of source depletion is a phony." With the exception of phosphorus, he tells us, "the most essential resources are virtually inexhaustible."[6] The physical chemist Eugene Rabinowitch, editor of the *Bulletin of Atomic Scientists*, is equally reassuring. Modern science, he informs his readers, is "showing us ways to create wealth from widely available raw materials—common minerals, air, sea water—with the aid of potentially unlimited sources of energy (fusion power, solar energy)."[7] If these scientists, these economists, are right, there simply is no "problem of conservation."

Very many scientists, of course, take the opposite view, especially if they are biologists. Expert committees set up by such scientific bodies as the American National Academy of Sciences have, in fact, been prepared to commit themselves to definite estimates of the dates at which this resource or that will be exhausted.[8] This is always, however, on certain assumptions. It makes a considerable difference whether one supposes or denies that rates of consumption will continue to increase exponentially as they have done since 1960; it makes a very great—in many cases overwhelming—difference whether one supposes or denies that substitutes will be discovered for our major resources. The Academy's extrapolations are best read as a *reductio ad absurdum* of the supposition that our present patterns of resource consumption can continue even over the next century.

The possibility that substitutes will be discovered introduces a note of uncertainty into the whole discussion, an uncertainty which cannot be simply set aside as irrelevant to our moral and political decisions about conservation, which it inevitably and properly influences. At the moment, for example, the prospect of developing a fuel-cell to serve as a substitute for petrol is anything but bright; confident predictions that by 1972 nuclear fission would be available as an energy source have proved to be unrealistic. But who can say what the situation will be in twenty years' time?[9] The now common-place comparison of earth to a space-ship is thus far misleading: the space-ship astronaut does not have the facilities to invent new techniques, nor can he fundamentally modify his habits of consumption. Any adequate extrapolation would also have to extrapolate technological advances. But by the nature of the case—although technologists have a bad habit of trying to persuade us otherwise—we cannot be at all certain when and whether those advances will take place, or what form they will assume, especially when, unlike the moonshots, they involve fundamental technological innovations such as the containing of nuclear fusion within a magnetic field.

No doubt, the space-ship analogy is justified as a protest against the pronouncements of nineteenth-century rhetoricians that the earth's resources are "limitless" or "boundless." (It was often supposed, one must recall, that oil was being produced underground as fast as it was being consumed.) Fuel-cells, nuclear fusion reactors, machinery for harnessing solar energy all have to be built out of materials, including, as often as not, extremely rare metals. Men can learn to substitute one source of energy or one metallic alloy for another, the more plentiful for the less plentiful. But that is the most they can do. They cannot harness energy without machines, without radiating heat, without creating wastes. Nor can they safely presume that no source of energy, no metal, is indispensable; there is nothing either in the structure of nature or in the structure of human intelligence to ensure that new resources will *always* be available to replace old resources. Think how dependent we still are on the crops our remote agriculture-creating forefathers chose to cultivate; we have not found substitutes for wheat, or barley, or oats, or rice. Nor have we domesticated new animals as beasts of burden. So, quite properly, the conservationist points out.

The uncertainties, however, remain. We can be confident that some day our society will run out of resources, but we do not know when it will do so or what resources it will continue to demand. The Premier of Queensland recently swept aside the protests of conservationists by arguing that Queensland's oil and coal resources should be fully utilized now, since posterity may have no need for them. This is not a wholly irrational attitude. One can readily see the force of an argument which would run thus: we are entitled, given the uncertainty of the future, wholly to ignore the interests of posterity, a posterity whose very existence is hypothetical—granted the possibility of a nuclear disaster—and whose needs, except for such fundamentals as air and water, we cannot possibly anticipate.

Let us begin by supposing, for the sake of argument, that the optimists are right, that we have no good ground for believing that any particular resource will still be in demand at any particular time in the future, that all we can say with certainty is that sooner or later, at some very remote epoch, civilization will run out of resources. That is as far as the space-ship analogy can carry us. Should we then set aside "the problem of resource depletion" as a mere pseudo-problem, on the ground that so distant a posterity is no concern of ours?

Kant thought it was *impossible* for us to do so. "Human nature," he once wrote, "is such that it cannot be indifferent even to the most remote epoch which may eventually affect our species, so long as this epoch can be expected with certainty."[10] And other philosophers have demonstrated a similar concern for their most distant descendants. That "the Universe is running down," as the theory of entropy was at one time interpreted as demonstrating, is a conclusion William James found intolerable; only a faith in God, he thought, could deliver men from "the nightmare of entropy."[11] John Stuart Mill was reduced to a state of profound melancholy by the reflection that since there is a limited number of musical combinations, music will some day come to a standstill.[12] (This is an illuminating example; Mill did not, of course, envisage electronic music.) Similarly, there are some who felt a sense of relief when astronauts reached the moon; it kept open the possibility, they thought, that men would one day be able to leave a planet made inhospitable by decay to continue the human race on some other shore. But for myself, I more than doubt whether a concern for the ultimate future of the human race forms an essential part of human nature. Kant notwithstanding, a man is not a *lusus naturae* or a moral monster because his sleep is undisturbed by the prospect that human beings will at some infinitely remote date run out of resources or that, however they behave, they will eventually be destroyed by cosmological convulsions.

But even if this is not so, even if men are inevitably perturbed by the reflection that in some remote epoch their race will be extinct, their perturbation does not, of itself, generate any sort of responsibility towards a posterity whose fate they may lament but cannot prevent. The case is no doubt rather indifferent when what is involved is the long-term exhaustion of resources rather than what James had in mind—the "running down of the Universe." For men can, in principle, so act as to delay that exhaustion whereas they cannot delay the earth's cosmological destruction. But if all that can be predicted—the hypothesis

I have for the moment adopted—is a very long-term exhaustion of resources, no *immediate* action on our part seems to be called for. Anything we can do would, over millions of years, be infinitesimal in its effects; not even by reducing our consumption of petrol to a thimbleful apiece could we ensure the availability of a similar quantity to our remotest descendants.

If the exhaustion of resources is really, as the more optimistic scientists assert, a problem only for a future so distant as to be scarcely imaginable, then I do not think there is any good reason for our troubling our heads about it.

Should we, then, move to the opposite extreme and leave the future to look after itself, concentrating all our efforts on making the best we can of today? Then it would not matter whether the scientists are right or wrong in predicting an early exhaustion of resources. For the future would be none of our business. This, according to Matthew (6:34), is what Jesus taught: "Take therefore no thought for the morrow; for the morrow shall take thought for the things of itself. Sufficient unto the day is the evil thereof." Nor is it by any means a preposterous attitude. We are confronted, in the present, by evils of every kind: in some of the developing countries by precisely the starvation, the illiteracy, the abysmal housing, the filth and disease which we fear for posterity; in many of our own cities by urban decay, impoverished schools, rising tides of crime and violence. It might well seem odd that the conservationist—and this is an argument not uncommonly directed against him—is so confident that he knows how to save posterity when he cannot even save his own contemporaries. Over a large part of the globe, too, the "needs of posterity" are already being used to justify not only tyranny but a conspicuous failure to meet the needs of the present. One can easily be led to the conclusion that it would be better to let the morrow look after itself and to concentrate, as more than sufficient, upon the evils of our own time.

The view that men ought to concern themselves about the fate of posterity as such—as distinct from the fate of their children, their reputation, their property—is a peculiarly Western one, characteristic, even then, only of the last two centuries. It arises out of a uniquely modern view about the nature of the world, man's place in it, and man's capacities. A Stoic confronting our present situation might well believe that we were nearing the end of a cycle, that there would shortly be a vast conflagration after which everything would begin again. Within such a cycle, men could help to make the world a better place, but they could not possibly, on the Stoic view, arrest its cyclical course. For theological reasons, Augustine rejected the cyclical conception of history; Christians, he was confident, are saved or damned for eternity, not only for a particular cycle of existence. It was he who introduced into Western thought the idea of "the future of mankind."[13] Augustine did not suggest, however, that human beings could in any way determine in what that future would consist: the future lay in God's hands. God, so Matthew (6:3) tells us, knows that men have need of food, of drink, of clothing; he will provide. Francis Bacon, writing late in the sixteenth century, was still of that opinion. "Men must pursue things which are just in [the] present," he wrote, "and leave the future to the divine Providence."[14] Although the lineal view of history left room for, it did not of itself entail, the doctrine that each generation should regard the future of mankind as its responsibility.

In Kant's philosophy, the idea of a duty to posterity assumes, perhaps for the first time, a central place. But although he exhorted men to sacrifice themselves for a posterity which would enjoy the fruits of their toil—thus initiating what Herzen was illuminatingly to call the "caryatid" theory of history, according to which man, like the figures on a baroque façade, bears the burden of the future on his shoulders—Kant had too little confidence in man to suggest that the future is entirely of his making.[15] Providence, working through laws of progress, is still for Kant the principal historical agent. And even when philosophers began to argue, against Kant, that secular, rather than supernatural, forces would determine the future of mankind, they still did not suggest that man's deliberate decisions could wholly determine the future course of history. Progress, Engels for example thought, was inevitable; men can co-operate with, or resist, the forces that will bring about a better society but they cannot in the end thwart its emergence.

Had Western man been able to continue to believe either that the future of the world lies in the hands of Providence or that progress is inevitable he would not feel his present qualms about the future; the problem of conservation would not exist for him. But while a belief in the inevitability of progress still affects the thinking of a great many of the world's inhabitants—not only, but most obviously, in Marxist countries—among Western intellectuals it tends now to be replaced by the quite opposite view that unless men change their ways, catastrophe is inevitable. Such prophecies are often conjoined with the assumption that simply by *deciding* to change their ways men can create a better world. This is the fullest expression of the activist side of the Western tradition, its Pelagianism, what its nineteenth-century critics called its "voluntarism." Men are now being called upon, entirely without help, to save the future. The future, it is presumed, lies entirely in their hands; tomorrow *cannot* take thought of itself; it is they, now, who have to save tomorrow, without any help either from Providence or from History. No previous generation has thought of itself as being confronted by so Herculean a task.

How far is such a picture of man's relation to the future acceptable? Obviously, this is a highly controversial matter, entangled, in the last resort, in the old dispute between determinism and free will. For my part, I am perfectly willing to accept what is negative in it: there is no guiding hand, secular or supernatural, which will ensure that man is bound to flourish, let alone survive. But it is a different matter to suppose that the future lies entirely in men's own hands, if this is taken to imply that, given sufficient goodwill, men need not fear for the future—or, more generally, that once men decide what they want the future to be like they can bring it to pass. . . .

Now to sum up. Whether there is a problem of conservation is warmly disputed; our conclusion on that point will depend on whether or not we are convinced that certain technological devices will be available for use in the very near future. And the grounds on which most of us have to come to a decision on this point are anything but adequate. We cannot be certain that posterity will need what we save—or on the other side that it will not need what we should not think of saving. There is always the risk, too, that our well-intentioned sacrifices will have the long-term effect of making the situation of posterity worse than it would otherwise be. That is the case for simply ignoring posterity, and doing what we can to repair present evils.

On the other wise, this is not our ordinary moral practice. We love, and in the virtue of that fact we are prepared to make sacrifices for the future and are not prepared to take risks, arising out of uncertainties, which would otherwise strike us as being rational. No doubt, we often make the wrong decisions; trying to protect what we love we in fact destroy it. Over-protection can be as damaging as neglect. But these uncertainties do not justify negligence. Furthermore, we now stand, if the more pessimistic scientists are right, in a special relationship to the future; unless we act, posterity will be helpless to do so. This imposes duties on us which would not otherwise fall to our lot.

But granted that it ought to do so, is any democratic-capitalist state likely to introduce effective measures to conserve resources? Conservation has not the same popular appeal as pollution; especially when it involves genuine sacrifices. Even those who are in favor of it in principle may still find it hard to accept the view that they must, for example, reduce the level of winter heating to which they have been accustomed. Indeed, if conservation were an isolated issue, one might be inclined to doubt whether the conservationist program would ever win widespread support. Admittedly it has already had its success. Afforestation and the control of soil erosion were both of them conservationist programs which have more than occasionally, if by no means universally, been implemented. But they involved far less, and less widespread, sacrifice than would the adoption of a policy of conserving fossil fuels. Furthermore, they were often accompanied by a degree of subsidization which more than compensated for temporary losses and which could not be matched in a thorough-going conservationist programme.

The fact is, however, that conservationism is not an isolated issue. By "doing what is just in the present," we may be doing what is best for posterity to a degree somewhat greater than is ordinarily allowed. If we were to concentrate on improving public facilities as distinct from private wealth, on diminishing noise and air pollution by substantially reducing automobile traffic, we might find that we have in the process decreased the level of industrial activity to a relatively harmless point. In general, people do not seem to find their present mode of life particularly enjoyable; we certainly need to experiment with alternatives which are at the same time less polluting and less wasteful of resources. The recycling of resources both benefits us, as helping to solve the problem of wastes, and would, we hope, also benefit posterity. The extension of public transport will help our cities as well as reducing the use of fossil fuels. Education, which certainly needs improvement, is not a great polluter, or a great user of scarce resources. In the uncertainties in which we find ourselves, it is perhaps on these double-benefit forms of action that we should concentrate our principal efforts. Certainly we shall have set up objectives which are hard enough to realize, but the benefits of which are immediately obvious—or obvious to those who are not enamored of squalor, decaying schools, over-crowded cities, inadequate hospitals and nursing homes, crime or official violence. If they are hard to realize, it is because the ownership of commodities, private affluence, is in fact generally preferred, when the crunch comes, to the improvement of the public conditions of life.

There is also the question of time. The degree of urgency, on the views of some scientists, is very great; political action is generally speaking slow, and in this case is subjected to an enormous range of special interests. In these circumstances there is a strong temptation to fall back on the ideal of the strong man, who could conserve by the direct exercise of coercion. I have refused to accept this as a "solution" for the conservation problem, partly because I do not think there is any good reason for believing that any "strong man" who is likely to emerge after the collapse of democracy would be primarily concerned with conservation and partly because I do not believe this to be the kind of cost we ought to be prepared to meet, for posterity's sake as well as for our own. Much the same is true of the suggestion that what we should work for is the collapse, as rapidly as possible, of our entire civilization, as the only way of conserving resources. The cost would be enormous; the benefit more than dubious.

My conclusions are limited and uncertain. That is how it should be; as Socrates liked to point out, confidence based on ignorance is not a virtue. Nor is there any point in turning to religion in an attempt to bolster up our confidence. Men do not need religion, or so I have argued against Montefiore, to justify their concern for the future. That concern arises out of their character as loving human beings; religion, indeed, often tells its adherents—whether in the accents of the East or the West—to set such concern aside, to "take no thought of the morrow." If Allah, as the proponents of stewardship now like to emphasize, is represented in the Koran as setting up men on earth as his "deputy," it by no means follows that a Muslim ruler can be rebuked for heresy when he sells off the oil which Allah has hidden beneath the desert sands. In short, the faithful cannot hope by recourse to Revelation, Christian or Muslim, to solve the problems which now confront them. Nor will mysticism help them. There is no substitute for hard thinking, thinking which cuts clean across the traditional disciplines. But intellectuals, no less than the religious, have their delusions of grandeur. Society, as much as nature, resists men's plans; it is not wax at the hands of the scientist, the planner, the legislator. To forget that fact, as a result of conservationist enthusiasm, is to provoke rather than to forestall disaster.

NOTES

1. Cf. O. C. Herfindahl, "What Is Conservation?," in *Three Studies in Minerals Economics* (Washington, DC: Resources for the Future, 1961), 1–12.

2. *The Declaration of Governors for Conservation of Natural Resources* is reprinted in I. Burton and R. W. Kates, eds., *Readings in Resource Management and Conservation* (Chicago: University of Chicago Press, 1965), 186–188. For the history of the conservationist movement see S. P. Hays, *Conservation and the Gospel of Efficiency* (Cambridge, MA: Harvard University Press, 1959). The arguments put forward by the leading conservationist Gifford Pinchot in his *The Fight for Conservation*, first published in New York in 1909, are remarkably like more recent arguments.

3. See the documents included in F. E. Smith, ed., *Conservation in the United States: A Documentary History* (New York: Chelsea House, 1971), especially vol. 2 on *Minerals*, ed. W. T. Doherty.

4. See Table 4 in D. H. Meadows, D. L. Meadows, J. Randers and W. W. Behrens III, *The Limits to Growth* (London: Universe Books, 1972), 56.

5. Norman McRae, "The Future of International Business," *Economist*, January 11, 1972.

6. "Prophet of Nuclear Power," Alvin Weinberg talks to Dr. Gerald Wick, *New Scientist*, January 20, 1972.

7. Eugene Rabinowitch, "Copping Out," *Encounter* 27, no. 3 (1971): 95.

8. See "The Energy Crisis," parts 1 and 2, in *Science and Public Affairs: Bulletin of the Atomic Scientists* 27 (September–October 1971): 8–9.

9. See M. K. Hubbert, "Energy Resources," in *Resources and Man: A Study and Recommendations* (San Francisco: W. H. Freeman, 1969), 157–242.

10. Immanuel Kant, "Idea for a Universal History with a Cosmopolitical Purpose," Proposition 8 in *Kant's Political Writings*, ed. H. Reiss, trans. H. B. Nisbet (Cambridge: Cambridge University Press, 1991), 50.

11. William James, *Pragmatism: A New Name for an Old Way of Thinking* (London: Longmans, Green, 1907), ch. 3.

12. J. S. Mill, *Autobiography* (London: Longmans, Green, Reader and Dyer, 1873), ch. 5.

13. For more detail see John Passmore, *The Perfectibility of Man* (London: Gerald Duckworth, 1972), 195–197.

14. Francis Bacon, *The Advancement of Learning* (Oxford, UK: Clarendon, 1926), bk. II, XXI.

15. Immanuel Kant, *Religion within the Bounds of Pure Reason* (bk. III, div. I, I) and *Idea for a Universal History*, Eighth Proposition. In his *The Contest of Faculties* (sect. 10) he ascribes even less importance to human goodwill as distinct from Providence. These last two works can now be conveniently read in *Kant's Political Writings*, ed. H. Reiss, trans. H. B. Nisbet (Cambridge: Cambridge University Press, 1991).

31

Sustainability

ALAN HOLLAND

Introduction: Birth of an Idea

The twentieth century saw unprecedented environmental change, much of it the cumulative and unintended result of human economic activity. In the judgment of many, this change—involving the exhaustion of natural resources and sinks, extensive pollution, and unprecedented impacts upon climate, life-forms, and life-sustaining systems—is undermining the conditions necessary for the economic activity to continue. In a word, present patterns of economic activity are judged to be "unsustainable."

An initial response was to suggest that human society would have to abandon the attempt to improve the human condition through economic growth, and settle instead for zero growth. The response was naturally unwelcome, both to political leaders anxious to assure voters of better times to come, and to business anxious to stay in business. Its logic, moreover, is open to question. For even if we accept that economic growth has been the chief cause of environmental degradation, it does not follow that abandoning growth is the remedy. If zero growth led to global war, for example, there would be environmental degradation, and zero growth to boot. And genetic technology holds out the hope, at least, that we might provide for human needs with decreasing impact on the natural environment, and even reverse some of the degradation that has already occurred.

This is the hope expressed by the idea of "sustainable development"—or "sustainability," for short. The origin of the idea is commonly dated to a report produced by the International Union for the Conservation of Nature in 1980. Over the ensuing years, and especially since the publication of the Brundtland Report,[1] it has come to dominate large areas of environmental discourse and policy-making. Replacing more confrontational discourse between advocates of economic development, and those increasingly concerned over its environmental consequences, Brundtland advanced the (conciliatory) proposition that the needs of the poor (among the declared aims of economic development) were best met by sustaining environmental capacity (among the declared aims of environmentalism). Development, it was suggested, could be pursued to the extent that it was compatible with sustaining environmental capacity. On the assumption that environmental capacity can be expressed in terms of the capacity to satisfy human needs, this was formulated as a principle of "sustainable development"—"development that meets the needs of the present without compromising the ability of future generation to meet their own needs."[2]

There are formal analogies between the principle of sustainability, so framed, and J. S. Mill's principle of liberty, which licenses the pursuit of liberty insofar as this is compatible with a similar liberty for all. The principle of sustainability in effect licenses the pursuit of quality of life insofar as this is compatible with a similar quality of life for all (including future people). Such a principle appears to safeguard the future of the environment, too. But far from abandoning the aim of economic growth, Brundtland foresees "a new era of economic growth" and believes such growth is "absolutely essential" for the relief of poverty.[3] On the other hand it holds that growth is not sufficient to relieve poverty, and sees the need for what it calls a "new development path,"[4] one that sustains environmental capacity. Its approach is notably human-centered, and aside from one reference to a "moral reason" for conserving wild beings,[5] the deep ecology perspective is absent. The loss of coral reefs, for example, is lamented simply on the grounds that they "have generated an unusual variety of toxins valuable in modern medicine";[6] and hope is expressed that "The Earth's endowment of species and natural ecosystems will soon be seen as assets to be conserved and managed for the benefit of all humanity."[7] Sustainable development is understood as development that sustains human progress into the distant future. At the same time, the human-centered reasons that are given are often as much moral as prudential. Poverty is declared to be an "evil in itself" and a strong thread of the argument centers on a concern that current economic activity imposes costs on the future: "We borrow environmental capital from future generations."[8] In this way Brundtland clearly links issues of intragenerational equity with those of intergenerational equity. The implications for environmental protection, however, are less clear. Environmentalists may welcome the recognition that it is not environmental protection that stands in the way of future development so much as the fallout from existing patterns of development. But the fundamental issue is whether the protection that the environment requires serves to determine the "new development path," or whether it is the new development path that dictates the nature of the environmental protection required. In the latter event, it makes a difference—or rather it makes all the difference—what conception of the human good is to govern the development path in question.

Reception of the Idea

In the world of policy, the concept of sustainability has been assimilated with remarkable speed, determining much of the agenda of the United Nations conference on Environment and Development held in Rio de Janeiro in 1992.

Agenda 21, a program to which governments all over the world have committed themselves, represents the agenda for putting sustainability into practice. At the local level, too, many sensible schemes are being put into operation in the name of sustainability, often under the auspices of Agenda 21, and the term figures increasingly as an overriding objective of environmental organizations, whether inside or outside the remit of government.

However, the appearance of consensus even at the highest policy levels continues to be accompanied by a sustained chorus of skepticism and suspicion. Environmentalists are suspicious that what is billed as a constraint on business as usual will turn out to be a cover for business as usual. Vandana Shiva comments on the paradox that development and growth—creatures of the market economy—are being offered as a cure for the very ecological crisis that they have served to bring about.[9] "Nature shrinks as capital grows," she writes: the market destroys both the economy of nature, and non-market "survival economies."[10] Business interests, on the other hand, are suspicious that the constraint on business as usual might be a cover for no business at all. The sustainability agenda is also a fertile source of suspicion between North and South—the poorer countries suspecting that the constraint on development now judged to be necessary as a result of patterns of economic activity from which they received little or no benefits is being used to justify a constraint on *their* development. This was already a key issue ahead of the first ever world conference on Environment and Development held in Stockholm in 1972.[11]

At the theoretical level, too, the principle of sustainability faces criticism. It is said to be muddled, even self-contradictory, and some question its moral stance if it means that the claims of the environment might override the claims of poor and hungry people.[12] Others argue that it encapsulates the very same values and assumptions as those it purports to challenge.[13] Many detect new managerial aspirations behind the sustainability agenda; others see merely the continuation of older aspirations—for control or domination of world resources. Last but not least, there is the—skeptical—view, advocated for example by Julian Simon,[14] that there never was an "environmental crisis" to begin with, and that a sustainability agenda is therefore otiose.

We shall first briefly consider two such skeptical challenges, before turning to consider (1) the objectives of the sustainability agenda, (2) the criteria for its achievement, and (3) issues of implementation.

Challenges

Although it tends to receive short shrift from environmentalists, there is more to Simon's case than meets the eye. He claims that almost every trend that affects human welfare, if we take a reasonably long-term view, "points in a positive direction."[15] While some of the evidence he cites appears to be chosen very selectively, and in fact to be extremely short-term, he is on stronger ground in challenging claims about the "exhaustion" of sinks and resources and "threats" to life-support systems. For nature does not present itself as either resource, sink, or life-support system. "Resource," and "sink," too, are relational terms; and precious little of nature presents itself conveniently for our use without an untold input of human effort and contrivance. The so-called environmental "services" or natural life-support systems are, of course, necessary, but they are in no way sufficient to support human life, unless humans also support themselves. For this reason, increasing encroachment on the natural world does not in itself mean the exhaustion of natural resources; indeed, the very availability of nature as a resource requires and presupposes much destruction of the natural world. It should be remarked, for example, that biodiversity can only be counted as a "resource" at all because of the technology made possible by its (partial) destruction, and that without the sustained efforts of the human race, major segments of the natural world would still threaten rather than support its continuance. Whilst Simon is rightly taken to task for seeming to suggest that resources are therefore "infinite" (i.e. not finite), he is right to infer that they can at least be said to be "indefinite" (i.e. as having no fixed point of termination). And since humans can only hope for an indefinite rather than an infinite future, it could be argued that this is enough. In response, and appealing to the second law of thermodynamics, ecological economists such as Herman Daly have argued that increasing entropy sets an inescapable limit to economic growth. However, this does not rule out indefinite growth, if the increase can be slowed down sufficiently; nor does it take into account the extent to which solar energy might be increasingly used to "subsidize" such growth. Some analysts argue that, thanks to technological advance, levels of welfare can be maintained or even increased on the basis of reduced resource consumption. Unfortunately, it does not follow that more resources will be left intact (because fewer are needed); on the contrary, fewer resources may be left intact, and for the very same reason—that fewer are needed.

Another challenge, raised by Wilfred Beckerman,[16] concerns the validity of pursuing a sustained path, as opposed to one that exhibits more variation. He rightly points out that in our individual lives we are perfectly ready to anticipate declining fortunes in our later years as more than compensated for by earlier flourishing. Now the argument is admittedly not straightforward, because in the intergenerational case those who experience the decline are different individuals from those who flourish, which raises issues of justice not raised by the intrapersonal case. But the point remains that the objective of a level path appears to privilege egalitarian conceptions of justice over more flexible ones, such as the notion of the political philosopher, John Rawls, that a situation is not unjust if everyone in it is no worse off than anyone in an alternative situation. It is conceivable than an uneven development path could leave everyone better off.

Objectives

For those who accept the need for a new, sustainable, development path, two immediate questions must be faced: (1) What values are affirmed by sustainability? (2) Are they all consistent?

The values most evident among the arguments advanced for sustainability are justice, well-being, and the value of nature "in its own right." The point of most interest to environmentalists is how far the claims of nature are going to be served by policies designed to secure human well-being and justice. For it is no foregone conclusion that some preordained "harmony" must obtain between these objectives. But equally of interest is the relation between sustainability and the pursuit of justice, both intragenerational and intergenerational. Sustainable development is sometimes defined as non-declining consumption per capita. However, non-declining per capita consumption is compatible with enormous per capita inequalities. And while it is true that a commitment to intergenerational equity appears logically to imply a commitment to intragenerational equity, realization of these two aims might in practice conflict. The opportunity costs of measures to "save" the environment that benefit future generations might happen to fall upon today's poor; and this just makes the point that although the poor suffer most from environmental degradation, because they depend more heavily than the rich on natural, as opposed to human-made resources, it does not follow that they benefit most from its restoration. But if environmental protection is not sufficient to ensure social justice, i.e. to help today's poor, is it not at least necessary? A skeptic might wonder whether environmental damage—e.g. damage that results from development—is not also at least as necessary a condition of social justice. The relation between environmental protection and the well-being of future humans is also not straightforward. Environmental protection does not guarantee the well-being of future humans, without many enabling factors being in place. Whether it is necessary depends on how environmental protection is understood. If it is explained as "maintaining the capacity of the environment to serve human interests," then indeed the connection is assured—by fiat. But if environmental protection is given a more radical slant—allowing nature to "go its own way," for example— then again, it might be environmental damage that is required to ensure the well-being of future humans.

Whether long-term human economic interests and the long-term integrity of the natural world really do coincide is one of the deep underlying problems of the sustainability debate. Of particular interest in this connection is the way that the sustainability debate itself has prompted searching critiques of the growth model, and questionings of the relation between growth and the human good. Increasingly, the distinction is drawn between economic growth, on the one hand, and development on the other, the latter denoting a richer register of human aspiration. Amartya Sen, for example, has explained development as "a process of expanding the capabilities of people," a notion that is intended to include expanded autonomy and greater access to justice.[17] There are in fact sound conceptual reasons for claiming that human well-being, if this is understood as implying a conscious state of sensitivity, cannot be intelligibly specified without reference to external circumstances, including states of the natural world. Peace of mind is simply not an option if a baby is crying, or people are starving. And there are conceptions of the human good that make a concern for nature for its own sake a contributing or even constitutive factor in human well-being. In this way human interest and ecological integrity do not just happen to coincide, but exhibit an interlocking conceptual relationship. This prompts a more hopeful line of thought—that human aspirations for a better quality of life might be met even though there is decreasing reliance upon economic growth as such, and therefore decreasing impact on non-human life-forms. At least on one scenario, then, the leading objectives of sustainability might after all be made consistent.

A further question is how sustainability compares with environmental objectives of longer standing that it has tended to supplant, such as (a) nature conservation, (b) land health, and (c) ecological integrity.

Nature conservation. Writing in the 1991 annual report of the Council for the Protection of Rural England, its Chairman David Astor voiced the opinion that "The great pioneers of CPRE in the 1920s did not use the term sustainable development but that is exactly what they stood for." But one might well take issue with this claim. For among these traditional nature conservationists one finds a concern for natural features that are indigenous, rare, venerable, fragile, and irreplaceable. It is not clear, however, that a single one of these categories would be guaranteed a place on a sustainability programme.

Land health. Aldo Leopold's notion of land health—"the capacity of the land for self-renewal"—on the other hand, comes much closer. It embraces cultivated as well as natural landscapes and, like sustainability, anticipates the idea of future capacity. But one difference is suggested by an analogy with the human body. We can function in pretty much the normal way (sustain our activities) with spectacles, dentures, and a walking-stick. But the conditions which these aids help us surmount are incapacities. The same goes for agricultural systems, which can continue to function provided humans continue to supply the necessary fertilizers, etc., but it does not exactly have the capacity for *self*-renewal (as specified in Leopold's definition of land health). It is not clear that sustainability requires land health as such.

Integrity. The contrast between the goal of sustainability and that of ecological integrity—understood as a condition of minimum human disturbance—appears at first sight more marked. But the prospects for compatibility depend on which of two questions are asked. The first question is: how much integrity do we need? There are those who believe that a tolerable human future actually requires the maintenance of substantial portions of the globe relatively free of human influence. But whether they are right depends upon the character of the human future being envisaged; and it is hard to avoid the suspicion that certain sustainable futures at any rate would feature a natural world containing only species that can abide, or avoid, the human footprint. The second question is: how much freedom from human disturbance do non-human species need? This way lies a brighter prospect for the compatibility of the sustainability and ecological agendas. But its realization depends on whether human society is prepared to ask the question, and act on the answer.

The Criteria

The issue of what will count as achieving sustainability, and what criterion will be used to map the new development path, is absolutely crucial. It will in effect determine the sustainable agenda. By far the most developed suggestions for measuring progress towards sustainability have come from economists, and in a number of countries many so-called "green accounting" schemes are in operation. At the heart of the economic approach to sustainability is the concept of capital. The modeling of nature on the analogy of financial "capital," capable of yielding "interest," arises readily from earlier notions of "sustainable yield," found in resource economics. Applied to nature generally, the proposed criterion states that if each generation bequeaths at least as much capital to the next generation as it receives, then this will constitute a sustained development path. (Economists appear not to be deterred from using the model despite there being no clear sense in which non-renewable resources can "yield" interest.)

The theory proceeds to mark a distinction between two kinds of capital—natural and human-made.[18] Human-made capital comprises all artifacts, as well as human and social capital—people, their skills, intelligence, virtues, and institutions. Natural capital comprises all naturally occurring organic and inorganic resources, including not just physical items but also genetic information, biodiversity, life-support systems, and sinks. This distinction in turn generates two versions of the criterion: so-called "weak sustainability," which stipulates an undiminished capital bequest irrespective of how it is composed; and so-called "strong sustainability," which stipulates an undiminished bequest of natural capital.

The distinguishing feature of weak sustainability is indifference between natural and human-made capital, provided only that human needs continue to be met. It is often alleged, but on no very good textual evidence, that advocates of weak sustainability are committed to unlimited sustainability between natural and human-made capital. What they are committed to, as economists, is that the value of all kinds of capital is comparable. But this is quite different from the claim that the valued *items* can be substituted: a visit to the cinema might cost as much as a good meal, but it doesn't follow that it can replace a good meal. Moreover, advocates of strong sustainability, if they are committed to the principles of neoclassical economics, must also hold that the value of human-made and natural capital is comparable in terms of economic value; for this is ultimately reducible to the preferences of rational economic agents, and therefore capable of being expressed in monetary terms. Indeed, it is just this feature of the account that makes it possible for the level of capital to be measured, and therefore possible for there to be a criterion of sustainability. The distinguishing feature of strong sustainability is that it is not indifferent between natural and human-made capital, but requires natural capital to be maintained. So it requires natural capital to be maintained not only where substitution by human-made capital is not possible (this much weak sustainability would agree with, since if natural capital cannot be substituted its loss would be a loss to total capital, which weak sustainability does not allow), but crucially, where it is possible.

The impact of weak sustainability is not insignificant. Although it allows environmental loss to be offset against other kinds of economic gain, it at least acknowledges that environmental loss harbors economic loss, and must therefore be taken into account rather than regarded as an "externality"—something which purely economic accounting can afford to ignore. From the environmentalist perspective, however, the most obvious defect of weak sustainability is that it appears to countenance severing the connection between sustainability and environmental protection. It permits the natural environment to be degraded provided that human well-being can still be secured. But it should be noted that if the claim made by some advocates of strong sustainability is correct, and the presence of substantial features of the natural world is indispensable for human well-being, then certainly it will justify the claim that the level of natural capital should be maintained. But, far from justifying strong sustainability as a distinct position from weak, it actually shows that it is redundant. The advantages of the economic approach are several:

1. It highlights the fact that environmental conservation carries economic costs and burdens; also—and crucially—that economic benefits carry environmental costs.
2. It offers a way of measuring the benefits of environmental conservation against those gained from other forms of expenditure, e.g. military, health.

3. It makes the case for compensating forgone income—e.g. "debt-for-nature" swaps (though this runs up against the problem that compensating "poor" countries may mean "compensating" individually rich people).

4. If offers a way of measuring the effectiveness of any conservation policy or program, which, however unclear its methods or contested its results, is vital for any durable program of implementation.

5. The concept of "natural capital" in particular makes the point that we should not regard our use of environmental resources as if we were living off *income*.

But how plausible is it to construe nature as capital? The account of the economists faces moral, methodological, ecological, and conceptual difficulties.

The moral objection, first, is that the natural world is not simply a resource but, for example, contains sentient creatures who have claims to moral consideration. An alternative, or additional, point is that the natural world embodies values other than its value to humans, values that are inherent rather than instrumental. A third objection stems from resistance to the idea that all values are commensurable, and especially resistance to the idea that they can be assessed according to a common economic numeraire.

Among methodological objections is the problem of how markets distorted by socio-political power structures can adequately reflect resource scarcity; and more generally, how the value of items and processes that span centuries can be adequately assessed according to the parochial and quotidian values of the day. Besides being presumptuous, the difficulty is that the slightest difference in evaluative criteria here and now—say a 1 per cent difference in the "discount rate"—will have enormous and amplified effects hereafter.

Among ecological objections is the difficulty of mapping ecological realities— involving processes that are episodic, non-linear, unstable, and unpredictable— with economic indicators and criteria. A further objection to the resource approach is that it suppresses recognition of the historical character of the biosphere. Even where there is recognition that the biosphere is the result of complex events spanning many centuries, this tends to be construed as merely a technical problem of restoration. Time and historical process is construed, in other words, as nothing more than a technical constraint on the preservation of natural capital—an approach which completely misses what is at stake in loss of biodiversity. Irreversibility, too, is seen as an annoying impediment to the maintenance of capital, instead of being seen for what it is—the very stuff of natural process.

There are also severe conceptual difficulties. First, the concepts "natural world" and "natural capital" have quite different connotations. Natural capital is the natural world construed as an "asset"; it is the natural world just insofar as it represents its capacity to service human needs. Large portions of the natural world therefore do not count as natural capital—for example, events such as volcanic eruptions and hurricanes, and species such as mosquitoes and locusts. On the one hand, degradation seems to abound; stunted tree development associated with deteriorating soil quality; premature leaf drop associated with air pollution; fish and aquatic mammals disappearing from rivers; thousands of miles of coral reef bleached or dying in the wake of 1998's highest sea temperatures on record. But the situation from the point of view of natural capital is far from clear. Is this a description of a decline in natural capital? The capacity of the natural world to serve human well-being appears unabated—perhaps *because* we have, for example, more efficient ways of destroying locusts. Hence, the economic criterion appears unable to identify what has gone wrong. If this is true, it cannot possibly function as a guide to the path that will put it right.

A second point is that the extent of natural capital depends on the availability of human-made capital. For our vulnerable ancestors, much of the natural world needed to be—and was—destroyed. Much else needed to be transformed—by fire, cultivation, and domestication. The implication is that natural capital relies so extensively on human-made capital for its capacity to be realized that it is by and large meaningless to talk of natural capital in abstraction from the human-made capital that mediates its use. Reflections of this kind lead Dale Jamieson to remark that "it is quite difficult to distinguish natural from human produced capital."[19] If this is true, then it will also be "quite difficult" to specify the requirements for strong sustainability. The amount of natural capital available simply cannot be judged independently of reference to human-made capital.

A third problem concerns the notion of substitution; levels of capital are supposed to be judged depending upon whether substitutions can or cannot be made. But the question whether A can or cannot be substituted for B is not intelligible in isolation; it depends on context—the purpose for which the substitution is made, the degree of precision required, and so forth.[20] Nor is it clear, in any case, how far these are empirical as opposed to normative questions. However, the real problem is not the proposed substitution between natural and human-made capital; it is the proposed substitution of capital for nature. Natural capital is not the yield of nature; on the contrary, it tends to be yielded through the destruction of nature.

One, final, difficulty with the economic approach is how we are to know, at any particular point along the "new development path," what the overall trajectory of the path is likely to be, and what "next step" will contribute to it. The source of the difficulty is both causal and epistemological, because each possible next step will have different outcomes, and these will rapidly become impossible to predict. It is hard to see how a sustainable path could be identified in any other way than retrospectively. This suggests that the condition or state of sustainability cannot be understood in terms of purely economic criteria, i.e. as measurable by any kind of efficiency or optimizing outcomes. It appears rather as an inter-temporal and path-dependent process which can only be maintained by procedures and traditions that are self-critical, self-renewing, and sensitive to distributional and historical concerns.

Implementation

A "new ethic," and new technologies, might both be seen as possible ingredients of a "new development path," but they cannot be realized in a vacuum. They require the presence of certain sorts of social and political institution, and above all a citizenry that is attuned to the demands of the new development path. Unfortunately, certain characteristics of the present global economy, for example, the mobility of labor and capital, and the centralization of knowledge and power, make transitions to environmental sustainability and to redistributive social and economic policies hard to envisage. The interests of multinational corporations, even though they put on a green costume and may sincerely speak of feeding the hungry, are not best served by widening access to natural resources and helping the hungry to feed themselves. Developments in science and technology, e.g. nuclear and genetic research programs, point to a similar conclusion. They are becoming less, rather than more, socially accountable, as their research agendas fall increasingly into private hands. The future looks set to belong to powerful and well-financed minorities rather than being held in common—the "common future" to which Brundtland aspired. For these reasons, the notion of a "technical solution" had best not be too readily dismissed, since from one perspective it appears the most likely eventuality. From the spinning Jenny to the information revolution, technology has shown a remarkable ability to lead society by the nose. In accordance with such reflections, it is now being increasingly recognized that "social" or "cultural" sustainability is likely to be crucial to the achievement of any kind of sustainability at all. Among the key elements of cultural sustainability that will need to be in place are: (a) resilient political institutions; (b) effective regulation; (c) appropriate social skills and habits; (d) accountable science and technology; and (e) a climate of trust.

There are signs that the sustainability agenda itself may be capable of provoking certain sorts of political institution and vision, just as much as it requires them for its implementation. The very idea of intergenerational justice, for example, at least complements and may well presuppose and inspire notions of cross-generational community that are by no means new, but have been somewhat muted in the face of prevailing utilitarian and individualistic ideologies. In modern times they reach back at least to the eighteenth-century Irish philosopher Edmund Burke's notion of a diachronic community, a "partnership . . . not only between those who are living, but between those who are living, those who are dead, and those who are to be born." Skeptics doubt whether such a community can enjoy reciprocal relations, or shared values. But such doubts appear misplaced. Later generations can reciprocate by honoring their predecessors or striving to fulfill their hopes. And the importance of shared values can be overrated: absence of dissent and lack of change over time are more symptomatic of a moribund than a healthy community.

Certain alternatives to the economic approach to sustainability gain relevance here, such as the idea of patrimonial management developed by several French researchers.[21] Inspired by relations between people and environment in nonmarket economics, the approach builds on gift and counter-gift exchanges that arise naturally among those whose livelihoods are invested in nature, and for whom, therefore, the environment itself is a medium of transaction and interaction over time. From this perspective, monetary transaction, far from being the key and measure of all environmental transactions, is a poor measure of the value of environmental resources, especially in the context of developing economies, besides also failing to capture the significance of transtemporal relations in developed market economies.

Conclusion

Sustainable development may be summarily defined as development of a kind that does not prejudice future development. It is intended to function essentially as a criterion for what is to count as acceptable environmental modification. Although we may not be able to predict future human needs with any precision, we can be sure that any future human development will require resources, sinks, and life-sustaining systems. We may be reasonably sure therefore that measures taken to minimize human impact on resources and sinks, and to minimize changes to life-support functions, will be steps toward sustainability thus understood. We can also be sure that measures taken to secure these environmental targets require a supporting social fabric. At the same time, it must not be forgotten that what any generation bequeaths to its successors is a package, including not only "costs," but also "benefits," such as technological expertise and other forms of human and social "capital," without which natural resources, sinks, and services would not have the value to humans that they do. But perhaps most crucial of all, its actions also shape and determine the very conditions under which succeeding generations will live. Hence, the question of whether any society is on a sustainable trajectory is at best an extremely complex one, and may well be in principle impossible to compute. What then, should we make of the attempts, especially by economists, to provide just such computations? It is true that environmental degradation has economic costs, and that these costs are a telling symptom of our environmental predicament. But it does not follow that these are the only costs, or that economic values can successfully measure these costs—for a variety of reasons. These include doubts about the various methodologies deployed, and difficulties of principle—about how to cost moral concerns, or how to cost features of the social fabric, such as a Royal Commission on Environmental Pollution, or the environmental habits of Swedish citizens. Above all, if our assessment of economic value is determined by the very economic system whose credentials are in question, it is hard to see how the translation of that value system into the environmental sphere can yield a just estimate of environmental value, or enforce the re-evaluation of environmental goods, as it is supposed to do.

The real importance of sustainability may lie in providing a new conceptual context within which issues of growth and environment can be debated, and in provoking us to reassess our notions of quality of life and environment. It answers also to a need, visceral as well as pragmatic, to do something in the face of loss. But as a guiding principle, it must be judged ultimately unsatisfying. It seems too closely locked in to conceptions of the world—a storehouse that must be kept filled, a machine that must be maintained—that are themselves no longer sustainable. In the wake of Darwin, the world looks much more like an open-ended historical process ill-suited for filling or maintaining. Our more modest task is how not to blight the interlocking futures of the human and the natural community that we have the power profoundly to affect but lack the capacity and the wisdom to manage.

NOTES

1. World Commission on Environment and Development (WCED), *Our Common Future* (Oxford: Oxford University Press, 1987). (Also called "The Brundtland Report," the classic text on sustainability.)
2. Ibid., 8.
3. Ibid., 1.
4. Ibid., 4.
5. Ibid., 13.
6. Ibid., 151.
7. Ibid., 160.
8. Ibid., 8.
9. V. Shiva, "Recovering the Real Meaning of Sustainability," in D. Cooper and J. Palmer, eds., *The Environment in Question* (London: Routledge, 1992), 187–93. (A brief but probing essay on sustainability.)
10. Ibid., 189.
11. H. H. Landsberg, "Looking Backward: Stockholm 1972," *Resources for the Future* 106 (1992): 2–3. (A brief, but useful backward glance at the Stockholm conference.)
12. W. Beckerman, "Sustainable Development: Is It a Useful Concept?," *Environmental Values* 3 (1994): 191–209. (A sharp critique of the sustainability agenda.)
13. M. Redclift, *On Ethics and Economics* (Oxford, UK: Blackwell, 1987). (A seminal work on quality-of-life issues.)
14. J. L. Simon, "Scarcity or Abundance?," in L. Westra and P. H. Werhane, eds., *The Business of Consumption* (Lanham, MD: Rowman and Littlefield, 1998), 237–45. (An outspoken critique of the sustainability agenda.)

15. Ibid., 237.
16. Beckerman, "Sustainable Development."
17. A. K. Sen, *On Ethics and Economics* (Oxford, UK: Blackwell, 1987).
18. D. Pearce, A. Markandya, and B. B. Barbler, *Blueprint for a Green Economy* (London: Earthscan, 1989), 34–35. (A classic of environmental economics.)
19. D. Jamieson, "Sustainability and Beyond," *Ecological Economics* 24 (1998): 185. (A sober assessment of the limitations of the sustainability discourse.)
20. A. Holland, "Substitutability; or, Why Strong Sustainability Is Weak and Absurdly Strong Sustainability Is Not Absurd," in J. Foster, ed., *Valuing Nature? Economics, Ethics and the Environment* (London: Routledge, 1997), 119–34. (Explores some conceptual difficulties with the notion of sustainability.)
21. G. Lescuyer, "Globalization of Environmental Monetary Valuation and Sustainable Development: An Experience in the Tropical Forest of Cameroon," *International Journal of Sustainable Development* 1 (1998): 115–33. (A timely reminder of the importance of non-market economies.)

32

Excerpts from "Theorising Environmental Justice: The Expanding Sphere of a Discourse"

DAVID SCHLOSBERG

Expanding the Space of Environmental Justice Discourse . . .

Environmental justice as a discourse has rapidly expanded its influence, and has been applied to both a broadening range of issues, and, increasingly, a global level. While these extensions are crucial, I also want to address the potential of extending the discourse beyond individual human beings, to conceptualisations of community-level justice and justice beyond the human.

Horizontal and Vertical Expansion

If there has been a single major development in the framing of environmental justice in the past decade, it has been the way the use of the concept, as an organising theme or value by a range of movements, has expanded spatially.[1] While there has been a continued focus on the original core of environmental justice issues in the distribution of toxins—or environmental bads more generally—in the United States, environmental justice discourse and literature has been extended in both topical and geographic scope. As Sze and London note in their important overview, environmental justice has seen the expansion into new issues and constituencies on the one hand, and new places and spatial analyses from the local to the global on the other. They celebrate this expansion, arguing that this attention to the expanding spatial realm of environmental justice has been the focus of many crucial researchers in the field, from politics to sociology to geography.[2] This expansion has been more than simply an exercise in academic interdisciplinarity—it has led to a broad extension of the foci of environmental justice scholarship.

Environmental justice may have been originally focused on the inequity of the distribution of toxics and hazardous waste in the United States, but it has moved far beyond this. Perhaps, however, such a broadening is not new, but a longstanding characteristic of the movement. Cole and Foster's[3] now classic study of the movement discussed the various "tributaries" that make up the environmental justice movement. They included the civil rights and anti-toxics movements, but also indigenous rights movements, the labour movement (including farm labour, occupational health and safety, and some industrial unions), and traditional environmentalists. Faber and McCarthy[4] added the solidarity movement and the more general social and economic justice movements. We could easily add immigrant rights groups and urban environmental and smart growth movements, as well as local foods and food justice movements, to the list. Environmental justice as an organising frame has been applied not only to the initial issues of toxins and dumps, but also analyses of transportation, access to countryside and green space, land use and smart growth policy, water quality and distribution, energy development and jobs, brownfields refurbishment, and food justice.[5] Questions of the role of scientific expertise, and the relationship between science and environmental justice communities, have also been examined.[6] There has also been more thorough examination of the roles of under-examined groups in the environmental justice movement, or exposed to environmental hazards—indigenous peoples, Asian and Latino workers, women and youth,[7] illustrating the broadening range of foci of environmental justice scholarship in the United States. I do not mean to imply that all of these studies offer similar or unproblematic analyses of the issues, but simply to note the longstanding and continuing trend of the expanding topical space of the environmental justice frame.

In addition to the expansion of issues, there has been a push to globalise environmental justice as an explanatory discourse. There are two distinct moments to this expansion: the application of the frame to movements in a variety of countries, and the examination of the globalised and transnational nature of environmental justice movements and discourse. Walker[8] sees this development as both a horizontal diffusion of environmental justice ideas, meanings, and framings, along with the vertical extension of an environmental justice frame beyond borders, and into relations between countries and truly global issues. As for the first, the applications of my own theoretical framework of environmental justice have been more broad than I would have imagined, including cases of postcolonial environmental justice in India, waste management in the United Kingdom, agrarian change in Sumatra, nuclear waste in Taiwan, salmon farming and First Nations in Canada, gold mining in Ghana, oil politics in Ecuador, indigenous water rights in Australia, wind farm development in Wales, pesticide drift in California, energy politics in Mexico, and many more.[9] In addition, there have been collections on environmental justice focused on issues and movements in Latin America, South Africa, Canada, and the ex-Soviet Union.[10] Walker[11] lists no fewer than 37 countries in which the environmental justice frame has been applied. Clearly, the discourse of environmental justice has expanded horizontally, and been engaged by both activists and academics involved in issues across the globe.

The vertical extension of an environmental justice framework is evidently illustrated by the use of environmental justice as an organising theme by a number of global movements, such as food security, indigenous rights, and anti-neoliberalism.[12] This global approach has been thoroughly analysed in Pellow's[13] enlightening work on the global toxics trade and both local community and global non-governmental organisation (NGO) resistance to it. Offering both a thorough analysis of the international production of waste, and keen observation of the transnational movement(s) that have risen in response, Pellow's work brings attention to the global potential of environmental justice analysis. The essence of transnational networks, he argues, is found in their critique of environmental inequities, the disruption of social relations that produce such inequities, and the articulation of ecologically sustainable and socially just institutions and practices.[14] Such an analysis focuses on both the nature of the injustice and the creative and crucially networked response on the part of movements. Mohai et al.[15] note a number of additional transnational issue networks that have environmental justice as an organising theme, from those concerned with e-waste to the movement for climate justice. Carmin and Agyeman[16] bring both of these elements of expansion together in a recent collection that focuses both on specific issues and movements and a larger global framework of analysis. Clearly, environmental justice analysis continues to expand in scope and scale.

Community

From my own perspective, environmental justice discourse has crossed or subverted two other spatial barriers. The first is the link between individual and community. While the traditional, liberal frame of reference for the conception of justice is purely individualist, environmental justice movements address injustice at both the level of the individual and the community. Just as the experience of environmental injustice pushed the reflection of the concept beyond a singular focus on equity, that experience also illustrates that the conception of justice used in the movements addresses impacts on, and limitations to, individuals and their communities simultaneously.

The case of Hurricane Katrina helps to illustrate this key point. The impacts of the disaster were, certainly, disproportionately experienced by poor African Americans. Many lost their homes, their jobs, their belongings; many were left behind or made invisible by the racism inherent in the city—and those that remained often experienced exclusion from the plans to rebuild. But the understanding of those impacts goes far beyond injustice to individuals. In the seminal set of reflections on the environmental justice implications of Katrina edited by Bullard and Wright,[17] a wide range of impacts is discussed. From transportation, employment, health, housing, and economic opportunities to broader issues of social disrespect, community diaspora, and political and economic participation, the authors reflecting on the disaster address a range of basic needs and functions that were undermined, and that would have to be restored in a just recovery. These needs are not simply about individuals, but neighbourhoods, communities, and the city itself. Ultimately, the question of environmental justice in the wake of the storm is about the very functioning of New Orleans, its neighbourhoods and communities.[18]

Environmental justice battles focusing on issues ranging from asthma in New York[19] to biopiracy against the San in Southern Africa[20] have never only been about individual illnesses or impacts, but always also about the impact on the social cohesion and functioning of the community. Movement groups frame their concerns in both individual and collective senses. This more communitarian conception of injustice is confronting to liberal individualist notions of justice, but it is a rather straightforward thing for environmental justice communities to experience, and to articulate. My own work has attempted to argue this point in relation to recognition and capabilities approaches to environmental and social justice. We can, for example, see community-based articulation in demands for environmental justice that emphasise and defend the basic needs and very functioning of indigenous communities.[21] Similarly, the protection of the ability of social reproduction—community functioning, not simply individual exposures—is central to many environmental justice movements.[22] The spatial barrier between the concern for individuals and that of communities has been thoroughly crossed, and expressed, in movements. This is an area ripe for further exploration.

Beyond Human

Finally, one of the remaining border challenges of environmental justice theory is to make important connections with the environment itself. There is a reason that we discuss *environmental* justice—the issues involved are about how, exactly, we are immersed in the environment, and the manipulation of nature, around us. Yes, most of the discussion is about environmental bads and injustices to human beings, but the origins of environmental injustices are as much in the treatment of the non-human realm as in relations among human beings. The shift suggested here is one from environmental conditions as an example or manifestation of social injustice to one where justice is applied to the treatment of the environment itself.

A number of analysts have made these connections. In the notion of just sustainability, Agyeman[23] insists on a conception of environmental justice that goes beyond socio-cultural impacts alone to the interactions between social and environmental communities. Post-Katrina, many reflections have involved not only the conditions of the people in the city, but also consideration of the ecological damage done to surrounding ecosystems that have led to greater vulnerabilities for both human and non-human communities.[24] Sze et al.[25] have continued their innovative work on environmental justice in the Sacramento (California) delta region by engaging this element of the socio-natural context. They see the examination of the relationship between the manipulation of nature and people for economic gain as a crucial component of an environmental justice analysis.

I have been making the argument that a capabilities approach to justice is a crucial tool for addressing the relationship between environment and human needs and, potentially, the functioning of ecosystems themselves.[26] A capabilities approach could enrich conceptions of environmental and climate justice by bringing recognition to the functioning of these systems, in addition to those who live within and depend on them. In this approach, the central issue continues to be the interruption of the capabilities and functioning of living systems—what keeps those living systems from transforming primary goods into the functioning, integrity, and flourishing of those that depend on them. When we interrupt, corrupt, or defile the potential functioning of ecological support systems, we do an injustice not only to human beings, but also to all of those non-humans that depend on the integrity of the system for their own functioning. It is the disruption and increasing vulnerability of the integrity of ecosystems that is at the heart of the injustice of climate change, for example, both in terms of its impact on vulnerable human communities and non-human nature. The treatment—or abuse—of human and non-human individuals and systems is based on the same loss of the ability to function.

This application of a capabilities approach to non-human nature brings both benefits and potential conflicts. The first benefit is that a focus on the needs of non-human systems would entail that human beings actually recognise the link between environmental conditions and the basic needs of both human beings and the non-human. In other words, extending a capabilities approach to non-human environments entails recognition of the value of the processes and provisions of natural systems. The second benefit is a discursive one, as a capabilities approach applied to both human and non-human can serve as a bridge between conceptions of social justice and a wide range of environmental concerns.

The main problem with this approach, of course, is the potential for conflict between the capabilities and functioning of human beings and those of the natural world.[27] Fully addressing this issue would take more space than is available here, but I would simply note that any conception of justice, as it is applied to actual issues and injustices, would entail potential conflict. One of the major problems of ideal justice theories is that they seek to eliminate the potential for conflict—at least in theory. But such theorists are mistaken to believe that the elimination of such conflict in theory makes for more harmonious application to social policy or practice. Conflicts of justice arise, whether in the human realm, or, in this example, between human beings and the nature in which they are immersed, no matter what the ideal. Actual problem solving entails the negotiation of different conceptions of (in)justice in and across different participants, from community or stakeholder groups to corporations or states; it requires recognition, conceptions of disadvantage, and political engagement. This is where potential conflicts can be addressed, and ways of life attentive to the creation and experience of disadvantage and disabled functioning—human and non-human alike—can be negotiated and designed.

One of the clear developments in the past decade, then, has been a thorough expansion of the scope of the environmental justice frame. Against the early warnings of some in the US environmental justice community that the term should remain limited to the experience of racial discrimination, my suggestion has always been that environmental justice has the potential to be an integrative and empowering framework for a variety of movements and concerns.[28] Likewise, Sze and London have insisted that "instead of imposing a restrictive boundary around the concepts of environmental justice, scholarship in this emerging field should embrace its wide-ranging and integrative character."[29] Clearly, the trend of environmental justice in both theory and practice has been this expansion of the discourse into new spaces, and across many boundaries. . . .

Theory and Movements: Environmental Justice Discourse and Praxis

All of this, to me, illustrates how environmental justice in practice offers a rich form of politics and practice—one that academics in the field would do well to engage. One of the signature characteristics in much environmental justice scholarship has been a relationship between academic work and movement groups. The original articulation of an environmental justice movement came out of academic studies and conferences. The early history of the academic side of the movement was based on the work of Robert Bullard[30] and early conferences, such as that organised by Bryant and Mohai,[31] that helped articulate and publicise findings of inequitable distribution of environmental goods and bads.[32] The relationship between academic studies and the environmental justice movement has been integral to its development and growth, and its discourses, in the past three decades. Sze and London[33] see this relationship as one of the continuing elements of reflection in the recent literature, and one of the promising trends in the field.

In part, this relationship is about the idea of praxis—that theory and practice must inform each other.[34] As Holifield et al. insist, there is "a need for environmental justice scholarship to actively work at its connections to activism and its engagement with those at the sharp end of injustice, however it is understood, and to bring theory to bear in meaningful ways into praxis and diverse forms of public engagement."[35] Theorising from movement experience works to expand our understanding of those movements; in return, those movements can and do inform theory in productive ways. There are numerous examinations of this intersection in the United States, from my own work on movement pluralism,[36] to Di Chiro's[37] work on social reproduction in environment/feminist coalitions, to Sze et al.'s[38] examination of water politics in California, to the range of responses to community organising after Katrina.[39] All of this illustrates the relationship between environmental justice as an academic idea and a social movement, to the benefit of each.

This focus on the relationship between practice and theory has also been central to my attempts to understand the "justice" of environmental justice.[40] Many attempts to define environmental or climate justice have been too detached from the actual demands of social movements that use the idea as an organising theme or identity. This does assume that there is a value to movement practice—that theory can, and should, actually learn from the language, demands, and action of movements. Why, the more purist academic or sceptic might ask, should we prioritise what activists believe or do? But the question should not be about who is the best judge of a conception of justice—activists or theorists. The point is that different discourses of justice, and the various experiences and articulations of injustice, inform how the concept is used, understood, articulated, and demanded in practice; the engagement with what is articulated on the ground is of crucial value to our understanding and development of the concepts we study. It continues to be unfortunate that there are those in the study of environmentalism, or in the theoretical realm, who simply cannot see the importance, and range, of these articulations at the intersection of theory and practice—especially when movement innovation is as broad and informative as it is in environmental justice.

Conclusion

There has always been something particularly salient about the term environmental justice. It simply fit the conditions many communities were subjected to, and expanded the conception of social justice into a whole new realm of inequity, misrecognition, and exclusion—that of environmental disadvantage. The idea of environmental justice reflected the lived experience of the reality of injustice on the ground, in the air, in one's food, at the workplace or school, and on the playground. It is the salience of those experiences that helped push the concept to be embraced more broadly, on an increasing array of issues across the globe. In doing so, environmental justice moved from being simply a reflection of social injustice generally to being a statement about the crucial nature of the relationship between environment and the provision of justice itself. The concept has pushed boundaries since its inception, and has expanded both spatially and conceptually. In its latest incarnation, environmental justice is now also about the material relationships between human disadvantage and vulnerability and the condition of the environment and natural world in which that experience is immersed. Like all iterations of environmental justice over the years, this focus has much to offer communities—both human and non-human—as well as academics.

NOTES

1. J. Sze and J. K. London, "Environmental Justice at the Crossroads," *Sociological Compass* 2, no. 4 (2008): 1331–1354; G. Walker, "Globalizing Environmental Justice," *Global Social Policy* 9, no. 3 (2009): 355–382.

2. In addition to Sze and London, see this spatial expansion in the work of, for example, J. T. Roberts and B. C. Parks, *A Climate of Injustice* (Cambridge, MA: MIT Press, 2007), D. Pellow, *Resisting Global Toxics: Transnational Movements for Environmental Justice* (Cambridge, MA: MIT Press, 2007); P. Mohai, D. Pellow, and J. Timmons Roberts, "Environmental Justice," *Annual Review of Environment and Resources* 34 (2009): 405–430; and Walker, "Globalizing Environmental Justice"; G. Walker, *Environmental Justice: Concepts, Evidence, and Politics* (London: Routledge, 2011).

3. L. W. Cole and S. R. Foster, *From the Ground Up: Environmental Racism and the Rise of the Environmental Justice Movement* (New York: NYU Press, 2001).

4. D. Faber and D. McCarthy, "Neo-liberalism, Globalization, and the Struggle for Ecological Democracy: Linking Sustainability and Environmental Justice," in J. Agyeman, R. D. Bullard, and B. Evans, eds., *Just Sustainabilities: Development in an Unequal World* (Cambridge, MA: MIT Press, 2003), 38–63.

5. See J. Agyeman, "Constructing Environmental (In)justice: Transatlantic Tales," *Environmental Politics* 11, no. 3 (2002): 31–35; G. T. Rowan and C. Fridgen, "Brownfields and Environmental Justice: The Threats and Challenges of Contamination," *Environmental Practice* 5, no. 1 (2003): 58–61; R. Bullard, *Growing Smarter: Achieving Livable Communities, Environmental Justice, and Regional Equity* (Cambridge, MA: MIT Press, 2007); V. Jones, *The Green Collar Economy* (New York: HarperOne, 2009); R. Gottlieb and A. Joshi, *Food Justice* (Cambridge, MA: MIT Press, 2010).

6. J. Corburn, *Street Science* (Cambridge, MA: MIT Press, 2005); G. Ottinger, B. Cohen, and K. Fortun, eds., *Technoscience and Environmental Justice: Expert Cultures in a Grassroots Movement* (Cambridge, MA: MIT Press, 2011).

7. G. Di Chiro, "Living Environmentalisms: Coalition Politics, Social Reproduction and Environmental Justice," *Environmental Politics* 17, no. 2 (2008): 276–298; J. Sze, J. London, F. Shilling, G. Gambirazzio, T. Filan, and M. Cadenasso, "Defining and Contesting Environmental Justice: Socio-Natures and the Politics of Scale in the Delta," in R. Holifield, M. Porter, and G. Walker, eds., *Spaces of Environmental Justice* (Chichester, UK: Wiley-Blackwell, 2010), 219–256; K. P. Whyte, "The Recognition Dimensions of Environmental Justice in Indian Country," *Environmental Justice* 4, no. 4 (2011): 199–205.

8. Walker, "Globalizing Environmental Justice."

9. In order, see G. Williams and E. Mawdsley, "Postcolonial Environmental Justice: Government and Governance in India," *Geoforum* 37, no. 5 (2006): 660–670; M. Watson and H. Bulkeley, "Just Waste? Municipal Waste Management and the Politics of Environmental Justice," *Local Environment* 10, no. 4 (2005): 411–426; J. F. McCarthy, "Processes of Inclusion and Adverse Incorporation: Oil Palm and Agrarian Change in Sumatra, Indonesia," *Journal of Peasant Studies* 37, no. 4 (2010): 821–850; M. Fan, "Environmental Justice and Nuclear Waste Conflicts in Taiwan," *Environmental Politics* 15, no. 3 (2006): 417–434; J. Page, "Salmon

Farming in First Nations' Territories: A Case of Environmental Injustice on Canada's West Coast," *Local Environment* 12, no. 6 (2007): 613–626; P. Tschakert, "Digging Deep for Justice: A Radical Re-imagination of the Artisanal Gold Mining Sector in Ghana," *Antipode* 41, no. 4 (2009): 706–740; P. Widener, "Benefits and Burdens of Transnational Campaigns: A Comparison of Four Oil Struggles in Ecuador," *Mobilization: An International Quarterly* 12, no. 1 (2007): 21–36; J. McLean, "Water Injustices and Potential Remedies in Indigenous Rural Contexts: A Water Justice Analysis," *Environmentalist* 27, no. 1 (2007): 25–38; R. Cowell, G. Bristow, and M. Munday, "Acceptance, Acceptability and Environmental Justice: The Role of Community Benefits in Wind Energy Development," *Journal of Environmental Planning and Management* 54, no. 4 (2011): 539–557; J. L. Harrison, *Pesticide Drift and the Pursuit of Environmental Justice* (Cambridge, MA: MIT Press, 2011); and D. Carruthers, "Environmental Justice and the Politics of Energy on the US-Mexico Border," *Environmental Politics* 16, no. 3 (2007): 394–413.

10. In order, see D. Carruthers, ed., *Environmental Justice in Latin America* (Cambridge, MA: MIT Press, 2008); D. McDonald, *Environmental Justice in South Africa* (Athens: Ohio University Press, 2002); J. Agyeman, *Speaking for Ourselves: Environmental Justice in Canada* (Seattle: University of Washington Press, 2010); and J. Agyeman and Y. Ogneva-Himmelberger, *Environmental Justice and Sustainability in the Former Soviet Union* (Cambridge, MA: MIT Press, 2009).

11. Walker, "Globalizing Environmental Justice," 361.

12. D. Schlosberg, "Reconceiving Environmental Justice: Global Movements and Political Theories," *Environmental Politics* 13, no. 3 (2004): 517–540.

13. Pellow, *Resisting Global Toxics*; D. Pellow, "Politics by Other Greens: The Importance of Transnational Environmental Justice Movement Networks," in J. Carmin and J. Agyeman, eds., *Environmental Inequalities beyond Borders* (Cambridge, MA: MIT Press, 2011), 247–266.

14. Pellow, "Politics," 248.

15. Mohai, Pellow, and Timmons Roberts, "Environmental Justice."

16. Carmin and Agyeman, *Environmental Inequalities beyond Borders*.

17. R. Bullard and B. Wright, eds., *Race, Place, and Environmental Justice after Hurricane Katrina* (Boulder, CO: Westview, 2009).

18. We will see the same set of community-based issues post-Sandy in New York and New Jersey.

19. J. Sze, *Noxious New York* (Cambridge, MA: MIT Press, 2006).

20. S. Vermeylen and G. Walker, "Environmental Justice, Values, and Biological Diversity: The San and the Hoddia Benefit-Sharing Agreement," in Carmin and Agyeman, *Environmental Inequalities beyond Borders*, 105–128.

21. D. Schlosberg and D. Carruthers, "Indigenous Struggles, Environmental Justice, and Community Capabilities," *Global Environmental Politics* 10, no. 4 (2010): 12–35.

22. G. Di Chiro, "Living Environmentalisms: Coalition Politics, Social Reproduction, and Environmental Justice," *Environmental Politics* 17, no. 2 (2008): 276–298; EcoEquity, "Greenhouse Development Rights," 2008, http://gdrights.org/.

23. J. Agyeman, *Sustainable Communities and the Challenge of Environmental Justice* (New York: NYU Press, 2005).

24. J. A. Ross and L. Zepeda, "Wetland Restoration, Environmental Justice and Food Security in the Lower 9th Ward," *Environmental Justice* 4, no. 2 (2011): 101–108.

25. Sze et al., "Defining and Contesting Environmental Justice."

26. D. Schlosberg, *Defining Environmental Justice* (Oxford: Oxford University Press, 2007); D. Schlosberg, "Justice, Ecological Integrity, and Climate Change," in A. Thompson and J.

Bendik-Keymer, eds., *Ethical Adaptation to Climate Change: Human Virtues of the Future* (Cambridge, MA: MIT Press, 2012), 165–184.

27. E. Cripps, "Saving the Polar Bear, Saving the World: Can the Capabilities Approach Do Justice to Humans, Animals and Ecosystems?," *Res Publica* 16, no. 1 (2010): 1–22.

28. D. Schlosberg, *Environmental Justice and the New Pluralism* (Oxford: Oxford University Press, 1999); Schlosberg, *Defining Environmental Justice*.

29. Sze and London, "Environmental Justice at the Crossroads," 1332.

30. R. Bullard, *Dumping in Dixie: Race, Class, and Environmental Quality* (Boulder, CO: Westview, 1990); R. Bullard, *Confronting Environmental Racism: Voices from the Grassroots* (Boston: South End, 1993).

31. B. Bryant and P. Mohai, eds., *Race and the Incidence of Environmental Hazards: A Time for Discourse* (Boulder, CO: Westview, 1992).

32. Then again, it is clear that there is a long and deep history of what we would now call environmental justice concerns in both race-based and environmental movements—environmental justice as a concern is not new, nor was it driven solely by academic concerns. See, for example, Dorceta Taylor's comprehensive histories. D. Taylor, "American Environmentalism: The Role of Race, Class and Gender in Shaping Activism, 1820–1995," *Race, Gender, Class* 5, no. 1 (1997): 16–62; D. Taylor, *The Environment and the People in American Cities, 1600s–1900s: Disorder, Inequality, and Social Change* (Durham, NC: Duke University Press, 2009).

33. Sze and London, "Environmental Justice at the Crossroads."

34. Ibid., 1347.

35. R. Holifield, M. Porter, and G. Walker, introduction to Holifield, Porter, and Walker, *Spaces of Environmental Justice*, 18.

36. Schlosberg, *Environmental Justice and the New Pluralism*.

37. Di Chiro, "Living Environmentalisms."

38. Sze et al., "Defining and Contesting Environmental Justice."

39. Bullard and Wright, *Race, Place, and Environmental Justice*.

40. Schlosberg, "Reconceiving Environmental Justice"; Schlosberg, *Defining Environmental Justice*.

Reading Questions and Further Readings

Reading Questions

1. How strong is the argument that untouched nature provides a "sense and pattern" (Goodin) that gives people context for their lives?
2. Do the readings give clear guidance on how to negotiate between conflicting values such as human welfare, environmental conservation, and future generations?
3. Does an individual emphasis on wildness (Thoreau) also help us to preserve nature?
4. If our responsibilities under global environmental change grow to include distant and far-off humans, animals, and nature, how does this impact our obligations near home?
5. Is a "new ethic" (Holland) required to account for the environment, or does the expansion of existing ones suffice?

Further Readings

Broome, John. *Climate Matters*. New York: Norton, 2012.

Carlson, Allen. *Aesthetics and the Environment: The Appreciation of Nature, Art and Architecture*. New York: Routledge, 2000.

Erikson, Kai, Robert Gramling, Shirley Laska, and William Freudenburg. *Catastrophe in the Making: The Engineering of Katrina*. Washington, DC: Island, 2010.

Shrader-Frechette, Kristen. *Environmental Justice: Creating Equality, Reclaiming Democracy*. Oxford: Oxford University Press, 2002.

Shue, Henry. *Climate Justice: Vulnerability and Protection*. Oxford: Oxford University Press, 2014.

Stone, Christopher. *Should Trees Have Standing? Toward Legal Rights for Natural Objects*. Wellsboro, PA: Tioga, 1988.

PART VI

Environmental Controversies

This part draws on the concepts and tools in previous parts to analyze three active controversies within Environmental Studies. These controversies are about urban and less dense places, industrial and nonindustrial modes of agriculture, and technical and nontechnical approaches to environmental problem solving.

City and Country

The first controversy is the tension between dense, efficient urbanization and immersion in wilderness. The controversy here surrounds the relationship between individual values and lifestyle and one's environmental impacts.

Edward Abbey, similar to Thoreau in part 5 (albeit with a different flair), sees an essential role for wilderness in self-realization. "We need wilderness because we are wild animals," Abbey writes. Wilderness helps to define one's role in the world, in this case a role that is not as a dominant species but rather one among many. David Owen counters that the most efficient human habitats are "more like Manhattan," meaning very dense and necessarily with less contact with nonhuman wilderness. The tension between these positions brings into view conflicts between lifestyles, comparative efficiency, and other environmental considerations.

Aldo Leopold offers us an opportunity to appreciate nature in all its forms and locations. Indeed, we are currently seeing the resurgence of red-tailed hawks in Washington Square Park and whales in New York Harbor, paired with the kind of appreciation Abbey and Leopold might approve of. In "Conservation Esthetic," Aldo Leopold wrote, "The weeds in a city lot convey the same lesson as the redwoods; the farmer may see in his cow-pasture what may not be vouchsafed to the scientist adventuring in the South Seas."[1]

NOTE
1. Aldo Leopold, *Sand Country Almanac* (1949; repr., New York: Oxford University Press, 1989).

33

Excerpts from "More like Manhattan"

DAVID OWEN

My wife and I got married right out of college, in 1978. We were young and naive and unashamedly idealistic, and we decided to make our first home in a utopian environmentalist community in New York State. For seven years, we lived, quite contentedly, in circumstances that would strike most Americans as austere in the extreme: our living space measured just seven hundred square feet, and we didn't have a dishwasher, a garbage disposal, a lawn, or a car. We did our grocery shopping on foot, and when we needed to travel longer distances we used public transportation. Because space at home was scarce, we seldom acquired new possessions of significant size. Our electric bills worked out to about a dollar a day.

The utopian community was Manhattan. (Our apartment was on Sixty-ninth Street, between Second and Third.) Most Americans, including most New Yorkers, think of New York City as an ecological nightmare, a wasteland of concrete and garbage and diesel fumes and traffic jams, but in comparison with the rest of America it's a model of environmental responsibility. By the most significant measures, New York is the greenest community in the United States, and one of the greenest cities in the world. The most devastating damage humans have done to the environment has arisen from the heedless burning of fossil fuels, a category in which New Yorkers are practically prehistoric. The average Manhattanite consumes gasoline at a rate that the country as a whole hasn't matched since the mid-nineteen-twenties, when the most widely owned car in the United States was the Ford Model T. Eighty-two percent of Manhattan residents travel to work by public transit, by bicycle, or on foot. That's ten times the rate for Americans in general, and eight times the rate for residents of Los Angeles County.[1] New York City is more populous than all but eleven states; if it were granted statehood, it would rank fifty-first in per-capita energy use, not only because New Yorkers drive less but because city dwellings are smaller than other American dwellings and are less likely to contain a superfluity of large appliances.[2] The average New Yorker (if one takes into consideration all five boroughs of the city) annually generates 7.1 metric tons of greenhouse gases, a lower rate than that of residents of any other American city, and less than 30 percent of the national average, which is 24.5 metric tons;[3] Manhattanites generate even less.

"Anyplace that has such tall buildings and heavy traffic is obviously an environmental disaster—except that it isn't," John Holtzclaw, a transportation consultant for the Sierra Club and the Natural Resources Defense Council, told me. "If New Yorkers lived at the typical American sprawl density of three households per residential acre, they would require many times as much land. They'd be driving cars, and they'd have huge lawns and be using pesticides and fertilizers on them, and then they'd be overwatering their lawns, so that runoff would go into streams." The key to New York's relative environmental benignity is its extreme compactness. Manhattan's population density is more than eight hundred times that of the nation as a whole. Placing one and a half million people on a twenty-three-square-mile island sharply reduces their opportunities to be wasteful, and forces the majority to live in some of the most inherently energy-efficient residential structures in the world: apartment buildings. It also frees huge tracts of land for the rest of America to sprawl into. . . .

Yet our move was an ecological catastrophe. Our consumption of electricity went from roughly four thousand kilowatt-hours a year, toward the end of our time in New York, to almost thirty thousand kilowatt-hours in 2003—and our house doesn't even have central air-conditioning. We bought a car shortly before we moved, and another one soon after we arrived, and a third one ten years later. (If you live in the country and don't have a second car, you can't retrieve your first car from the mechanic after it's been repaired; the third car was the product of a mild midlife crisis, but soon evolved into a necessity.) My wife and I both work at home, but we manage to drive thirty thousand miles a year between us, mostly doing ordinary errands.[4] Nearly everything we do away from our house requires a car trip. Renting a movie and later returning it, for example, consumes almost two gallons of gasoline, since the nearest Blockbuster is ten miles away and each transaction involves two round trips. When we lived in New York, heat escaping from our apartment helped to heat the apartment above ours; nowadays, many of the BTUs produced by our brand-new, extremely efficient oil-burning furnace leak through our two-hundred-year-old roof and into the dazzling star-filled winter sky above.

*　*　*

The history of civilization is a chronicle of destruction: people arrive, eat anything slow enough to catch, supplant indigenous flora with species bred for exploitation, burn whatever can be burned, and move on or spread out. No sensitive modern human can contemplate that history without a shudder. Everywhere we look, we see evidence of our recklessness, as well as signs that our destructive reach is growing. For someone standing on the North Rim of the Grand Canyon on a moonless night, the brightest feature of the sky is no longer the Milky Way but the glow of Las Vegas, 175 miles away.[5] Tap water in metropolitan Washington, D.C., has been found to contain trace amounts of caffeine, ibuprofen, naproxen sodium, two antibiotics, an anticonvulsive drug used to treat seizures and bipolar disorder, and the antibacterial compound triclocarban, which is an ingredient of household soaps and cleaning agents.[6] Modern interest in environmentalism is driven by a yearning to protect what we haven't ruined already, to conserve what we haven't used up, to restore as much as possible of what we've destroyed, and to devise ways of reconfiguring our lives so that civilization as we know it can be sustained through our children's lifetimes and beyond.

To the great majority of Americans who share these concerns, densely populated cities look like the end of the world. Because such places concentrate high levels of human activity, they seem to manifest nearly every distressing symptom of the headlong growth of civilization—the smoke, the filth, the crowds, the cars—and we therefore tend to think of them as environmental crisis zones. Calculated by the square foot, New York City generates more greenhouse gases, uses more energy, and produces more solid waste than any other American region of comparable size. On a map depicting negative environmental impacts in relation to surface area, therefore, Manhattan would look like an intense hot spot, surrounded, at varying distances, by belts of deepening green.

But this way of thinking obscures a profound environmental truth, because if you plotted the same negative impacts by resident or by household the color scheme would be reversed. New Yorkers, individually, drive, pollute, consume, and throw away much less than do the average residents of the surrounding suburbs, exurbs, small towns, and farms, because the tightly circumscribed space in which they live creates efficiencies and reduces the possibilities for reckless consumption. Most important, the city's unusually high concentration of population enables the majority of residents to live without automobiles—an unthinkable deprivation almost anywhere else in the United States, other than in a few comparably dense American urban cores, such as the central parts of San Francisco and Boston. The scarcity of parking spaces in New York, along with the frozen snarl of traffic on heavily traveled streets, makes car ownership an unbearable burden for most, while the compactness of development, the fertile mix of commercial and residential uses, and the availability of public transportation make automobile ownership all but unnecessary in most of the city. A pedestrian crossing Canal Street at rush hour can get the impression that New York is the home of every car ever built, but Manhattan actually has the lowest car-to-resident ratio of anyplace in America. . . .

New York City is by no means the world's only or best example of the environmental benefits of concentrating human populations. Almost any large old city in Europe—where the main population centers arose long before the automobile, and therefore evolved to be served by less environmentally disastrous means of getting around—is both denser and less wasteful than New York. The most energy-efficient and least automobile-dependent city in the world is almost certainly Hong Kong, whose overall density greatly exceeds even that of Manhattan. But New York is a useful example for Americans, both because it is familiar and because it proves that affluent people are capable of living comfortably while consuming energy and inflicting environmental damage at levels well below U.S. averages. And—as is the case with all dense cities—New York's efficiencies are built-in and, therefore, don't depend on a total, sudden transformation of human nature. Even for people who live in sparsely populated areas far from urban centers, dense cities like New York offer important lessons about how to permanently reduce energy use, water consumption, carbon output, and many other environmental ills.

Thinking of crowded cities as environmental role models requires a certain willing suspension of disbelief, because most of us have been accustomed to viewing urban centers as ecological nightmares. New York is one of the most thoroughly altered landscapes imaginable, an almost wholly artificial environment, in which the terrain's primeval contours have long since been obliterated and most of the parts that resemble nature (the trees on side streets, the rocks in Central Park) are essentially decorations. Quite obviously, this wasn't always the case. When Europeans first began to settle Manhattan, in the early seventeenth century, a broad salt marsh lay where the East Village does today, the area now occupied by Harlem was flanked by sylvan bluffs, and Murray Hill and Lenox Hill were hills. Streams ran everywhere, and beavers built dams near what is now Times Square. One early European visitor described Manhattan as "a land excellent and agreeable, full of noble forest trees and grape vines," and another called it a "terrestrial Canaan, where the Land floweth with milk and honey."[7]

But then, across a relatively brief span of decades, Manhattan's European occupiers leveled the forests, flattened the hills, filled the valleys, buried the streams, and superimposed an unyielding, two-dimensional grid of avenues and streets, leaving virtually no hint of what had been before. The earliest outposts of metropolitan civilization, such as it was, were confined to the island's southern tip, but in the eighteenth and nineteenth centuries settlement spread northward at an accelerating pace. In 2007, Eric Sanderson, a landscape ecologist who was completing a three-dimensional computer re-creation of pre-European Manhattan, told Nick Paumgarten of the *New Yorker*, "It's hard to think of any place in the world with as heavy a footprint, in so short a time, as New York. It's probably the fastest, biggest land-coverage swing in history."[8] Picturing even a small part of that long-lost world requires a heroic act of the imagination—or, as in Sanderson's case, a vast database and complex computer-modeling software.

Given the totality of what has been erased, contemplation of New York's evolution into a megalopolis inspires mainly a sense of loss, and ecology-minded discussions of the city tend to have a forlorn air. Nikita Khrushchev, who visited New York in the fall of 1960, found the scarcity of foliage in the city depressing, by comparison with Moscow, saying, "It is enough to make a stone sad."[9] In environmental triage, New York is usually consigned to the hopeless category, worthy of palliative care only. Environmentalists tend to focus on a handful of ways in which the city might be made to seem somewhat less oppressively man-made: by easing the intensity of development; by creating or enlarging open spaces around structures; by relieving traffic congestion and reducing the time that drivers spend aimlessly searching for parking spaces; by increasing the area devoted to parks, greenery, and gardening; by incorporating vegetation into buildings themselves.

But such discussions miss the point, because in most cases changes like these would actually undermine the features that create the city's extraordinary efficiency and keep the ecological impact of its residents small. Spreading buildings out enlarges the distance between local destinations, thereby limiting the utility of walking and public transportation; making automobile traffic move more efficiently enhances the allure of owning cars and, inevitably, reduces ridership on the subway. Because urban density, in itself, is such a powerful generator of environmental benefits, the most critical environmental issues in dense urban cores tend to be seemingly unrelated matters like law enforcement and public education, because anxieties about crime and school quality are among the strongest forces motivating flight to the suburbs. By comparison, popular feel-good urban eco-projects like adding solar panels to the roofs of apartment buildings are decidedly secondary. Planting trees along city streets, always a popular initiative, has high environmental utility, but not for the reasons that people usually assume: trees are ecologically important in dense urban areas not because they provide temporary repositories for atmospheric carbon—the usual argument for planting more of them—but because their presence along sidewalks makes city dwellers more cheerful about dwelling in cities. Unfortunately, much conventional environmental activism has the opposite effect, since it reinforces the view that urban life is artificial and depraved, and makes city residents feel guilty about living where and how they do.

A dense urban area's greenest features—its low per-capita energy use, its high acceptance of public transit and walking, its small carbon footprint per resident—are not inexplicable anomalies. They are the direct consequences of the very urban characteristics that are the most likely to appall a sensitive friend of the earth. Yet those qualities are ones that the rest of us, no matter where we live, are going to have to find ways to emulate, as the world's various ongoing energy and environmental crises deepen and spread in the years ahead. In terms of sustainability, dense cities have far more to teach us than solar-powered mountainside cabins or quaint old New England towns. . . .

*　*　*

. . . "Sustainable living" is actually much harder in small, far-flung places than it is in dense cities. [Ben] Jervey [author of *The Big Green Apple*] cites New Yorkers' "overactive dependence" on fresh water as an example of their supposed wastefulness, and he marvels that the city's total use "amounts to well over one billion gallons per day."[10] A billion is a big number, to be sure, but New York City's population is more than thirteen times that of the entire state of Vermont, so the city's total consumption figures in any category will appear overwhelming in any direct comparison. It's per-capita consumption that is telling, though, and by that measure Vermonters use more water than New Yorkers do. They also use more than three and a half times as much gasoline—545 gallons per person per year versus 146 for all New York City residents and just 90 for Manhattan residents—with the result that, among the fifty states, pastoral Vermont ranks eleventh-highest in per-capita gasoline consumption while New York State, thanks entirely to New York City, ranks last. The average Vermonter also consumes more than four times as much electricity as the average New York City resident, has a larger carbon footprint, and generates more solid waste, backyard compost bins notwithstanding.[11]

Jervey is by no means alone. The prominent British environmentalist Herbert Girardet—who is an author, a documentary filmmaker, and a cofounder of the World Fortune Council—treats large cities mainly as environmental catastrophes. "The bulk of the world's energy consumption is *within* cities," he has written, "and much of the rest is used for producing and transporting goods and people *to and from* cities."[12] He proposes dramatically reducing urban energy consumption and making city dwellers less dependent on agricultural and other inputs from outlying areas, while improving overall energy efficiency through technological innovation. He has observed that cities cover just 3 or 4 percent of the earth's land area while accounting for 80 percent of the world's consumption of natural resources—as though population density were an ecological negative, and as though there were no meaningful distinction to be made between dense urban cores and lightly populated suburbs. Urban dwellers, by his way of thinking, are environmental freeloaders, parasitically drawing sustenance from the countryside, while people living at lower densities are more nearly at harmony with nature.[13] Girardet is a victim (and perpetuator) of the same optical illusion as Jervey.

New Yorkers themselves seldom fully appreciate the environmental virtues of their own way of living. On Earth Day 2007, the city announced an ambitious two-decade environmental initiative, called PlaNYC, which includes dozens of far-reaching proposals, among them the planting of more than a million trees, the collection of tolls from most private and commercial vehicles using the most traffic-clogged parts of Manhattan during the busiest times of the day, the imposition of a surcharge on the bills of the city's electrical customers, and other measures.[14] Actually implementing the plan has encountered the usual difficulties (shortly before Earth Day 2008, the state legislature killed the toll-collection scheme, which is known as "congestion pricing"), but one of the most striking features of the entire plan is how little recognition it gives to the numerous ways in which New York City's environmental performance is already exemplary, even extraordinary, at least in comparison with the rest of the United States. Shortly before the plan was made public, the mayor's office released a study showing that the city's buildings are responsible for 79 percent of its greenhouse-gas emissions—an ominous statistic, the study suggested, since the national average for buildings is just 32 percent. Daniel L. Doctoroff, a deputy mayor and the city official in charge of the plan, said, "We know we have to dramatically rethink the way we work with buildings"—probably an understatement, since the mayor's announced goal was to cut greenhouse-gas emissions by 30 percent by 2030.[15]

Cutting greenhouse-gas emissions is a fine idea, but in the case of the city's buildings the mayor's office obscured a far more important point. The proportion of emissions attributable to buildings in New York City is high because the number of cars, which are the main source of greenhouse emissions in the rest of the country, is extremely low in relation to the city's population: it's a sign of environmental success, not failure. Thinking in terms of proportions can only be misleading, since there's no way to decrease the percentage attributable to one element without increasing the percentage attributable to others: they're pieces of the same pie. Bringing down overall emissions levels is a worthy goal, but the mayor's emphasis was misplaced. The proportion of greenhouse-gas emissions attributable to buildings is higher in energy-efficient old European cities, too....

*   *   *

The hostility of many environmentalists toward densely populated cities is a manifestation of a much broader phenomenon, a deep antipathy to urban life which has been close to the heart of American environmentalist since the beginning. Henry David Thoreau, who lived in a cabin in the woods near Concord, Massachusetts, between 1845 and 1847, established an image, still potent today, of the sensitive nature lover living simply, and in harmony with the environment, beyond the edge of civilization. Thoreau wasn't actually much of an outdoorsman, and his cabin was closer to the center of Concord than to any true wilderness, but for many Americans he remains the archetype—the natural philosopher guiltlessly living off the grid. John Muir, who was born twenty years after Thoreau and founded the Sierra Club in 1892, viewed city living as toxic to both body and soul.[16] The National Park Service, established by Congress in 1916, was conceived as an increasingly necessary corrective to urban life, and national parks were treated in large measure as sanctuaries from urban depravity. The modern environmental movement arose, in the 1960s and 1970s, when a growing sense of ecological crisis, first inspired nationwide by Rachel Carson's extraordinarily influential book *Silent Spring*,[17] combined with other social forces, including the civil rights movement, opposition to the Vietnam War, and the power of OPEC, to create a sense among large numbers of mainly young people that just about everything wrong with the United States was urban in essence, and could be combated only by establishing, or reestablishing, a direct connection to "the land." American environmentalists in every age have tended to agree with Thomas Jefferson, who, in 1803, dismissed "great cities" as "pestilential to the morals, the health and the liberties of man."[18]

Jefferson made that disparaging remark in a letter to Dr. Benjamin Rush, a fellow signer of the Declaration of Independence. Daniel Lazare, in *America's Undeclared War: What's Killing Our Cities and How We Can Stop It*, cites that letter as a key document in the history of what he identifies as an enduring national antagonism toward urban life. Recently, I asked Lazare whether he detected that same antagonism in the modern American environmental movement. "Unquestionably," he said.

> Green ideology is a rural, agrarian ideology. It seeks to integrate man into nature in a very kind of direct, simplistic way—scattering people among the squirrels and the trees and the deer. To me, that seems mistaken, and it doesn't really understand the proper relationship between man and nature. Cities are much more efficient, economically, and also much more benign, environmentally, because when you concentrate human activities in confined spaces you reduce the human footprint, as it were. That is why the disruption of nature is much less in Manhattan than it is in the suburbs. The environmental movement is deeply stained with a sort of Malthusian current. It's anti-urban, anti-industrial, agrarian, primitivist. Manhattan seems to be a supremely unnatural place because of all the concrete and glass and steel, but the paradox is that it's actually more harmonious and more benign, in terms of nature, than ostensibly greener human environments, which depend on huge energy inputs, mainly in the form of fossil fuels. In order to surround ourselves with nature, we get in our cars and drive long distances, and then build silly pseudo-green houses in the middle of the woods—which are actually extremely disruptive, and very, very wasteful.

To be sure, there has always been plenty to loathe about urban living. The history of large cities all over the world is a history of filth and squalor and disease. Benjamin Rush placed himself at tremendous personal risk in 1793, a decade before Jefferson's letter, while attempting to combat a yellow fever epidemic in Philadelphia, which was then both the nation's capital and, with a population of 55,000, its largest city. No one in those days knew how yellow fever was transmitted, but there was no local shortage of plausible explanations. The streets of Philadelphia, like the streets of most cities, were reeking, open sewers, and that particular summer the air had been made especially rank by the arrival from the Caribbean of a large shipment of spoiled coffee beans, which had been left to rot on the wharf and seemed to Rush to be the most likely cause of the disease.[19] Jefferson's letter made specific reference to that epidemic, which killed 4,000 Philadelphians (and caused Jefferson himself to flee the city, along with many other government officials and most of the city's wealthier inhabitants, including most of its physicians). "When great evils happen," Jefferson wrote to Rush, "I am in the habit of looking out for what good may arise from them as consolations to us, and Providence has in fact so established the order of things, as that most evils are the means of producing some good. The yellow fever will discourage the growth of great cities in our nation"—a providential result, in his view. He acknowledged that cities "nourish some of the elegant arts," but stated that "the useful ones can thrive elsewhere, and less perfection in the others, with more health, virtue & freedom, would be my choice."[20] New York City, he wrote twenty years later, "seems to be a Cloacina[21] of all the depravities of human nature."[22] . . .

Anti-urbanism still animates American environmentalism, and is evident in the technical term that is widely used for sprawl: "urbanization." This is unfortunate, because thinking of freeways and strip malls as "urban" phenomena obscures the ecologically monumental difference between Manhattan and Phoenix, or between Copenhagen and Kansas City, and fortifies the perception that population density is an environmental ill. In 2006, Melissa Holbrook Pierson, a writer who lives in a smallish town in the Hudson River Valley, in upstate New York, published a book called *The Place You Love Is Gone*, a deeply felt paean to the lost American landscape, the one obliterated by sprawl. At one point, driven by what she refers to as "lacerating nostalgia," she describes the nightmare transformation of Akron, Ohio, where she grew up in the late 1950s and early 1960s. "I can't help it if I want to live in the past!" she writes. "It's *my* past, the time forty years ago when there was still some wide-open space into which to insert some dreaming, and still some darkness at night over it." She even manages to weep a little over Hoboken, New Jersey, where she lived, mostly unhappily, as a young adult. Her bitterest emotions, though, she reserves for New York City, which she accuses of having destroyed a pastoral paradise in order to create the extensive upstate reservoir system that supplies its drinking water—of "rubbing its chin in contemplation of turning faraway valleys into pipes to service its water closets." The city's early-twentieth-century planners, anticipating the population growth to come, condemned farms and rural hamlets far from the city in order to build the extraordinary chain of reservoirs without which New York City could not exist, and Pierson describes this massive engineering project as "larceny."[23] Her arguments persuaded Anthony Swofford, who reviewed the book in the *New York Times*. He wrote, "The story of New York City's water grab is astonishing, nearly unbelievable in its scope and greed," and he describes the creation of the city's water system, as recounted by Pierson, as "rural slaughter for the survival of the city."[24]

But this is wrong. If New York City could somehow be dismantled and its residents dispersed across the state at the density of Pierson's current hometown, what remains today of pastoral New York State would vanish under a tide of asphalt. Dense urban concentrations of people, along with the freshwater reservoirs and other infrastructure necessary to support them, are not the enemies of the images she clings to. It is the existence of Manhattan, not the nostalgia of Baby Boomers, that makes the Catskills possible, and it's small-town residents, not subway-riding apartment dwellers, who foster strip malls. You create open spaces not by spreading people out but by moving them closer together. Pierson does write, near the end of her book, that "it is the thousands of acres of uninhabited, forested land in the buffer zones of the New York City watershed that have preserved wilderness in the midst of an inexorably creeping urbanization."[25] But she doesn't acknowledge the role of her own form of nostalgia in the creation of the thing she hates. Many more acres of upstate pastoral paradise were destroyed by the steady spread of towns like hers than by the creation of the water supply system that makes it possible for New York City to exist. Building the city didn't fill the Hudson Valley with parking lots; fleeing the city did.

NOTES

1. Mark Ginsburg and Mark Strauss, "New York City—A Case Study in Density after 9/11," paper presented at the Density Conference, Boston, 2003, http://www.architects.org/emplibrary/C6_b.pdf.

2. American Council for an Energy-Efficient Economy.

3. Mayor's Office of Operations, "Inventory of New York City Greenhouse Gas Emissions 2007," City of New York, Office of Long-Term Planning and Sustainability, 2007, http://www.nyc.gov/html/om/pdf/ccp_report041007.pdf.

4. Our annual mileage, which has gone down since our kids went off to college, is fairly close to the national average. In 2001, the most recent year for which complete national data are available, the average American household consisted of 2.58 people, of whom 1.77 were licensed drivers, and put 21,187 miles on their 1.89 cars. The average American in 2001 traveled 35,244 miles by car—a bigger number because on some trips cars have more than one occupant. These statistics are from the "2001 National Household Travel Survey," published in 2004 by the Federal Highway Administration of the U.S. Department of Transportation. You can find a comprehensive "Summary of Travel Trends" here: http://nhts.ornl.gov/2001/pub/STT.pdf

5. David Owen, "The Dark Side," New Yorker, August 20, 2007.

6. Carol D. Leonnig, "Area Tap Water Has Traces of Medicines," Washington Post, March 10, 2008.

7. The first quotation is from the website of the Mannahatta Project, http://www.wcs.org/sw-high_tech_tools/landscapeecology/mannahatta; the second is from Daniel Denton, an English essayist, as quoted in Edwin G. Burrows and Mike Wallace, Gotham: A History of New York City to 1898 (New York: Oxford University Press, 1999), 3.

8. Nick Paumgarten, "The Mannahatta Project," New Yorker, October 1, 2007. Sanderson has published his findings in a book: Mannahatta: A Natural History of New York City (New York: Abrams, 2009).

9. Quoted in Fred R. Shapiro, ed., The Yale Book of Quotations (New Haven, CT: Yale University Press, 2006), 426.

10. Ben Jervey, The Big Green Apple (Guilford, CT: Globe Pequot, 2006), 39.

11. California Energy Commission, "U.S. Gasoline Per Capita Use by State, 2004" and "U.S. Per Capita Electricity Use by State, 2003," http://www.energy.ca.gov, based on consumption data from the Energy Information Administration and population data from the U.S. Census Bureau and other sources.

12. Herbert Girardet, *Creating Sustainable Cities* (Dartington, UK: Green Books, 1999), 42; also in Herbert Girardet, "The Metabolism of Cities," in Stephen M. Wheeler and Timothy Beatley, eds., *The Sustainable Urban Development Reader* (London: Routledge, 2004), 130 (italics in original).

13. For a brief profile of Girardet, see Bonnie Alter, "Herbert Girardet: Reluctant Optimist," Treehugger, May 8, 2007, http://www.treehugger.com/files/2007/05/herbert_girarde.php.

14. The PlaNYC website can be found at http://www.nyc.gov/html/planyc2030/html/home/home.shtml.

15. See Diane Cardwell, "Buildings Called Key Source of City's Greenhouse Gases," *New York Times*, April 11, 2007.

16. Robert Gottlieb, *Forcing the Spring: The Transformation of the American Environmental Movement* (Washington, DC: Island, 1993), 29–30. Gottlieb writes, "Although he wrote for urban, cosmopolitan publications that allowed him to establish urban support for wilderness protection, Muir was nevertheless especially hostile to urban living."

17. Rachel L. Carson, *Silent Spring* (New York: Houghton Mifflin, 1962).

18. Thomas Jefferson, letter to Dr. Benjamin Rush, April 21, 1803.

19. See Jim Murphy, *An American Plague: The True and Terrifying Story of Yellow Fever Epidemic of 1793* (New York: Clarion Books, 2003); and J. H. Powell, *Bring Out Your Dead: The Great Plague of Yellow Fever in Philadelphia in 1793* (Philadelphia: University of Pennsylvania Press, 1993). Murphy's book is intended for younger readers, but it is concise and well researched, and it contains many old prints.

20. Thomas Jefferson, letter to Dr. Benjamin Rush, April 21, 1803.

21. Cloacina was the goddess of the Roman sewer system. The name comes from the Latin word for "sewer" or "drain."

22. Thomas Jefferson, letter to William Short, September 8, 1823.

23. Melissa Holbrook Pierson, *The Place You Love Is Gone* (New York: Norton, 2006), 39 ("lacerating nostalgia"), 43 ("I can't help it"), 126 ("water closets"), 133 ("larceny").

24. Anthony Swofford, "A Meditation on Change," *New York Times*, January 15, 2006.

25. Pierson, *The Place You Love Is Gone*, 164.

34

Excerpts from "Freedom and Wilderness,
Wilderness and Freedom"

EDWARD ABBEY

When I lived in Hoboken, just across the lacquered Hudson from Manhattan, we had all the wilderness we needed. There was the waterfront with its decaying piers and abandoned warehouses, the jungle of bars along River Street and Hudson Street, the houseboats, the old ferry slips, the mildew-green cathedral of the Erie-Lackawanna Railway terminal. That was back in 1964–65: then came Urban Renewal, which ruined everything left lovable in Hoboken, New Jersey.

What else was there? I loved the fens, those tawny marshes full of waterbirds, mosquitoes, muskrats, and opossums that intervened among the black basaltic rocks between Jersey City and Newark, and somewhere back of Union City on the way to gay, exotic, sausage-packing, garbage-rich Secaucus. I loved also and finally and absolutely, as a writer must love any vision of eschatological ultimates, the view by twilight from the Pulaski Skyway (Stop for Emergency Repairs Only) of the Seventh Circle of Hell. Those melancholy chemical plants, ancient as acid, sick as cyanide, rising beyond the cattails and tules; the gleam of oily waters in the refineries' red glare; the desolation of the endless, incomprehensible uninhabitable (but inhabited) slums of Harrison, Newark, Elizabeth; the haunting and sinister odors on the wind. Rust and iron and sunflowers in the tangled tracks, the great grimy sunsets beyond the saturated sky. . . . It will all be made, someday, a national park of the mind, a rigid celebration of industrialism's finest frenzy.

We tried north too, up once into the Catskills, once again to the fringe of the Adirondacks. All I saw were Private Property Keep Out This Means You signs. I live in a different country now. Those days of longing, that experiment in exile, are all past. The far-ranging cat returns at last to his natural, native habitat. But what wilderness there was in those bitter days I learned to treasure. Foggy nights in greasy Hoboken alleyways kept my soul alert, healthy and aggressive, on edge with delight.

The other kind of wilderness is also useful. I mean now the hardwood forests of upper Appalachia, the overrated mountains of Colorado, the burnt sienna hills of South Dakota, the raw umber of Kansas, the mysterious swamps of Arkansas, the porphyritic mountains of purple Arizona, the mystic desert of my own four-cornered country—this and 347 other good, clean, dangerous places I could name.

Science is not sufficient. "Ecology" is a word I first read in H. G. Wells twenty years ago and I still don't know what it means. Or seriously much care. Nor am I primarily concerned with nature as living museum, the preservation of spontaneous plants and wild animals. The wildest animal I know is you, gentle reader, with this helpless book clutched in your claws. No, there are better reasons for keeping the wild wild, the wilderness open, the trees up and the rivers free, and the canyons uncluttered with dams.

We need wilderness because we are wild animals. Every man needs a place where he can go to go crazy in peace. Every Boy Scout troop deserves a forest to get lost, miserable, and starving in. Even the maddest murderer of the sweetest wife should get a chance for a run to the sanctuary of the hills. If only for the sport of it. For the terror, freedom, and delirium. Because we need brutality and raw adventure, because men and women first learned to love in, under, and all around trees, because we need for every pair of feet and legs about ten leagues of naked nature, crags to leap from, mountains to measure by, deserts to finally die in when the heart fails.

The prisoners in Solzhenitsyn's labor camps looked out on the vast Siberian forests—within those shadowy depths lay the hope of escape, of refuge, of survival, of hope itself—but guns and barbed wire blocked the way. The citizens of our American cities enjoy a high relative degree of political, intellectual, and economic liberty; but if the entire nation is urbanized, industrialized, mechanized, and administered, then our liberties continue only at the sufferance of the technological megamachine that functions both as servant and master, and our freedoms depend on the pleasure of the privileged few who sit at the control consoles of that machine. What makes life in our cities at once still tolerable, exciting, and stimulating is the existence of an alternative option, whether exercised or not, whether even appreciated or not, of a radically different mode of being out there, in the forests, on the lakes and rivers, in the deserts, up in the mountains.

Who needs wilderness? Civilization needs wilderness. The idea of wilderness preservation is one of the fruits of civilization, like Bach's music, Tolstoy's novels, scientific medicine, novocaine, space travel, free love, the double martini, the secret ballot, the private home and private property, the public park and public property, freedom of travel, the Bill of Rights, peppermint toothpaste, beaches for nude bathing, the right to own and bear arms, the right not to own and bear arms, and a thousand other good things one could name, some of them trivial, most of them essential, all of them vital to that great, bubbling, disorderly, anarchic, unmanageable diversity of opinion, expression, and ways of living which free men and women love, which is their breath of life, and which the authoritarians of church and state and war and sometimes even art despise and always have despised. And feared.

The permissive society? What else? I love America because it is a confused, chaotic mess—and I hope we can keep it this way for at least another thousand years. The permissive society is the free society, the open society. Who gave us permission to live this way? Nobody did. We did. And that's the way it should be—only more so. The best cure for the ills of democracy is more democracy.

The boundary around a wilderness area may well be an artificial, self-imposed, sophisticated construction, but once inside that line you discover the artificiality beginning to drop away; and the deeper you go, the longer you stay, the more interesting things get—sometimes fatally interesting. And that too is what we want: Wilderness is and should be a place where, as in Central Park, New York City, you have a fair chance of being mugged and buggered by a shaggy fellow in a fur coat—one of Pooh Bear's big brothers. To be alive is to take risks; to be always safe and secure is death.

Enough of these banalities—no less true anyhow—which most of us embrace. But before getting into the practical applications of this theme, I want to revive one more argument for the preservation of wilderness, one seldom heard but always present, in my own mind at least, and that is the political argument.

Democracy has always been a rare and fragile institution in human history. Never was it more in danger than now, in the dying decades of this most dangerous of centuries. Within the past few years alone we have seen two more relatively open societies succumb to dictatorship and police rule—Chile and India. In all of Asia there is not a single free country except Israel—which, as the Arabs say, is really a transplanted piece of Europe. In Africa, obviously going the way of Latin America, there are none. Half of Europe stagnates under one-man or one-party domination. Only Western Europe and Britain, Australia and New Zealand, perhaps Japan, and North America can still be called more or less free, open, democratic societies.

As I see it, our own nation is not free from the danger of dictatorship. And I refer to internal as well as external threats to our liberties. As social conflict tends to become more severe in this country—and it will unless we strive for social justice—there will inevitably be a tendency on the part of the authoritarian element—always present in our history—to suppress individual freedoms, to utilize the refined techniques of police surveillance (not excluding torture, of course) in order to preserve—not wilderness!—but the status quo, the privileged positions of those who now so largely control the economic and governmental institutions of the United States.

If this fantasy should become reality—and fantasies becoming realities are the story of the twentieth century—then some of us may need what little wilderness remains as a place of refuge, as a hideout, as a base from which to carry on guerrilla warfare against the totalitarianism of my nightmares. I hope it does not happen; I believe we will prevent it from happening; but if it should, then I, for one, intend to light out at once for the nearest national forest, where I've been hiding cases of peanut butter, home-brew, ammunition, and C-rations for the last ten years. I haven't the slightest doubt that the FBI, the NSA, the CIA, and the local cops have dossiers on me a yard thick. If they didn't, I'd be insulted. Could I survive in the wilderness? I don't know—but I do know I could never survive in prison.

Could we as a people survive without wilderness? To consider that question we might look at the history of modern Europe, and of other places. As the Europeans filled up their small continent, the more lively among them spread out over the entire planet, seeking fortune, empire, a new world, a new chance—but seeking most of all, I believe, for adventure, for the opportunity of self-testing. Those nations that were confined by geography, bottled up, tended to find their outlet for surplus energy through war on their neighbors; the Germans provide the best example of this thesis. Nations with plenty of room for expansion, such as the Russians, tended to be less aggressive toward their neighbors. . . .

As we return to a happier equilibrium between industrialism and a rural-agrarian way of life, we will of course also encourage a gradual reduction of the human population of these states to something closer to the optimum: perhaps half the present number. This would be accomplished by humane social policies, naturally, by economic and taxation incentives encouraging birth control, the single-child family, the unmarried state, the community family. Much preferable to war, disease, revolution, nuclear poisoning, etc., as population control devices.

What has all this fantasizing to do with wilderness and freedom? We can have wilderness without freedom; we can have wilderness without human life at all; but we cannot have freedom without wilderness, we cannot have freedom without leagues of open space beyond the cities, where boys and girls, men and women, can live at least part of their lives under no control but their own desires and abilities, free from any and all direct administration by their fellow men. "A world without wilderness is a cage," as Dave Brower says.

I see the preservation of wilderness as one sector of the front in the war against the encroaching industrial state. Every square mile of range and desert saved from the strip miners, every river saved from the dam builders, every forest saved from the loggers, every swamp saved from the land speculators means another square mile saved for the play of human freedom.

All this may seem utopian, impossibly idealistic. No matter. There comes a point at every crisis in human affairs when the ideal must become the real—or nothing. It is my contention that if we wish to save what is good in our lives and give our children a taste of a good life, we must bring a halt to the ever-expanding economy and put the growth maniacs under medical care.

Let me tell you a story.

A couple of years ago I had a job. I worked for an outfit called Defenders of Fur Bearers (now known as Defenders of Wildlife). I was caretaker and head janitor of a 70,000-acre wildlife refuge in the vicinity of Aravaipa Canyon in southern Arizona. The Whittell Wildlife Preserve, as we called it, was a refuge for mountain lion, javelina, a few black bear, maybe a wolf or two, a herd of whitetail deer, and me, to name the principal fur bearers.

I was walking along Aravaipa Creek one afternoon when I noticed fresh mountain lion tracks leading ahead of me. Big tracks, the biggest lion tracks I've seen anywhere. Now I've lived most of my life in the Southwest, but I am sorry to admit that I had never seen a mountain lion in the wild. Naturally I was eager to get a glimpse of this one.

It was getting late in the day, the sun already down beyond the canyon wall, so I hurried along, hoping I might catch up to the lion and get one good look at him before I had to turn back and head home. But no matter how fast I walked and then jogged along, I couldn't seem to get any closer; those big tracks kept leading ahead of me, looking not five minutes old, but always disappearing around the next turn in the canyon.

Twilight settled in, visibility getting poor. I realized I'd have to call it quits. I stopped for a while, staring upstream into the gloom of the canyon. I could see the buzzards settling down for the evening in their favorite dead cottonwood. I heard the poor-wills and the spotted toads beginning to sing, but of that mountain lion I could neither hear nor see any living trace.

I turned around and started home. I'd walked maybe a mile when I thought I heard something odd behind me. I stopped and looked back—nothing; nothing but the canyon, the running water, the trees, the rocks, the willow thickets. I went on and soon I heard that noise again—the sound of footsteps.

I stopped. The noise stopped. Feeling a bit uncomfortable now—it was getting dark—with all the ancient superstitions of the night starting to crawl from the crannies of my soul, I looked back again.

And this time I saw him. About fifty yards behind me, poised on a sand bar, one front paw still lifted and waiting, stood this big cat, looking straight at me. I could see the gleam of the twilight in his eyes. I was startled as always by how small a cougar's head seems but how long and lean and powerful the body really is. To me, at that moment, he looked like the biggest cat in the world. He looked dangerous. Now I know very well that mountain lions are supposed almost never to attack human beings. I knew there was nothing to fear—but I couldn't help thinking maybe this lion is different from the others. Maybe he knows we're in a wildlife preserve, where lions can get away with anything. I was not unarmed; I had my Swiss army knife in my pocket with the built-in can opener, the corkscrew, the two-inch folding blade, the screwdriver. Rationally there was nothing to fear; all the same I felt fear.

And something else too: I felt what I always feel when I meet a large animal face to face in the wild: I felt a kind of affection and the crazy desire to communicate, to make some kind of emotional, even physical contact with the animal. After we'd stared at each other for maybe five seconds—it seemed at the time like five minutes—I held out one hand and took a step toward the big cat and said something ridiculous like, "Here, kitty, kitty." The cat paused there on three legs, one paw up as if he wanted to shake hands. But he didn't respond to my advance.

I took a second step toward the lion. Again the lion remained still, not moving a muscle, not blinking an eye. And I stopped and thought again and this time I understood that however the big cat might secretly feel, I myself was not yet quite ready to shake hands with a mountain lion. Maybe someday. But not yet. I retreated.

I turned and walked homeward again, pausing every few steps to look back over my shoulder. The cat had lowered his front paw but did not follow me. The last I saw of him, from the next bend of the canyon, he was still in the same place, watching me go. I hurried on through the evening, stopping now and then to look and listen, but if that cat followed me any further I could detect no sight or sound of it.

I haven't seen a mountain lion since that evening, but the experience remains shining in my memory. I want my children to have the opportunity for that kind of experience. I want my friends to have it. I want even our enemies to have it—they need it most. And someday, possibly, one of our children's children will discover how to get close enough to that mountain lion to shake paws with it, to embrace and caress it, maybe even teach it something, and to learn what the lion has to teach us.

Reading Questions and Further Readings

Reading Questions

1. Does one of Abbey's lifestyle virtues like hiking benefit the environment, or is it primarily of personal importance?
2. How do we reconcile the high efficiency and density of urban areas with the importance of connecting with nature?

Further Readings

Cronon, William. *Nature's Metropolis: Chicago and the Great West*. New York: Norton, 1991.

Sax, Joseph. *Mountains without Handrails: Reflections on the National Parks*. Ann Arbor: University of Michigan Press, 1980.

Agrarian and Industrial Agriculture

As with the prior readings in this part, the tension between agrarian and industrial models of production arises in the context of industrial society. Both conceptual and empirical debates arise in the case of genetic technology in agriculture. Norman Borlaug, a Nobel Laureate, defends the use of industrial pesticides, fertilizer, and breeding pioneered in the Green Revolution. This is justified on humanitarian grounds, mainly to stave off famine in a rapidly growing population.

In contrast, the poet and social critic Wendell Berry expresses skepticism about agricultural philosophies and techniques not rooted in tradition, locality, and humility. He defends a place-based approach to agriculture, based on familiarity with a piece of land over multiple generations, and with a strong skepticism about industrial technology, corporate power, and centralized government.

The anti-industrial ideal is a common framework for understanding environmental impact, evident in such influential books as Michael Pollan's *Omnivore's Dilemma* and popular press depictions of agricultural problems (think of the environmental movement's strong negative response to genetically engineered crops, for instance). One of the twists to this story is that a large percentage of environmental impacts attributable to agriculture originate with animals. This includes both industrial and nonindustrial modes of animal agriculture and is constituted by land-use change (for grazing land and feed crops), methane emissions from ruminants, and waste management. While there are environmentally harmful agricultural systems as well as environmentally preferable ones, the common industrial/nonindustrial distinction does not offer a reliable correlation to environmental impact.

In addition to the debates about industrial agriculture, this controversy also draws on disagreements about the effectiveness of technical and nontechnical approaches to solving environmental problems (also discussed later in this part in the context of geoengineering).

35

Excerpts from "The Green Revolution Revisited and the Road Ahead"

NORMAN BORLAUG

Introduction

It is a great pleasure to be here in Oslo, nearly 30 years after I was awarded the Nobel Peace Prize. I wish to thank the Norwegian Nobel Institute and the U.S. Embassy in Norway for arranging this lecture. Today, I am here to take stock of the contributions of the so-called "Green Revolution," and explore the role of science and technology in the coming decades to improve the quantity, quality, and availability of food for all of the world's population.

Although I am an agricultural scientist, my work in food production and hunger alleviation was recognized through the Nobel Peace Prize because there is no Nobel Prize for food and agriculture. I have often speculated that if Alfred Nobel had written his will to establish the various prizes and endowed them fifty years earlier, the first prize established would have been for food and agriculture. However, by the time he wrote his will in 1895 establishing the prize, the horrors of the widespread potato famine that had swept across western Europe in 1845–51—taking the lives of untold millions—had been forgotten.[1] The subsequent migration of millions of western Europeans to the Americas during 1850–60 restored a reasonable, yet still tenuous balance in the land-food-population equation. Moreover, the European food supply was further greatly increased during the last three decades of the 19th century through the application of improved agricultural technology developed earlier in the century (i.e., restoration of soil fertility, better control of diseases, and use of improved varieties and breeds of crops and animals). Hence, when Alfred Nobel wrote his will, there was no serious food production problem haunting Europe.

I am now in my 56th year of continuous involvement in agricultural research and production in the low-income, food-deficit developing countries. I have worked with many colleagues, political leaders, and farmers to transform food production systems. Despite the successes of the Green Revolution, the battle to ensure food security for hundreds of millions of miserably poor people is far from won.

Mushrooming populations, changing demographics and inadequate poverty intervention programs have eaten up many of the gains of the Green Revolution.

This is not to say that the Green Revolution is over. Increases in crop management productivity can be made all along the line—in tillage, water use, fertilization, weed and pest control, and harvesting. However, for the genetic improvement of food crops to continue at a pace sufficient to meet the needs of the 8.3 billion people projected in 2025, both conventional breeding and biotechnology methodologies will be needed.

Dawn of Modern Agriculture

Science-based agriculture is really a 20th century invention. Until the 19th century, crop improvement was in the hands of farmers, and food production grew largely by expanding the cultivated land area. As sons and daughters of farm families married and formed new families, they opened new land to cultivation. Improvements in farm machinery expanded the area that could be cultivated by one family. Machinery made possible better seedbed preparation, moisture utilization, and improved planting practices and weed control, resulting in modest increases in yield per hectare.

By the mid-1800s, German scientist Justus von Leibig and French scientist Jean-Baptiste Boussingault had laid down important theoretical foundations in soil chemistry and crop agronomy. Sir John Bennett Lawes produced super phosphate in England in 1842, and shipments of Chilean nitrates (nitrogen) began arriving in quantities to European and North American ports in the 1840s. However, the use of organic fertilizers (animal manure, crop residues, green manure crops) remained dominant into the early 1900s.

Groundwork for more sophisticated genetic crop improvement was laid by Charles Darwin in his writings on the variation of life species (published in 1859) and by Gregor Mendel through his discovery of the laws of genetic inheritance (reported in 1865). Darwin's book immediately generated a great deal of interest, discussion and controversy. Mendel's work was largely ignored for 35 years. The rediscovery of Mendel's work in 1900 provoked tremendous scientific interest and research in plant genetics.

The first decade of the 20th century brought a fundamental scientific breakthrough, followed by the rapid commercialization of that breakthrough. In 1909, Nobel Laureate in Chemistry (1918) Fritz Haber demonstrated the synthesis of ammonia from its elements. Four years later, in 1913, the company BASF, thanks to the innovative solutions of Carl Bosch, began operation of the world's first ammonia plant. The expansion of the fertilizer industry was soon arrested by WWI (ammonia used to produce nitrate for explosives), then by the great economic depression of the 1930s, and then by the demand for explosives during WWII. However, after the war, rapidly increasing amounts of nitrogen became available and contributed greatly to boosting crop yields and production.

It is only since WWII that fertilizer use, and especially the application of low-cost nitrogen derived from synthetic ammonia, has become an indispensable component of modern agricultural production (nearly 80 million nutrient tonnes consumed annually). It is estimated that 40% of today's 6 billion people are alive thanks to the Haber-Bosch process of synthesizing ammonia (Vaclav Smil, University Distinguished Professor, University of Manitoba).

By the 1930s, much of the scientific knowledge needed for high-yield agricultural production was available in the United States. However, widespread adoption was delayed by the great economic depression of the 1930s, which paralyzed the world agricultural economy. It was not until WWII brought about a much greater demand for food to support the Allied war effort that the new research findings began to be applied widely, first in the United States and later in many other countries.

Maize cultivation led the modernization process. In 1940, U.S. farmers produced 56 million tons of maize on roughly 31 million hectares, with an average yield of 1.8 t/ha. In 1999, U.S. farmers produced 240 million tons of maize on roughly 29 million hectares, with an average yield of 8.4 t/ha. This more than four-fold yield increase is the impact of modern hybrid seed–fertilizer–weed control technology!

Following WWII, various bilateral and multilateral agencies, led by the United States and the Food and Agriculture Organization (FAO) of the United Nations, initiated technical-agricultural assistance programs in a number of countries in Europe, Asia, and Latin America. In the beginning, there was considerable naiveté especially about the transferability of modern production technology from the industrialized temperate zones to the tropics and subtropics. Most varieties from the United States, for example, were not well suited in the environments in which they were introduced.

There was another model of technical assistance that preceded these public sector foreign technical assistance programs, which ultimately proved to be superior. This was the Cooperative Mexican Government–Rockefeller Foundation agricultural program, which began in 1943. This foreign assistance program initiated research programs in Mexico to improve maize, wheat, beans, and potato technology. It also invested significantly in human resource development, training scores of Mexican scientists and helping to establish the national agricultural research system.

Green Revolution

The breakthrough in wheat and rice production in Asia in the mid-1960s, which came to be known as the Green Revolution, symbolized the process of using agricultural science to develop modern techniques for the Third World. It began

TABLE 35.1. Cereal Production in Asia, 1961–99

		Milled rice	Wheat	All cereals
		(million tonnes)		
China	1961	48	14	91
	1970	96	29	163
	1999	170	114	390
India	1961	46	11	70
	1970	54	20	93
	1999	112	71	186
Developing Asia	1961	155	44	248
	1970	233	71	372
	1999	449	242	809

Source: FAO AGROSTAT, April 2000

TABLE 35.2. Changes in Factors of Production in Developing Asia

	Irrigation (million ha)	Fertilizer nutrient consumption (million tonnes)	Tractors (millions)
1961	87	2	0.2
1970	106	10	0.5
1980	129	29	2.0
1990	158	54	3.4
1998	176	70	4.5

Source: FAO ARGOSTAT, April 2000

in Mexico with the "quiet" wheat revolution in the late 1950s. During the 1960s and 1970s, India, Pakistan, and the Philippines received world attention for their agricultural progress (Table 35.1). Since 1980, China has been the greatest success story. Home to one-fifth of the world's people, China today is the world's biggest food producer. With each successive year, its cereal crop yields approach that of the United States.

Over the past four decades FAO reports that in Developing Asia, the irrigated area has more than doubled—to 176 million hectares. Fertilizer consumption has increased more than 30-fold, and now stands at about 70 million tonnes of nutrients, and the number of tractors in use has increased from 200,000 to 4.6 million (Table 35.2).

I often ask the critics of modern agricultural technology what the world would have been like without the technological advances that have occurred, largely during the past 50 years. For those whose main concern is protecting the "environment," let's look at the positive impact that the application of science-based technology has had on land use.

Had the global cereal yields of 1950 still prevailed in 1999, we would have needed nearly 1.8 billion ha of additional land of the same quality—instead of the 600 million that was used—to equal the current global harvest (see Figure 35.1 at the end of text). Obviously, such a surplus of land was not available, and certainly not in populous Asia, where the population has increased from 1.2 to 3.8 billion over this time period. Moreover, if more environmentally fragile land had been brought into agricultural production, think of the impact on soil erosion, loss of forests and grasslands, biodiversity and extinction of wildlife species that would have ensued. . . .

Water Resources

. . . Proven technologies, such as drip irrigation, which saves water and reduces soil salinity, are suitable for much larger areas than currently used. Various new precision irrigation systems are also on the horizon, which will supply water to plants only when they need it. There is also a range of improved small-scale and supplemental irrigation systems to increase the productivity of rainfed areas, which offer much promise for smallholder farmers.

Clearly, we need to rethink our attitudes about water, and move away from thinking of it as nearly a free good, and a God-given right. Pricing water delivery closer to its real costs is a necessary step to improving use efficiency. Farmers and irrigation officials (and urban consumers) will need incentives to save water. Moreover, management of water distribution networks, except for the primary canals, should be decentralized and turned over to the farmers. Farmers' water user associations in the Yaqui valley in northwest Mexico, for example, have done a much better job of managing the irrigation districts than did the Federal Ministry of Agriculture and Water Resources previously.

In order to expand food production for a growing world population within the parameters of likely water availability, the inevitable conclusion is that humankind in the 21st century will need to bring about a "Blue Revolution" to complement the "Green Revolution" of the 20th century. In the new Blue Revolution, water-use productivity must be wedded to land-use productivity. New science and technology must lead the way. . . .

Genetic Improvement

Continued genetic improvement of food crops—using both conventional as well as biotechnology research tools—is needed to shift the yield frontier higher and to increase stability of yield. While biotechnology research tools offer much promise, it is also important to recognize that conventional plant breeding methods are continuing to make significant contributions to improved food production and enhanced nutrition. In rice and wheat, three distinct, but inter-related strategies are being pursued to increase genetic maximum yield poten-tial: changes in plant architecture, hybridization, and wider genetic resource utilization.[2] Significant progress has been made in all three areas, although widespread impact on farmers' fields is still probably 10–12 years away. IRRI [International Rice Research Institute] claims that the new "super rice" plant type, in association with direct seeding, could increase rice yield potential by 20–25 percent.[3]

In wheat, new plants with architecture similar to the "super rices" (larger heads, more grains, fewer tillers) could lead to an increase in yield potential of 10–15 percent.[4] Introducing genes from related wild species into cultivated wheat can introduce important sources of resistance for several biotic and abi-otic stresses, and perhaps for higher yield potential as well, especially if the trans-genic wheats are used as parent material in the production of hybrid wheats.[5]

The success of hybrid rice in China (now covering more than 50 percent of the irrigated area) has led to a renewed interest in hybrid wheat, when most research had been discontinued for various reasons, mainly low heterosis while trying to exploit cytoplasmic male sterility, and high seed production costs. However, recent improvements in chemical hybridization agents, advances in biotechnology, and the emergence of the new wheat plant type have made an assessment of hybrids worthwhile. With better heterosis and increased grain fill-ing, the yield frontier of the new plant material could be 25–30 percent above the current germplasm base.

Maize production has really begun to take off in many Asia countries, espe-cially China. It now has the highest average yield of all the cereals in Asia, with much of the genetic yield potential yet to be exploited. Moreover, recent devel-opments in high-yielding quality protein maize (QPM) varieties and hybrids using conventional plant breeding methods stand to improve the nutritional quality of the grain without sacrificing yields. This research achievement offers important nutritional benefits for livestock and humans. With biotechnology tools, it is likely that we will see further nutritional "quality" enhancements in the cereals in years to come.

The recent development of high-yielding sorghum varieties and hybrids with resistance to the heretofore-uncontrollable parasitic weed *Striga* spp., by researchers at Purdue University in the USA is an important research break-through for many areas of Asia and Africa.

There is growing evidence that genetic variation exists within most cereal crop species for developing genotypes that are more efficient in the use of nitro-gen, phosphorus, and other plant nutrients than are currently available in the best varieties and hybrids. In addition, there is good evidence that further heat and drought tolerance can be built into high-yielding germplasm. . . .

Educating Urbanites about Agriculture

The current backlash against agricultural science and technology evident in some industrialized countries is hard for me to comprehend. How quickly humankind becomes detached from the soil and agricultural production! Less than 4 percent of the population in the industrialized countries (less than 2 percent in the USA) is directly engaged in agriculture. With low-cost food supplies and urban bias, is it any wonder that consumers don't understand the complexities of re-producing the world food supply each year in its entirely, and expanding it further for the nearly 85 million new mouths that are born into this world each year. I believe we can help address this "educational gap" in industrialized urban nations by making it compulsory in secondary schools and universities for students to take courses on biology and science and technology policy.

As the pace of technological change has accelerated the past 50 years, the fear of science has grown. Certainly, the breaking of the atom and the prospects of a nuclear holocaust added to people's fear, and drove a bigger wedge between the scientist and the layman. Rachel Carson's book *Silent Spring*, published in 1962, which reported that poisons were everywhere, also struck a very sensitive nerve. Of course, this perception was not totally unfounded. By the mid 20th century, air and water quality had been seriously damaged through wasteful industrial production systems that pushed effluents often literally into "our own backyards."

We all owe a debt of gratitude to environmental movement in the industrialized nations, which has led to legislation over the past 30 years to improve air and water quality, protect wildlife, control the disposal of toxic wastes, protect the soils, and reduce the loss of biodiversity.

However, I agree also with environmental writer Gregg Easterbrook, who argues in his book *A Moment on the Earth* that "In the Western world the Age of Pollution is nearly over . . . Aside from weapons, technology is not growing more dangerous and wasteful but cleaner and more resource-efficient. Clean technology will be the successor to high technology."

However, Easterbrook goes on to warn that, "As positive as trends are in the First World, they are negative in the Third World. One reason why the West must shake off its instant-doomsday thinking about the United States and Western Europe is so that resources can be diverted to ecological protection in the developing world."

In his writings, U.S. Professor Robert Paarlberg, who teaches at Wellesley College and Harvard University, sounded the alarm about the deadlock between agriculturalists and environmentalists over what constitutes "sustainable agriculture" in the Third World. This debate has confused—if not paralyzed—many in the international donor community who, afraid of antagonizing powerful environmental lobbying groups, have turned away from supporting science-based agricultural modernization projects still needed in much of smallholder Asia, sub-Saharan Africa, and Latin America.

This deadlock must be broken. We cannot lose sight of the enormous job before us to feed 10–11 billion people, 90 percent of whom will begin life in a developing country, and probably in poverty. Only through dynamic agricultural development will there be any hope to alleviate poverty and improve human health and productivity and reduce political instability.

Closing Comments

Thirty years ago, in my acceptance speech for the Nobel Peace Prize, I said that the Green Revolution had won a temporary success in man's war against hunger, which if fully implemented, could provide sufficient food for humankind through the end of the 20th century. But I warned that unless the frightening power of human reproduction was curbed, the success of the Green Revolution would only be ephemeral.

I now say that the world has the technology—either available or well advanced in the research pipeline—to feed on a sustainable basis a population of 10 billion people. The more pertinent question today is whether farmers and ranchers will be permitted to use this new technology? While the affluent nations can certainly afford to adopt ultra low-risk positions, and pay more for food produced by the so-called "organic" methods, the one billion chronically undernourished people of the low-income, food-deficit nations cannot.

It took some 10,000 years to expand food production to the current level of about 5 billion tons per year. By 2025, we will have to nearly double current production again. This cannot be done unless farmers across the world have access to current high-yielding crop-production methods as well as new biotechnological breakthroughs that can increase the yields, dependability, and nutritional quality of our basic food crops.

Moreover, higher farm incomes will also permit small-scale farmers to make added investments to protect their natural resources. As Kenyan archeologist Richard Leakey likes to remind us, "you have to be well-fed to be a conservationist!" We need to bring common sense into the debate on agricultural science and technology and the sooner the better!

Most certainly, agricultural scientists have a moral obligation to warn the political, educational, and religious leaders about the magnitude and seriousness of the arable land, food, population and environmental problems that lie ahead. These problems will not vanish by themselves. Unless they are addressed in a forthright manner future solutions will be more difficult to achieve.

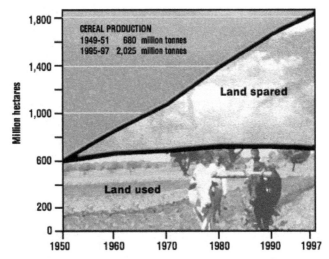

Figure 35.1. World cereal production—area saved through
improved technology, 1950–1998

NOTES

1. Douglas C. Daly, "The Leaf That Launched a Thousand Ships," *Natural History* 1 (1996): 24–25.
2. S. Rajaram and N. E. Borlaug, "Approaches to Breeding for Wide Adaptation, Yield Potential, Rust Resistance and Drought Tolerance," paper presented at Primer Simposio Internacional de Trigo, Ciudad Obregon, Mexico, April 7–9, 1997; P. L. Pingali and S. Rajaram, "Technological Opportunities for Sustaining Wheat Productivity Growth," paper presented at the World Food and Sustainable Agriculture Program Conference: Meeting the Demand for Food in the 21st Century: Challenges and Opportunities for Illinois Agriculture, Urbana, Illinois, May 27, 1997.
3. G. S. Khush, "Modern Varieties—Their Real Contribution to Food Supply and Equity," *Geojournal* 35 (1995): 275–284.
4. Rajaram and Borlaug, "Approaches to Breeding."
5. A. Mujeeb-Kazi and G. P. Hettel, eds., *Utilizing Wild Grass Biodiversity in Wheat Improvement—15 Years of Research in Mexico for Global Wheat Improvement*, Wheat Special Report 29 (Mexico, DF: International Maize and Wheat Improvement Center, 1995).

36

The Agrarian Standard

WENDELL BERRY

The Unsettling of America was published twenty-five years ago; it is still in print and is still being read. As its author, I am tempted to be glad of this, and yet, if I believe what I said in that book, and I still do, then I should be anything but glad. The book would have had a far happier fate if it could have been disproved or made obsolete years ago.

It remains true because the conditions it describes and opposes, the abuses of farmland and farming people, have persisted and become worse over the last twenty-five years. In 2002 we have less than half the number of farmers in the United States that we had in 1977. Our farm communities are far worse off now than they were then. Our soil erosion rates continue to be unsustainably high. We continue to pollute our soils and streams with agricultural poisons. We continue to lose farmland to urban development of the most wasteful sort. The large agribusiness corporations that were mainly national in 1977 are now global, and are replacing the world's agricultural diversity, which was useful primarily to farmers and local consumers, with bioengineered and patented monocultures that are merely profitable to corporations. The purpose of this now global economy, as Vandana Shiva has rightly said, is to replace "food democracy" with a worldwide "food dictatorship."

To be an agrarian writer in such a time is an odd experience. One keeps writing essays and speeches that one would prefer not to write, that one wishes would prove unnecessary, that one hopes nobody will have any need for in twenty-five years. My life as an agrarian writer has certainly involved me in such confusions, but I have never doubted for a minute the importance of the hope I have tried to serve: the hope that we might become a healthy people in a healthy land.

We agrarians are involved in a hard, long, momentous contest, in which we are so far, and by a considerable margin, the losers. What we have undertaken to defend is the complex accomplishment of knowledge, cultural memory, skill, self-mastery, good sense, and fundamental decency—the high and indispensable art—for which we probably can find no better name than "good farming." I mean farming as defined by agrarianism as opposed to farming as defined by industrialism: farming as the proper use and care of an immeasurable gift.

I believe that this contest between industrialism and agrarianism now defines the most fundamental human difference, for it divides not just two nearly opposite concepts of agriculture and land use, but also two nearly opposite ways of understanding ourselves, our fellow creatures, and our world.

* * *

The way of industrialism is the way of the machine. To the industrial mind, a machine is not merely an instrument for doing work or amusing ourselves or making war; it is an explanation of the world and of life. Because industrialism cannot understand living things except as machines, and can grant them no value that is not utilitarian, it conceives of farming and forestry as forms of mining; it cannot use the land without abusing it.

Industrialism prescribes an economy that is placeless and displacing. It does not distinguish one place from another. It applies its methods and technologies indiscriminately in the American East and the American West, in the United States and in India. It thus continues the economy of colonialism. The shift of colonial power from European monarchy to global corporation is perhaps the dominant theme of modern history. All along, it has been the same story of the gathering of an exploitive economic power into the hands of a few people who are alien to the places and the people they exploit. Such an economy is bound to destroy locally adapted agrarian economies everywhere it goes, simply because it is too ignorant not to do so. And it has succeeded precisely to the extent that it has been able to inculcate the same ignorance in workers and consumers.

To the corporate and political and academic servants of global industrialism, the small family farm and the small farming community are not known, not imaginable, and therefore unthinkable, except as damaging stereotypes. The people of "the cutting edge" in science, business, education, and politics have no patience with the local love, local loyalty, and local knowledge that make people truly native to their places and therefore good caretakers of their places. This is why one of the primary principles in industrialism has always been to get the worker away from home. From the beginning it has been destructive of home employment and home economies. The economic function of the household has been increasingly the consumption of purchased goods. Under industrialism, the farm too has become increasingly consumptive, and farms fail as the costs of consumption overpower the income from production.

The idea of people working at home, as family members, as neighbors, as natives and citizens of their places, is as repugnant to the industrial mind as the idea of self-employment. The industrial mind is an organizational mind, and I think this mind is deeply disturbed and threatened by the existence of people who have no boss. This may be why people with such minds, as they approach the top of the political hierarchy, so readily sell themselves to "special interests." They cannot bear to be unbossed. They cannot stand the lonely work of making up their own minds.

The industrial contempt for anything small, rural, or natural translates into contempt for uncentralized economic systems, any sort of local self-sufficiency in food or other necessities. The industrial "solution" for such systems is to increase the scale of work and trade. It brings Big Ideas, Big Money, and Big Technology into small rural communities, economies, and ecosystems—the brought-in industry and the experts being invariably alien to and contemptuous of the places to which they are brought in. There is never any question of propriety, of adapting the thought or the purpose or the technology to the place.

The result is that problems correctable on a small scale are replaced by large-scale problems for which there are no large-scale corrections. Meanwhile, the large-scale enterprise has reduced or destroyed the possibility of small-scale corrections. This exactly describes our present agriculture. Forcing all agricultural localities to conform to economic conditions imposed from afar by a few large corporations has caused problems of the largest possible scale, such as soil loss, genetic impoverishment, and groundwater pollution, which are correctable only by an agriculture of locally adapted, solar-powered, diversified small farms—a correction that, after a half century of industrial agriculture, will be difficult to achieve.

The industrial economy thus is inherently violent. It impoverishes one place in order to be extravagant in another, true to its colonialist ambition. A part of the "externalized" cost of this is war after war.

* * *

Industrialism begins with technological invention. But agrarianism begins with givens: land, plants, animals, weather, hunger, and the birthright knowledge of agriculture. Industrialists are always ready to ignore, sell, or destroy the past in order to gain the entirely unprecedented wealth, comfort, and happiness supposedly to be found in the future. Agrarian farmers know that their very identity depends on their willingness to receive gratefully, use responsibly, and hand down intact an inheritance, both natural and cultural, from the past.

I said a while ago that to agrarianism farming is the proper use and care of an immeasurable gift. The shortest way to understand this, I suppose, is the religious way. Among the commonplaces of the Bible, for example, are the admonitions that the world was made and approved by God, that it belongs to Him, and that its good things come to us from Him as gifts. Beyond those ideas is the idea that the whole Creation exists only by participating in the life of God, sharing in His being, breathing His breath. "The world," Gerard Manley Hopkins said, "is charged with the grandeur of God." Some such thoughts would have been familiar to most people during most of human history. They seem strange to us, and what has estranged us from them is our economy. The industrial economy could not have been derived from such thoughts any more than it could have been derived from the Golden Rule.

If we believed that the existence of the world is rooted in mystery and in sanctity, then we would have a different economy. It would still be an economy of use, necessarily, but it would be an economy also of return. The economy would have to accommodate the need to be worthy of the gifts we receive and use, and this would involve a return of propitiation, praise, gratitude, responsibility, good use, good care, and a proper regard for the unborn. What is most conspicuously absent from the industrial economy and industrial culture is this idea of return. Industrial humans relate themselves to the world and its creatures by fairly direct acts of violence. Mostly we take without asking, use without respect or gratitude, and give nothing in return. Our economy's most voluminous product is waste—valuable materials irrecoverably misplaced, or randomly discharged as poisons.

To perceive the world and our life in it as gifts originating in sanctity is to see our human economy as a continuing moral crisis. Our life of need and work forces us inescapably to use in time things belonging to eternity, and to assign finite values to things already recognized as infinitely valuable. This is a fearful predicament. It calls for prudence, humility, good work, propriety of scale. It calls for the complex responsibilities of caretaking and giving-back that we mean by "stewardship." To all of this the idea of the immeasurable value of the resource is central.

*   *   *

We can get to the same idea by a way a little more economic and practical, and this is by following through our literature the ancient theme of the small farmer or husbandman who leads an abundant life on a scrap of land often described as cast-off or poor. This figure makes his first literary appearance, so far as I know, in Virgil's Fourth Georgic:

> I saw a man,
> An old Cilician, who occupied
> An acre or two of land that no one wanted,
> A patch not worth the ploughing, unrewarding
> For flocks, unfit for vineyards; he however
> By planting here and there among the scrub
> Cabbages or white lilies and verbena
> And flimsy poppies, fancied himself a king
> In wealth, and coming home late in the evening
> Loaded his board with unbought delicacies.

Virgil's old squatter, I am sure, is a literary outcropping of an agrarian theme that has been carried from earliest times until now mostly in family or folk tradition, not in writing, though other such people can be found in books. Wherever found, they don't vary by much from Virgil's prototype. They don't have or require a lot of land, and the land they have is often marginal. They practice subsistence agriculture, which has been much derided by agricultural economists and other learned people of the industrial age, and they always associate frugality with abundance.

In my various travels, I have seen a number of small homesteads like that of Virgil's old farmer, situated on "land that no one wanted" and yet abundantly productive of food, pleasure, and other goods. And especially in my younger days, I was used to hearing farmers of a certain kind say, "They may run me out, but they won't starve me out" or "I may get shot, but I'm not going to starve." Even now, if they cared, I think agricultural economists could find small farmers who have prospered, not by "getting big," but by practicing the ancient rules of thrift and subsistence, by accepting the limits of their small farms, and by knowing well the value of having a little land.

How do we come at the value of a little land? We do so, following this strand of agrarian thought, by reference to the value of *no* land. Agrarians value land because somewhere back in the history of their consciousness is the memory of being landless. This memory is implicit, in Virgil's poem, in the old farmer's happy acceptance of "an acre or two of land that no one wanted." If you have no land you have nothing: no food, no shelter, no warmth, no freedom, no life. If we remember this, we know that all economies begin to lie as soon as they assign a fixed value to land. People who have been landless know that the land is invaluable; it is worth everything. Pre-agricultural humans, of course, knew this too. And so, evidently, do the animals. It is a fearful thing to be without a "territory." Whatever the market may say, the worth of the land is what it always was: It is worth what food, clothing, shelter, and freedom are worth; it is worth what life is worth. This perception moved the settlers from the Old World into the New. Most of our American ancestors came here because they knew what it was to be landless; to be landless was to be threatened by want and also by enslavement. Coming here, they bore the ancestral memory of serfdom. Under feudalism, the few who owned the land owned also, by an inescapable political logic, the people who worked the land.

Thomas Jefferson, who knew all these things, obviously was thinking of them when he wrote in 1785 that "it is not too soon to provide by every possible means that as few as possible shall be without a little portion of land. The small landholders are the most precious part of a state. . . ." He was saying, two years before the adoption of our constitution, that a democratic state and democratic liberties depend upon democratic ownership of the land. He was already anticipating and fearing the division of our people into settlers, the people who wanted "a little portion of land" as a home, and, virtually opposite to those, the consolidators and exploiters of the land and the land's wealth, who would not be restrained by what Jefferson called "the natural affection of the human mind." He wrote as he did in 1785 because he feared exactly the political theory that we now have: the idea that government exists to guarantee the right of the most wealthy to own or control the land without limit.

In any consideration of agrarianism, this issue of limitation is critical. Agrarian farmers see, accept, and live within their limits. They understand and agree to the proposition that there is "this much and no more." Everything that happens on an agrarian farm is determined or conditioned by the understanding that there is only so much land, so much water in the cistern, so much hay in the barn, so much corn in the crib, so much firewood in the shed, so much food in the cellar or freezer, so much strength in the back and arms—and no more. This is the understanding that induces thrift, family coherence, neighborliness, local economies. Within accepted limits, these become necessities. The agrarian sense of abundance comes from the experienced possibility of frugality and renewal within limits.

This is exactly opposite to the industrial idea that abundance comes from the violation of limits by personal mobility, extractive machinery, long-distance transport, and scientific or technological breakthroughs. If we use up the good possibilities in this place, we will import goods from some other place, or we will go to some other place. If nature releases her wealth too slowly, we will take it by force. If we make the world too toxic for honeybees, some compound brain, Monsanto perhaps, will invent tiny robots that will fly about pollinating flowers and making honey.

* * *

To be landless in an industrial society obviously is not at all times to be jobless and homeless. But the ability of the industrial economy to provide jobs and homes depends on prosperity, and on a very shaky kind of prosperity too. It depends on "growth" of the wrong things—on what Edward Abbey called "the ideology of the cancer cell"—and on greed with purchasing power. In the absence of growth, greed, and affluence, the dependents of an industrial economy too easily suffer the consequences of having no land: joblessness, homelessness, and want. This is not a theory. We have seen it happen.

I don't think that being landed necessarily means owning land. It does mean being connected to a home landscape from which one may live by the interactions of a local economy and without the routine intervention of governments, corporations, or charities.

In our time it is useless and probably wrong to suppose that a great many urban people ought to go out into the countryside and become homesteaders or farmers. But it is not useless or wrong to suppose that urban people have agricultural responsibilities that they should try to meet. And in fact this is happening. The agrarian population among us is growing, and by no means is it made up merely of some farmers and some country people. It includes urban gardeners, urban consumers who are buying food from local farmers, consumers who have grown doubtful of the healthfulness, the trustworthiness, and the dependability of the corporate food system—people, in other words, who understand what it means to be landless.

*   *   *

Apologists for industrial agriculture rely on two arguments. In one of them, they say that the industrialization of agriculture, and its dominance by corporations, has been "inevitable." It has come about and it continues by the agency of economic and technological determinism. There has been simply nothing that anybody could do about it.

The other argument is that industrial agriculture has come about by choice, inspired by compassion and generosity. Seeing the shadow of mass starvation looming over the world, the food conglomerates, the machinery companies, the chemical companies, the seed companies, and the other suppliers of "purchased inputs" have done all that they have done in order to solve "the problem of hunger" and to "feed the world."

We need to notice, first, that these two arguments, often used and perhaps believed by the same people, exactly contradict each other. Second, though supposedly it has been imposed upon the world by economic and technological forces beyond human control, industrial agriculture has been pretty consistently devastating to nature, to farmers, and to rural communities, at the same time that it has been highly profitable to the agribusiness corporations, which have submitted not quite reluctantly to its "inevitability." And, third, tearful over human suffering as they have always been, the agribusiness corporations have maintained a religious faith in the profitability of their charity. They have instructed the world that it is better for people to buy food from the corporate global economy than to raise it for themselves. What is the proper solution to hunger? Not food from the local landscape, but industrial development. After decades of such innovative thought, hunger is still a worldwide calamity.

The primary question for the corporations, and so necessarily for us, is not how the world will be fed, but who will control the land, and therefore the wealth, of the world. If the world's people accept the industrial premises that favor bigness, centralization, and (for a few people) high profitability, then the corporations will control all of the world's land and all of its wealth. If, on the contrary, the world's people might again see the advantages of local economies, in which people live, so far as they are able to do so, from their home landscapes, and work patiently toward that end, eliminating waste and the cruelties of landlessness and homelessness, then I think they might reasonably hope to solve "the problem of hunger," and several other problems as well.

But do the people of the world, allured by TV, supermarkets, and big cars, or by dreams thereof, *want* to live from their home landscapes? *Could* they do so, if they wanted to? Those are hard questions, not readily answerable by anybody.

Throughout the industrial decades, people have become increasingly and more numerously ignorant of the issues of land use, of food, clothing, and shelter. What would they do, and what *could* they do, if they were forced by war or some other calamity to live from their home landscapes?

It is a fact, well attested but little noticed, that our extensive, mobile, highly centralized system of industrial agriculture is extremely vulnerable to acts of terrorism. It will be hard to protect an agriculture of genetically impoverished monocultures that is entirely dependent on cheap petroleum and long-distance transportation. We know too that the great corporations, which now grow and act so far beyond the restraint of "the natural affections of the human mind," are vulnerable to the natural depravities of the human mind, such as greed, arrogance, and fraud.

The agricultural industrialists like to say that their agrarian opponents are merely sentimental defenders of ways of farming that are hopelessly old-fashioned, justly dying out. Or they say that their opponents are the victims, as Richard Lewontin put it, of "a false nostalgia for a way of life that never existed." But these are not criticisms. They are insults.

For agrarians, the correct response is to stand confidently on our fundamental premise, which is both democratic and ecological: The land is a gift of immeasurable wealth. If it is a gift, then it is a gift to all the living in all time. To withhold it from some is finally to destroy it for all. For a few powerful people to own or control it all, or decide its fate, is wrong.

From that premise we go directly to the question that begins the agrarian agenda and is the discipline of all agrarian practice: What is the best way to use land? Agrarians know that this question necessarily has many answers, not just one. We are not asking what is the best way to farm everywhere in the world, or everywhere in the United States, or everywhere in Kentucky or Iowa. We are asking what is the best way to farm in each one of the world's numberless places, as defined by topography, soil type, climate, ecology, history, culture, and local need. And we know that the standard cannot be determined only by market demand or productivity or profitability or technological capability, or by any other single measure, however important it may be. The agrarian standard, inescapably, is local adaptation, which requires bringing local nature, local people, local economy and local culture into a practical and enduring harmony.

Reading Questions and Further Readings

Reading Questions

1. If industrial agriculture both was efficient and had a low environmental impact, would there still be arguments against its practice?
2. In a modern democratic society, what is the role of the agrarian culture proposed by Berry?

Further Readings

McWilliams, James E. *Just Food: Where Locavores Get It Wrong and How We Can Truly Eat Responsibly*. New York: Little, Brown, 2009.

Ronald, Pamela C., and Raoul W. Adamchak, *Tomorrow's Table: Organic Farming, Genetics, and the Future of Food*. Oxford: Oxford University Press, 2010.

Managing Nature versus Stewardship

The human influence on the planet is becoming increasingly obvious and systematic. Some environmentalists have responded by telling us to lighten up and let nature be. The lawyer and engineer Brad Allenby tells us that it is too late for this response, and anyway the traditional environmentalist notion of nature is a fiction. It is time for us to recognize our impact on the planet and accept our responsibilities as planetary managers. He urges us to take the joystick in hand and manage the planet rationally according to our values.

Not so fast, says the Harvard professor David W. Keith. The earth is not yet a human artifact, and nature is not a fiction. In fact, according to Keith, nature has intrinsic value. Keith admits that the human impact is growing and perhaps irresistible; but it is a matter of degree, and there are still lightly influenced landscapes that have aesthetic values and history that cannot be replaced or duplicated by the products of clever management. Keith thinks we should imagine ourselves as stewards of values that should be treasured rather than as managers of substitutable commodities.

This is an ancient debate but one that has gained greater urgency as the human colonization of nature has grown. Will our environmentalist impulses be satisfied by rational management of earth's resources, or must the autonomy and spontaneity of nature be respected if it is to exist at all? This remains an open question as we hurtle toward an ever more human-dominated planet.

37

Excerpts from "Earth Systems Engineering and Management"

BRAD ALLENBY

The impact of human activities on natural systems of all kinds has grown to the point that we need to engage consciously in earth systems engineering and management. I address why this is the case, and what I mean by such a provocative term. In addition, I explore what we can learn from relevant experience, and how this daunting task should be approached.

False Dichotomy

Technology is the means by which human cultures interact with the physical, chemical, and biological world. It is through technology that a human imprint has been made on the physics and chemistry of every cubic meter of air and of water. Critical dynamics of heavy metals flows and grand elemental cycles—nitrogen, carbon, sulfur, phosphorus, hydrologic—are increasingly dominated by the (usually unintended and sometimes unforeseen) byproducts of the technological activities of our species. The biosphere itself, at levels from the genetic to the landscape, is increasingly a human product. Few biological communities can be found that do not reflect human predation, management, or consumption. As Gallagher and Carpenter remark in introducing a special issue of *Science* on human-dominated ecosystems, the concept of a pristine ecosystem, untouched by human activity, "is collapsing in the wake of scientists' realization that there are no places left on Earth that don't fall under humanity's shadow."[1] Even those considered "natural" almost inevitably contain invasive species, frequently in dominant roles: "The world's ecosystems will never revert to the pristine state they enjoyed before humans began to routinely criss-cross the globe. . . ."[2] While it may be true that we as individuals did not deliberately set out to dominate the carbon or nitrogen cycle, or to create a planet of mixmastered species, our economic, technology, energy, and transportation systems—the linked technological, economic, and population growth characteristic of the industrial revolution—have had precisely that effect.[3] In short, the earth has become a human artifact. Science fiction books have long spoken of terraforming Mars; ironically, we have all the while been terraforming earth.

The increasingly tight coupling between predominantly human systems and predominantly natural systems is not a sudden phenomenon. Greenland ice deposits reflect copper production during the Sung Dynasty in ancient China (ca. 1000 B.C.), as well as by the ancient Greeks and Romans; spikes in lead concentrations in the sediments of Swedish lakes reflect production of that metal in ancient Athens, Rome, and medieval Europe.[4] Contrary to popular belief, anthropogenic carbon dioxide buildup in the atmosphere began not with the industrial revolution with its reliance on fossil fuel, but with the deforestation of Eurasia and Africa over the past millennia.[5] Human impacts on ecosystems have similarly been going on for centuries, from the probable role of humans in eliminating megafauna in Australia and North America, to the clear role of human transportation systems in supporting invasive species around the world.[6] In this sense, recent books lamenting the "end of nature"[7]—"nature" taken as pre-human and pristine wilderness—obviously miss the mark by centuries. It is not that nonhuman "nature" has ended; rather, what such often anguished commentary represents is that the gap between the reality of the world as human artifact and the cultural construct of "nature" has grown so large that the fiction of "nature as other" can no longer be maintained.[8]

Thus, in a real sense there are no "natural" systems anymore, and the distinction between "human" and "natural" systems is somewhat misleading in its superficial clarity. Clearly there are phenomena, primarily geologic—volcanoes and plate tectonics come to mind—that human activities do not affect. But against this must be balanced the reality that the dynamics of most fundamental systems—the nitrogen, carbon, phosphorous, sulfur, and hydrologic cycles; the biosphere at various scales from genetic to regional; and the climate and oceanic circulation systems—are increasingly dominated by anthropogenic activities. Of course, there are degrees of influence: there are obviously systems whose dominant dimensions are nonanthropogenic, such as perhaps an isolated tundra biological community, and those whose dominant dimensions are anthropogenic, such as industrial or economic systems (and it is in this sense that this paper will refer to "natural" and "human" systems throughout). But the reality of tight coupling between the two domains must always be borne in mind. This is of particular importance given the well-known difficulty that social scientists and physical scientists have in communicating with each other, and the institutional realities of differing languages, communities, and worldviews between the two groups which exacerbate the natural–human dichotomy, and make integrated study of such systems extremely challenging.

Key Themes

There are three important themes that weave through the evolution of the terraformed earth. The first is the relationship of humans to the planet. Initially, humans, then in a hunter-gatherer culture, were like any other species embedded in the biosphere, and their activities were endogenous to it. With the evolution of agriculture, however, the species began to evolve to dominate local natural systems and, with the deforestation of Europe and North Africa during the 10th to 14th centuries, began the process of climatic change as well.[9] At this point, elements of human culture grew to be exogenous to the natural systems they worked with, and the mental model of stewardship evolved: humans as caretakers for the relevant aspects of nature (primarily, as the stewardship image implies, pastoral). This model is still prevalent today, but it has been rendered obsolete by the industrial revolution and concomitant growth in human population, technology, and economic systems. Humans are now once more endogenous to fundamental earth systems, but in a different relationship than originally: they now increasingly define the dynamics and behavior of those systems. The journey has been from "being natural" (e.g., hunter/gatherer society), to opposing and controlling nature (the Enlightenment, settlement of the American West),[10] to absorbing nature into the human experience. Perhaps it is best, if somewhat simplistically, summarized by saying that human experience was endogenous to nature; now much of nature is endogenous to human experience.

A second important question involves the level at which subjectivity—independent status as a free, rational, and moral agent—is posited. The reason this issue arises is because, especially when looking at the effects of fundamental technological systems over long timeframes, the idea that people "engineered" the results seems somewhat of a stretch. Thus, for example, virtually all island ecologies have been significantly impacted by invasive species in the last few centuries.[11] These species arrived as a result of the expansion of Eurocentric civilization, a cultural phenomenon, which was made possible by a complex evolution of technological capabilities (especially in transportation and navigation systems) and economic interests. If the concept of "engineering" is limited to the individual engineer and a particular artifact (e.g., an accurate timepiece), it does not make sense to say that the global restructuring of biota as a result of this process was "engineered." On the other hand, it is apparent that, although each engineering advance, and each voyage, may be taken individually, taken as a whole the processes involved were so ubiquitous and pervasive that they resulted in an anthropogenic restructuring of island biota (as well as a lot of other effects). Thus, what was not engineered at the level of the individual was arguably engineered at the level of society (in this case, European). Thus, the appropriate level of subjectivity in this case may be cultural, rather than individual.[12]

The reason this matters is because ethical responsibility vests at the level of conscious choice. If the scope of engineering, invention, and design is limited to individual artifacts and projects, then it enables refusal to take operational and ethical responsibility for the systems effects. Thus, for example, it is apparent that human activities over the past millennium have contributed significantly to climatic change, and that shifts in climate, especially if accompanied by changes in oceanic circulation patterns, will have dramatic effects on biodiversity and human society (including, quite probably, mortality rates). But limiting the concept of human engineering only to the individual, artifactual level means that, despite current activity, there is no sense in which human society, taken as a whole, is yet being tasked with the moral obligation to respond rationally and constructively to its clear impacts on the carbon cycle and climate system.

It is therefore necessary to expand the definition of engineering, design, and management to the scale of the technological and cultural systems that are, in fact, now beginning to dominate the dynamics of many natural systems. This is a principal rationale for earth systems engineering and management (ESEM), which can be seen as the acceptance of responsibility for the ethical and operational implications of what the human species has already been doing for centuries, and is continuing to do at a rapidly increasing rate. It is not traditional engineering and project management—but neither is it a new and completely foreign concept. It is, in a sense, the denouement of the evolution of applied science and technology (techne) that began in ancient Greece and Rome, and has been globalized by the industrial revolution.[13]

Another important theme is complexity, but here also a certain caution is necessary to avoid superficiality. For example, there has been a considerable literature recently implicitly drawing on the analogy between natural and ecological systems, and human systems. This can be useful, as the development of the field of industrial ecology demonstrates.[14] Indeed, both human and natural systems are similar in that they are technically complex; the learnings from the latter can indeed inform our understanding of the former. But the relationship is one of analogy. Failure to also understand the profound differences can lead to superficial reasoning or even nonsense—the burgeoning literature that suggests restructuring global capitalism or transnational corporations to resemble gardens is ample evidence of that. In particular, it is important to understand that human systems are of a different, and higher, class of complexity than natural systems. Human systems and human history are strongly affected by unpredictable contingency, partially as a result of the exercise of (bounded) free will, and the nature of humans as relatively autonomous, moral agents.[15] Moreover, human systems are characterized by a powerful reflexivity: a natural system such as a salt marsh is not changed by what a scientist may learn about it, but human systems, which internalize knowledge as it is developed, and thus change continually in an accelerating process of reflexive growth, are.[16] Additionally, the evolutionary processes of culture, technology, and social knowledge are uniquely human projects, with their own dynamics and timeframes, which have no parallel in traditional natural systems.[17]

Context

Thus, the context of ESEM requires that one comprehend not just the scientific and technological domains, but the social science domains—culture, religion, politics, institutional dynamics—as well. The human systems implicated in ESEM are extraordinarily powerful, with huge inertia and resistance to change built into them, and ESEM will fail as a response to the conditions of our modern world unless these are respected and understood. The evolution of eurocentric Judeo-Christian capitalist and technology systems has swept the globe.[18] Relevant implications of this historical process include commoditization of the world, including nature, which in developed countries is increasingly purchased at stores in upscale malls, in theme parks (reflecting not "natural" ecological dynamics but late 20th century ideology, including, of course, corporate sponsorship), and as eco-tour packaged "experiences."[19] The process of commoditization, the globalization of culture through movements such as, e.g., postmodernism,[20] and urbanization have had several fundamental effects.

First, the concepts of "nature," "wilderness," and related terms are fundamentally changing for many people. Second—and contrary to the oversimplistic and mistaken idea that the world is, in globalizing, becoming more homogeneous and simple—the complexity of society, and of the couplings between human and natural systems, is increasing radically. A superficial homogeneity (e.g., proliferation of U.S. fast food options around the world) is more than overcome by the increased access to global cultures and artifacts of all places, kinds, and times enabled by world transportation and information networks (indeed, this cultural heterogeneity—pastiche—is the very foundation of postmodernism). Finally, this increased complexity is reflexively affecting the governance structures within which environmental and technological issues have traditionally been addressed: the absolute primacy of the nation-state is being replaced by a far more fluid, and complex, dynamic structure involving a number of stakeholders, including private firms, NGOs, and communities of all kinds.

The implications of this evolution are not just of academic interest, but are absolutely pivotal to understanding the challenges faced by ESEM. Because of the increasingly tight coupling between human and natural systems, we are now reaching a point where the dynamics of natural systems are no longer governed simply by their internal structures and dynamics, but can only be understood in relationship to the human systems with which they are coupled. The systems implications of this are profound in two ways. First, it means that, from a systems perspective, the contingency and reflexivity inherent in human systems is imported into natural systems. Think, for example, of the carbon cycle and the way it is now being affected by the Kyoto process, and its associated scientific, institutional, and cultural elements. Trying to understand this cycle without understanding its human elements would be foolish. Or consider the ecology of the Hawaiian Islands: it simply cannot be studied without dealing with invasive species of all kinds—and they, in turn, reflect the expansion of the eurocentric Judeo-Christian civilization, and associated technologies, especially transportation modes. The natural history of the Hawaiian Islands is a human history, written into its ecology.

The "humanization" of natural systems raises another absolutely critical point: the pivotal role of values in determining the structure of the external world. Because virtually all natural systems of any consequence have already been inescapably altered by human activity, there can be no question of returning to "pristine" nature (if such a state ever existed). Whether it is the Western U.S. or Hawaii with invasive species; Australia with the ancient extinction of magafauna; the Great Lakes, Baltic Sea, or Everglades; or just rebuilding a river ecology severely impacted by industrial activity—there is no "natural" state to return to. First, of course all these systems have been evolving to respond to different conditions over geologic time; there is no particular "base state." With human intervention, they have been further modified—and any attempt to "restore" them is, in effect, a further modification to suit human cultural and ideological norms. Difficult as it is to accept—there is no natural history anymore: there is only human history. What these systems will be in the future is a human decision, a human choice. Having realized this, we cannot escape the ethical responsibility for that choice. But whose values will determine that choice— what religion, what culture, what subgroup? The centrality of values and ethical choice to environmental decision-making around the world is apparent.[21]

Definition and Case Studies

Given this context, a more precise definition of ESEM can now be presented: ESEM is the capability to rationally engineer and manage human technology systems and related elements of natural systems in such a way as to provide the requisite functionality while facilitating the active management of strongly coupled natural systems. ESEM also aims to minimize the risk and scale of unplanned or undesirable perturbations in coupled human or natural systems. "Technology systems" in this sense is to be read broadly, as the means by which human beings interact with their environment; it thus includes not just artifacts, but the economic, cultural, and ideological context within which they are used.[22] As the examples below illustrate, ESEM in many cases will deal with large, complex, evolving projects and technologies, with complicated governance, ethical, scientific, cultural, and religious dimensions and uncertainties. It does not replace traditional disciplines such as political science, economics, sociology, engineering, or physical and biological sciences: rather, it draws upon and integrates them in the context of ESEM applications, and thus expands them as well. ESEM augments, but does not replace, existing fields.

In most cases, the social and physical scientific and technical knowledge necessary to support ESEM approaches is weak or nonexistent, and the evolution of the institutional and ethical capacity necessary to complement ESEM is, if anything, in an even more primitive state. Accordingly, ESEM is best thought of as a capability that must be developed over a period of decades, rather than something to be implemented in the short term. It is also important to note that ESEM is not "new" in that it argues that humans as a species should now begin to engineer the world. That is already happening. What is new about ESEM is the assumption of responsibility for what we as a species are already doing, and the determination to develop the capability to do so more rationally and ethically.

ESEM is a new field of study and practice, but it does not spring from nothingness. Rather, it builds on practices and activities that are already being explored, some newer and less developed than others. In this, it is like industrial ecology, which is an umbrella area of study including both applied methodologies (such as life cycle assessment, or LCA, and design for environment, or DFE), and research methods (such as materials flow analysis, or MFA). Thus, the first case study explores an ESEM methodology, "adaptive management," which has been developed in the context of resource management.

The second looks at urban areas as ESEM projects. There is obviously a huge body of knowledge and several disciplines that deal with this subject matter—much of which is informative for the development of ESEM principles. What the ESEM approach adds in this instance is a new envisioning of urban centers (particularly megacities, defined as over ten million population) as nodes in energy, material, ecological, and knowledge systems scaled from the local to the global, across many time scales, and involving many different kinds of human systems. The third case study looks at carbon cycle and climate system management efforts, an area where the ESEM approach is already latent (in geoengineering proposals, for example), but where the rational, comprehensive, systems-based ESEM approach has yet to be adopted. Finally, the last case study involves scenarios drawn from the information revolution: these illustrate both the complexity of the issues with which ESEM must deal, and the almost complete ignorance which characterizes current understanding. . . .

ESEM and Earth as Artifact

Earth systems engineering and management may be defined as the capability to rationally engineer and manage human technology systems and related elements of natural systems in such a way as to provide the requisite functionality while facilitating the active management of strongly coupled natural systems. The need for ESEM arises because, as a result of the industrial revolution and concomitant changes in agriculture, population levels, culture, and human systems, the world has become a human artifact. Partially because this process has occurred over time frames that are longer than individual time horizons, and has involved institutions and technology systems rather than conscious individual decisions, recognition of this phenomenon, and appropriate responses, have yet to occur. Indeed, it is apparent that the science and technology, institutional, and ethical infrastructures necessary to support such a response have not yet been developed. The issue is not, however, whether the earth will be engineered by the human species: that has been and is already occurring. The issue is whether humans will do so rationally, intelligently, and ethically.

NOTES

1. R. Gallagher and B. Carpenter, "Human-Dominated Ecosystems: Introduction," *Science* 277 (1997): 485.
2. J. Kaiser, "Stemming the Tide of Invading Species," *Science* 285 (1999): 1836–1841.
3. B. R. Allenby, "Earth Systems Engineering: The Role of Industrial Ecology in an Engineered World," *Journal of Industrial Ecology* 2 (1999): 73–93; "Human Dominated Ecosystems," *Science* 277 (1997): 485–525; V. Smil, *Cycles of Life: Civilization and the Biosphere* (New York: Scientific American Library, 1997); P. M. Vitousek, H. A. Mooney, J. Lubchenco, and J. M. Melillo, "Human Domination of Earth's Ecosystems," *Science* 277 (1997): 494–499; B. L. Turner, W. C. Clark, R. W. Kates, J. F. Richards, J. T. Mathews, and W. B. Meyer, eds., *The Earth as Transformed by Human Action* (Cambridge: Cambridge University Press, 1990).
4. S. Hong, J. Candelone, C. C. Patterson, and C. F. Boutron, "History of Ancient Copper Smelting Pollution during Roman and Medieval Times Recorded in Greenland Ice," *Science* 272 (1996): 246–249; I. Renberg, M. W. Persson, and O. Emteryd, "Preindustrial Atmospheric Lead Contamination in Swedish Lake Sediments," *Nature* 363 (1994): 323–326.
5. J. Jager and R. G. Barry, "Climate," in Turner et al., *Earth as Transformed by Human Action*.
6. Kaiser, "Stemming the Tide"; D. Jablonski, "Extinctions: A Paleontological Perspective," *Science* 253 (1991): 754–757; C. Barlow, *Green Space, Green Time* (New York: Springer-Verlag, 1997).
7. B. McKibben, *The End of Nature* (New York: Random House, 1989).
8. Although it is a basic and well-established principle of modern sociology that many concepts that people regard as self-evidently "real" are in fact cultural constructs, many people have trouble accepting this. This is particularly true in the environmental arena, where advocates, who frequently are personally committed to concepts such as "wilderness" and "nature," view them as purely objective and self-evidently worth protecting as such. That different cultures may construe these constructs differently, or that they may be transitory reflections of a particular stage of Western thought as opposed to transcendent verities, is accordingly not accepted by many Western, especially American, environmentalists.
9. A. Grubler, *Technology and Global Change* (Cambridge: Cambridge University Press, 1998).
10. W. Cronon, *Uncommon Ground: Rethinking the Human Place in Nature* (New York: Norton, 1995).
11. "Human Dominated Ecosystems."
12. Kaiser, "Stemming the Tide."
13. M. Heidegger, *The Question Concerning Technology and Other Essays*, trans. W. Lovitt (New York: Harper Torchbooks, 1977).
14. R. Socolow, C. Andrews, F. Berkhout, and V. Thomas, *Industrial Ecology and Global Change* (Cambridge: Cambridge University Press, 1994); T. E. Graedel and B. R. Allenby, *Industrial*

Ecology (Upper Saddle River, NJ: Prentice-Hall, 1995); B. R. Allenby, *Industrial Ecology: Policy Framework and Implementation* (Upper Saddle River, NJ: Prentice-Hall, 1999).

15. D. Harvey, *Justice, Nature and the Geography of Difference* (Cambridge, MA: Blackwell, 1996); D. S. Landes, *The Wealth and Poverty of Nations* (New York: Norton, 1998); J. C. Scott, *Seeing like a State* (New Haven, CT: Yale University Press, 1998); I. Hacking, *The Social Construction of What?* (Cambridge, MA: Harvard University Press, 1999).

16. A. Giddens, *The Constitution of Society* (Berkeley: University of California Press, 1984).

17. Grubler, *Technology and Global Change*; Heidegger, *The Question Concerning Technology and Other Essays*; D. F. Noble, *The Religion of Technology* (New York: Knopf, 1998).

18. Landes, *The Wealth and Poverty of Nations*; Harvey, *Justice, Nature and the Geography of Difference*; Noble, *The Religion of Technology*; J. Diamond, *Guns, Germs and Steel* (New York: Norton, 1997).

19. Cronon, *Uncommon Ground*.

20. Terms such as "high modernism" and "postmodernism" are not well defined, even in the relevant literature. "High modernism" approaches have tended to be elitist and technocratic. Think, for example, of Robert Moses and the way he drove his expressways and constructions through existing neighborhoods in New York City, destroying them in the process; or the way the old Soviet Union diverted the flows of two major rivers feeding the Aral Sea, the Amu Dar'ya and the Syr Dar'ya, resulting in changes in climate across Asia, extinction of some 85% of the fish species in the Sea; and a loss of 75% of the Sea's volume in a few short years. Postmodernism is generally atemporal and ageographical, characterized by pastiches of cultures, times, places, and ideas; it also tends towards consumerism. Think of Disney's Epcot Center, with (primarily plastic) reproductions of bits and buildings from various times and cultures, from all around the world. A second principle characteristic of postmodernity is its insistence on moral and cultural relativism: if taken as supportive of multiculturalism and tolerance, this arguably informs an appropriate governance structure for ESEM. If taken at its extreme, of course, it is patently ridiculous.

21. It is also extensively ignored, for many reasons. First, many of the stakeholders— environmentalists, technologists, scientists, industrialists—are naïve positivists, and don't recognize the value-laden mental models with which they approach these issues and conflicts. Second, most people of any stripe are much more comfortable speaking the "language of science" rather than the "language of values," even when it is clearly values that are at issue. Thus, discussions around the Kyoto Accord are often framed in technical terms, when in fact what is at issue are very different concepts of the world.

22. As Grubler comments, "technology cannot be separated from the economic and social context out of which it evolves, and which is responsible for its production and its use. In turn, the social and economic context is shaped by the technologies that are produced and used" (*Technology and Global Change*, 21). Heidegger in his essay "The Question Concerning Technology" takes the more extreme position that "the essence of technology is nothing technological," in that technology reflects the essence of being human, and cannot be separated out from the project of human evolution (*Question Concerning Technology*, 35).

38

The Earth Is Not Yet an Artifact

DAVID W. KEITH

In the previous article in this issue, Brad Allenby argues for a more active style of planetary management. He notes, "the biosphere itself, at levels from the genetic to the landscape, is increasingly a human product."[1]

While the earth is undoubtedly being transformed by human action,[2] the implications of the transformation for environmental management are in dispute. Allenby suggests that the scale of human impact makes it appropriate to treat all landscapes as human products, subject to active management aimed at goals we collectively define. I question Allenby's claim that "there are no 'natural' systems anymore,"[3] and question the utility of engineering-based management methods as tools for planetary management. In contrast to Allenby, I suggest that we learn to control before embarking on active planetary management.

Allenby defines "Earth Systems Engineering and Management" (ESEM) as "the capacity to rationally engineer and manage human technology systems and related elements of natural systems in such a way as to provide the requisite functionality."[4] The goal is active environmental management as opposed to minimization of environmental impact.[5] In advocating ESEM, Allenby stresses the extent of human transformation of the environment, arguing, "the earth has become a human artifact."[6] He implies that, if the world is an artifact, then it is naïve to try to maintain a landscape in its natural state, and naïve to practice stewardship.[7] I argue that while one may accept that artificiality demands active management, one ought to reject the claim that the world is now artificial.

Is There Any Nature Left?

It is undeniable that the boundary between natural and human systems is fuzzy. Likewise, it is fair to characterize the world's landscapes as ranging from heavily to lightly managed, and to be wary of unreflective characterization of any landscape as wild. Yet the claim that the earth has become a human artifact seems far too strong. Most landscapes are in the middle ground, and the details of this middle ground matter in disputes about environmental management.

In arguing for the world's artificiality Allenby notes that some degree of human influence can be detected everywhere; for example, that trace metals mobilized by bronze age smelting can be detected in the Greenland icesheet.

The argument that any trace of humanity makes a landscape unnatural assumes that true nature must be pristine; or, equivalently that there is a sharp dichotomy between nature and culture. Allenby rightly questions this dichotomy, yet his analysis proceeds by claiming that all is culture, that "there are no 'natural' systems anymore," that, "difficult as it is to accept—there is no natural history anymore: there is only human history."[8] Allenby critiques McKibben's "anguished commentary" about *The End of Nature*, but the essence of the critique is merely that McKibben misread the date of nature's demise. Excepting the date, Allenby's argument that the ubiquity of human imprint makes the world artificial is very similar to McKibben's statement that "By changing the weather, we make every spot on the earth man-made and artificial."[9]

Many landscapes bear no visible trace of humanity yet are nevertheless modified by human action. At the northern limit of trees in the central Canadian Northwest Territories, for example, one may travel for days and see no sign of humanity except satellites in the evening sky. Yet even here, human imprint is ubiquitous from the top of the food chain where the absence of Pleistocene megafauna such as mammoths is (arguably) due to human caused extinction[10] to the lowest trophic levels where primary productivity is (likely) changing due to fertilization by anthropogenic nitrogen and carbon dioxide. An absence of visible signs is thus an unreliable test for the absence of impact. How then should we assess human influence? Consider some more technical measures.

In contrast to visible signs, the presence of a detectable trace of human action seems a reassuringly scientific measure, yet it tells us little because it hangs on the question of detectability and thus on the detection technology employed. Suppose that some future technology could detect signs of early Eurasian agriculture in North American ice-cores at a date prior to the first human arrival in the Americas. Would we conclude that the Americas had ceased to be natural at that date? I think not. This measure of human influence is irrelevant to the arguments for the world's artificiality both because it depends on the detection technology, and, as discussed above, because it assumes a sharp dichotomy between nature and culture.

The disruption of biochemistry[11] and biodiversity[12] provide quantitative measures of the degree of human imprint that are more relevant to debates about environmental management that concern Allenby. On a global scale it is clear that some—but not all—such measures suggest pervasive anthropogenic disruption. The nitrogen cycle and the extinction rate are cases for which the anthropogenic perturbation is as large or (for extinction rate) much larger than the natural background. Despite these global aggregates, there are many areas where human influence remains slight.

Landscapes have aesthetic values that are not captured in the technical measures outlined above. One such value rests on a landscape's history. Consider, for example, a project aimed at restoring an environment that has been destroyed, say the restoration of a prairie after strip-mining. Such an artificially created environment could be pristine by all the measures discussed above, yet we would likely value it less because of its very artificiality. There is a strong analogy with the values we ascribe to works of art. We value a fake painting less than an original. We would likely value a fake even less if its creation necessitated the destruction of the original, as is the case with a prairie restored after strip-mining. One may ascribe a similar value to landscapes, a value tied to their historic origin: "What is significant about wilderness is its causal continuity with the past."[13]

Now reconsider the claim "there are no 'natural' systems anymore" in light of the measures of human impact outlined above. While the world in aggregate is heavily influenced by humanity, there nevertheless exist many landscapes in which human influence is light. Specifically, there are still many areas where there is essentially no visible human imprint; where the majority of species have evolved in situ, largely unaffected by human influence;[14] where the biochemical perturbations are small; and finally, where there is historic continuity with a pre-human landscape. Illustrative examples of lightly influenced landscapes are found in the arctic and in the interior of tropical rainforests. Such landscapes exist, they have a value, and they are not artificial.

Why does the existence of lightly influenced landscapes matter? The heart of the ESEM prescription for environmental management is the goal of "providing the requisite functionality."[15] The joint claim that there are no natural systems anymore and that environmental management should aim only at functionality denies much of the traditional justification for the preservation of nature. Since the 19th century, the politics of wilderness preservation have been driven, in part, by convictions about the intrinsic value of nature.[16] Allenby urges us to dismiss these romantic ideals, and replace them with functional measures of nature's utility and with rational environmental management. In part, Allenby's comments draw from so-called postmodern efforts to demonstrate that nature is a cultural construct, and often a physical product of human action.[17] Interestingly, this assault on the reality of nature has been used by advocates from across the political spectrum, ranging from the pro-development Wise Use Movement to People for the Ethical Treatment of Animals, to argue against the protection of natural ecosystems.[18]

Allenby would have us justify the preservation of nature by assessing the functional values that it provides. While such utilitarian arguments are important, I doubt that they will prove sufficient. I suggest that natural landscapes have some form of intrinsic rights[19] or moral value,[20] and that these values are not simply tradable in a management regime aimed at achieving "the requisite functionality."

Artificial World?

Would an expansion of human influence make the world artificial? Not necessarily. Artificiality is a much stronger claim than influence. Artificial systems involve human agency in a fundamental way. The distinction is clear when we study a system's behavior. For example, one may characterize engineering and the social sciences as "the sciences of the artificial,"[21] and may distinguish them from the natural sciences by the kind of explanation that is sought. Natural science explains events as the results of simple laws acting without human intent, whereas engineering and the social sciences are concerned with design and with the consequences of human intentions.

Similarly, one would only call a landscape artificial if resolving central questions about its behavior demanded an understanding of human agency. What would an artificial world look like? We may one day arrive at a world where climate and weather are actively controlled, where new genetically engineered cultivars and new fauna are common in every landscape, where the human gene pool has entered a period of rapid divergence.[22] In such a world, attributes ranging from the global concentration of CO_2 to the local distribution of biota would be determined more by human decision making than by nature. Even if ecosystems closely resembled their pre-industrial counterparts, one would ask not how they evolved but why they were put there. Such a world would justify the term artificial in that geochemistry would be a form of engineering rather than a science.

The conclusion that the world is now artificial arises in part from a failure to distinguish between deliberately engineered environments, such as gardens, and the landscapes that are marked by what Allenby calls the "unintended byproducts of human engineering." This description covers the many landscapes that bear the scars of our industrial system without playing an economically important role in that system. Pollution, however, is not engineering. Intent matters. Intent is central to our understanding of engineering.[23] To claim that: "Throughout history, humans have continually designed and engineered the carbon cycle. That is, of course, what agriculture is all about" is to equate to affect with to design. On this reading of design, the following claim is justified: beavers design hydrological cycle, that is what dam building is all about. As another critic of Allenby puts it, "This is not engineering, just making a mess."[24] The distinction matters: If we are just making a mess, then minimizing our mess makes sense. But if we are engineering the world, then ESEM—management not minimization—makes sense.

Implications for Climate Policy

What would an ESEM climate policy look like? It would be systematic climate management aimed at providing "the requisite functionality." It would entail the engineering of climate, or at least of CO_2 concentration, so as to provide the required meteorological conditions as well as the required ecosystem services. Allenby makes it clear that ESEM would draw from a large toolkit, ranging from source abatement to geoengineering. In addition to controlling industrial CO_2 emissions, an ESEM climate policy might employ countervailing measures such as the use of ocean fertilization to enhance oceanic carbon sinks, or alteration of the effective solar flux (via the use of space-based solar shields) to offset the climatic effects of increased CO_2.[25]

ESEM would build on the best examples of modern-engineering project management and would incorporate features such as distributed decision making, rigorous performance measurement, and extensive dialog with stakeholders. All these are fine features of project management once a project's overall goal is reasonably well defined. Yet for the climate problem the choice of goals is the crux of debate. Put most simply, what should the CO_2 concentration target be? Any choice has weighty consequences. There are winners and losers in any climate scenario, and trade-offs both between people and between people and natural systems.

Climate policy is now a contentious subject of international politics. While climate policy could use some improvement, it is unclear how ESEM would help. Allenby's ESEM includes a coherent vision of management process, but it is much less clear about goals and governance. The tools that Allenby draws on (e.g., industrial ecology and life cycle analysis) are intended to evaluate the environmental consequences of policy choices, but these tools have very little to say about the formation of environmental values and the resolution of conflicts between them.[26] Yet ESEM is essentially a proposal for planetary governance built from a project-engineering mold.

An ESEM climate policy is doubly challenging because Allenby makes clear that climate policy ought to be unmoored from its preindustrial reference. The goal of ESEM is to find the climate that best enhances "functionality" rather than to minimize impacts so as to return us to the pre-industrial climate. Allenby presents two arguments for this unmooring. First, the artificiality argument: we have already messed with the climate; thus, the climate is an artifact; thus, the goal of climate policy should be management not minimization. Second, an argument in the language of postmodern analysis: "A critical point implicit in many of the global environmental discourses is whether it is ethical to privilege the present. Thus, for example, the global climate change negotiations seek to stabilize current climatic conditions . . . [and] thus seek to remove an important driver of biological evolution."[27]

The second argument depends on an odd conflation of timescales. Climatic fluctuations, such as the glacial cycles that drive biological evolution, have timescales that exceed 10,000 years; concern about removing these fluctuations seems irrelevant given the overwhelming biological impact arising from the much more rapid human transformation of the planet. Before we concern ourselves with managing the planet on the glacial timescale, we need to learn to mitigate the immediate impacts of our industrial system.

ESEM is a radical prescription for climate policy. It suggests that we cast off our ties to pre-industrial climate and embark on a broad program of planetary engineering aimed at managing (improving?) the functionality of tightly coupled biogeochemical systems. One may accept that the world may become an artifact in the strong sense defined above; further, one may accept that wise management in such a world might bear some resemblance to ESEM, that, "not minimization but management" would be a fine maxim. But we do not yet live in such a world. Allenby urges us to punt the problem of learning to mitigate industrial emissions of CO_2 and move directly to a regime of global planetary management in which the mitigation of industrial emissions is just one of many tools. I disagree. We would be wise to learn to walk before we try to run, to learn to mitigate before we try to manage.

NOTES
1. B. Allenby, "Earth Systems Engineering and Management," *IEEE Technology & Society Magazine* (2000): 10–24.
2. G. P. Marsh, *Man and Nature; or, The Earth as Modified by Human Action* (Cambridge, MA: Belknap Press of Harvard University Press, 1965), originally published 1864; B. L. I. Turner, W. C. Clark, R. W. Kates, J. F. Richards, J. T. Mathews, and W. B. Meyer, eds., *The Earth as Transformed by Human Action* (New York: Cambridge University Press, 1990); W. L. Thomas, *Man's Role in Changing the Face of the Earth* (Chicago: University of Chicago Press, 1956).
3. Allenby, "Earth Systems Engineering and Management."
4. Ibid.
5. B. Allenby, "Earth Systems Engineering: The Role of Industrial Ecology in an Engineered World," *Journal of Industrial Ecology* 2 (1999): 76.
6. Allenby, "Earth Systems Engineering and Management."
7. Allenby, "Earth Systems Engineering: The Role of Industrial Ecology," 84.
8. Allenby, "Earth Systems Engineering and Management.
9. B. McKibben, *The End of Nature* (New York: Random House, 1989), 215.
10. P. S. Martin and R. G. Klein, eds., *Quaternary Extinctions: A Prehistoric Revolution* (Tucson: University of Arizona Press, 1984).
11. V. Smil, *Carbon Nitrogen Sulfur: Human Interference in Grand Biospheric Cycles* (New York: Plenum, 1985); P. M. Vitousek, P. R. Ehrlich, A. H. Ehrlich, and P. A. Matson, "Human Appropriation of the Products of Photosynthesis," *Bioscience* 36 (1986): 368–373.
12. E. O. Wilson, "Threats to Biodiversity," *Scientific American* 261 (1989): 108–116.
13. R. Elliot, "Faking Nature," in L. P. Pojman, ed., *Environmental Ethics: Readings in Theory and Application* (Boston: Jones and Bartlett, 1994), 230.

14. M. E. Soulé, "The Social Siege of Nature," in M. E. Soulé and G. Lease, eds., *Reinventing Nature? Responses to Postmodern Deconstruction* (Washington, DC: Island, 1995), 158.

15. Allenby, "Earth Systems Engineering and Management."

16. R. Nash, *Wilderness and the American Mind*, 3rd ed. (New Haven, CT: Yale University Press, 1982).

17. W. Cronon, ed., *Toward Reinventing Nature* (New York: Norton, 1995).

18. Soulé and Lease, *Reinventing Nature?*

19. P. W. Taylor, *Respect for Nature: A Theory of Environmental Ethics* (Princeton, NJ: Princeton University Press, 1986).

20. H. Rolston, *Environmental Ethics* (Philadelphia: Temple University Press, 1988).

21. H. A. Simon, *The Sciences of the Artificial*, 3rd ed. (Cambridge, MA: MIT Press, 1996).

22. L. M. Silver, *Remaking Eden* (New York: Avon, 1997).

23. Simon, *The Sciences of the Artificial*.

24. R. M. Friedman, "When You Find Yourself in a Hole, Stop Digging," *Journal of Industrial Ecology* 3 (2000): 17.

25. Allenby, "Earth Systems Engineering and Management"; Allenby, "Earth Systems Engineering: The Role of Industrial Ecology"; D. W. Keith, "Geoengineering the Climate: History and Prospect," *Annual Review of Energy and Environment* 25 (2000): 245–284.

26. G. Morgan, M. Kandlikar, J. Risbey, and H. Dowlatabadi, "Why Conventional Tools for Policy Analysis Are Often Inadequate," *Climatic Change* 41 (1999): 271–281.

27. Allenby, "Earth Systems Engineering and Management."

Reading Questions and Further Readings

Reading Questions

1. Has the earth become a human artifact?
2. What is the difference between seeing humans as planetary managers (Allenby's view) and humans as planetary stewards (Keith's view)?

Further Readings

Allenby, Braden, and Daniel Sarewitz. *The Techno-Human Condition*. Cambridge, MA: MIT Press, 2011.

Moore, Kathleen Dean, and Michael Nelson. *Moral Ground: Ethical Action for a Planet in Peril*. San Antonio, TX: Trinity, 2010.

INDEX

ABOUT THE EDITORS

Maria Damon analyzes policy instruments for sustainable development as well as capacity building in response to climate change. She was previously Assistant Professor of Environmental Studies and Public Policy at New York University.

Dale Jamieson is Professor of Environmental Studies and Philosophy and Chair of Environmental Studies at New York University. His most recent books are *Love in the Anthropocene* (2015, with Bonnie Nadzam) and *Reason in a Dark Time* (2014).

Colin Jerolmack is Associate Professor of Environmental Studies and Sociology at New York University. His most recent book is *The Global Pigeon* (2013). He is currently writing a book on how hydraulic fracturing impacts communities in central Pennsylvania.

Anne Rademacher is Associate Professor of Environmental Studies and Anthropology at New York University. Her recent books include *Reigning the River* (2011) and (as coeditor) *Ecologies of Urbanism in India* (2013). She is currently researching green design expertise in a transnational context.

Christopher Schlottmann is Clinical Associate Professor and Associate Chair in the Department of Environmental Studies at New York University. His recent books include *Reflecting on Nature* (2012, coedited with Lori Gruen and Dale Jamieson) and *Conceptual Challenges for Environmental Education* (2012). He is currently writing two books on the ethics of food and the environment.

Lightning Source UK Ltd.
Milton Keynes UK
UKHW032107290921
391396UK00009B/434